Physics: Advanced Principles and Applications

Physics: Advanced Principles and Applications

Editor: Tom Gladstone

NY RESEARCH
P R E S S

New York

Published by NY Research Press
118-35 Queens Blvd., Suite 400,
Forest Hills, NY 11375, USA
www.nyresearchpress.com

Physics: Advanced Principles and Applications
Edited by Tom Gladstone

International Standard Book Number: 978-1-63238-582-6 (Hardback)

Cataloging-in-Publication Data

Physics : advanced principles and applications / edited by Tom Gladstone.
    p. cm.
Includes bibliographical references and index.
ISBN 978-1-63238-582-6
1. Physics. I. Gladstone, Tom.
QC21.3 .P49 2018
530--dc23

Contents

Preface

Physics is a natural science which focuses on concepts like matter, motion, energy, space and time. New developments in the field of physics, especially electromagnetism has drastically changed the modern world with products like television, computers or nuclear weapons. This book unfolds the innovative aspects of physics which will be crucial for the progress of this field in the future. A number of latest researches have been included to keep the readers up-to-date with the global concepts in this area of study.

All of the data presented henceforth, was collaborated in the wake of recent advancements in the field. The aim of this book is to present the diversified developments from across the globe in a comprehensible manner. The opinions expressed in each chapter belong solely to the contributing authors. Their interpretations of the topics are the integral part of this book, which I have carefully compiled for a better understanding of the readers.

At the end, I would like to thank all those who dedicated their time and efforts for the successful completion of this book. I also wish to convey my gratitude towards my friends and family who supported me at every step.

Editor

Structured Laser Illumination Planar Imaging Based Classification of Ground Coffee Using Multivariate Chemometric Analysis

Olivier K. Bagui[1], Kenneth A. Kaduki[2], Edouard Berrocal[3] & Jeremie T. Zoueu[1]

[1] Laboratoire d'Instrumentation Image et Spectroscopie, Institut National Polytechnique Felix Houphouet-Boigny, BP 1093 Yamoussoukro, Cote d'Ivoire

[2] Laser Physics and Spectroscopy Group, Department of Physics, University of Nairobi, P. O. Box 30197-00100, Nairobi, Kenya

[3] Department of Physics, Division of Combustion Physics, Lund Institute of Technology, Box 118, Lund 221 00, Sweden

Correspondence: J. T. Zoueu, Laboratoire d'Instrumentation Image et Spectroscopie, Institut National Polytechnique Felix Houphouet-Boigny, BP 1093 Yamoussoukro, Cote d'Ivoire. E-mail:Jeremie.zoueu@nphb.edu.ci

Abstract

Most commercially available ground coffees are processed from Robusta or Arabica coffee beans. In this work, we report on the potential of Structured Laser Illumination Planar Imaging (SLIPI) technique for the classification of five types of Robusta and Arabica commercial ground coffee samples (Familial, Belier, Brazil, Colombia and Malaga). This classification is made, here, from the measurement of the extinction coefficient μ_e and of the optical depth OD by means of SLIPI. The proposed technique offers the advantage of eliminating the light intensity from photons which have been multiply scattered in the coffee solution, leading to an accurate and reliable measurement of μ_e. Data analysis uses the chemometric techniques of Principal Component Anaysis (PCA) for variable selection and Hierarchical Cluster Analysis (HCA) for classification. The chemometric model demonstrates the potential of this approach for practical assessment of coffee grades by correctly classifying the coffee samples according to their species.

Keywords: extinction coefficient, optical density, structured illumination, SLIPI, PCA, HCA

1. Introduction

Coffee is a popular beverage used throughout the world (Oder, 2015). Though not widely consumed on the continent, African countries make a significant contribution to world coffee trade. Ethiopia was the third largest coffee producer in the world in 2015 with Cote d'Ivoire being number twelve (International Coffee organization, 2015). The quality of coffee depends on many factors, such as the growth environment and processing techniques (Carelli et al., 2006; Clifford & Willson, 1985). Arabica and Robusta varieties represent over 90% of the world production of coffee. Arabica coffee, originally from Ethiopia, is now widely cultivated in South America. Robusta, which is a variety of the canephora species, is grown in Africa (mainly in Cote d'Ivoire) and in the Far East (Vietnam in particular). Robusta coffee is richer in caffeine than arabica (from 2% to 3% against 1.3%), (National Coffee Association of USA, 2015). The arabica variety fetches higher prices on the world market.

In order to distinguish between Robusta and Arabica varieties on the market, non-specialists rely on information provided on the packaging. As it is the case with other high value agricultural products, there have been increased incidences of counterfeit coffee on sale in the market. Chemical and genetic procedures for identification of the origin of ground coffee exist but they are time consuming. Another method is sensory evaluation but it is not appropriate for accurate and repeatable classification. Development of measurement techniques and technologies that can objectively discriminate between Robusta and Arabica ground coffee are therefore highly desirable. Nowadays, the use of optical laser techniques for various quantitative measurements is commonly applied for a large number of applications. Here, we aim at measuring the extinction coefficient (e.g. Berrocal et al., 2012; Kristensson et al., 2011, Kristensson et al., 2012) in different coffee solutions and analyze the results by means of chemometrics.

Chemometrics is the use of mathematical and statistical methods to extract chemical information and to correlate quality parameters or physical properties from an experimental data-set. The basic process involves modeling of patterns in the data. The models are then routinely applied to future data in order to predict quality parameters. The chemometrics approach has been gaining interest in assessing product quality. The only requirements are the extraction of reliable measurements and adequate software to interpret the patterns in the data.

In this article, the extinction coefficient of various coffee solutions is measured using a recent approach called single-phase Structured Laser Illumination Planar Imaging (SLIPI) (Berrocal et al., 2012). While transmission measurement records the light intensity of a single beam crossing the sample of interest (where the initial and final intensities are recorder) SLIPI is based on imaging a spatially modulated light sheet from the side (at 90° angle). The main advantage of SLIPI over conventional transmission measurements is its efficient capability in rejecting the light intensity from multiply scattered photons, allowing more accurate measurement of the extinction coefficient in turbid media (such as coffee solutions). We aim, then, at combining SLIPI measurements with a data analysis based on chemometrics to make a reliable classification of various types of coffee.

2. Methods

2.1 Sample Preparation

The coffee samples were prepared following the same procedure. For each type of coffee, the solutions were prepared by weighting 4g of coffee using a Satorius VIC-303 balance with 1 mg resolution and dissolving it in 100 mL of boiled distilled water. We stirred the water and coffee mixture for 15 seconds to get a homogeneous suspension, filtered the suspension into a sealed glass flask and left it to cool to 20°C before starting measurements.

2.2 Experimental Setup

The SLIPI technique was first created and applied in 2008 for imaging spray systems typically used in combustion engines (Berrocal et al., 2008). A description of it, in its various configurations, can be found in the doctoral thesis of E. Kristensson (Kristensson et al., 2012). We employ, here, the single-phase SLIPI approach for the measurements of the extinction coefficients and optical depths. The method has been presented and fully described in (Berrocal et al., 2012). Figure 1 shows a schematic of our experimental setup.

In the experiment, a coffee solution is illuminated in a cuvette with a spatially modulated laser sheet constructed using by a 5 lp/mm Ronchi grating and shaped by spherical and cylindrical lenses. The incident light is produced by three diode lasers emitting at 450 nm, 532 nm and 638 nm, respectively. A 650 nm high pass filter is positioned in front of the camera to only detect the laser induced fluorescence signal from the coffee.

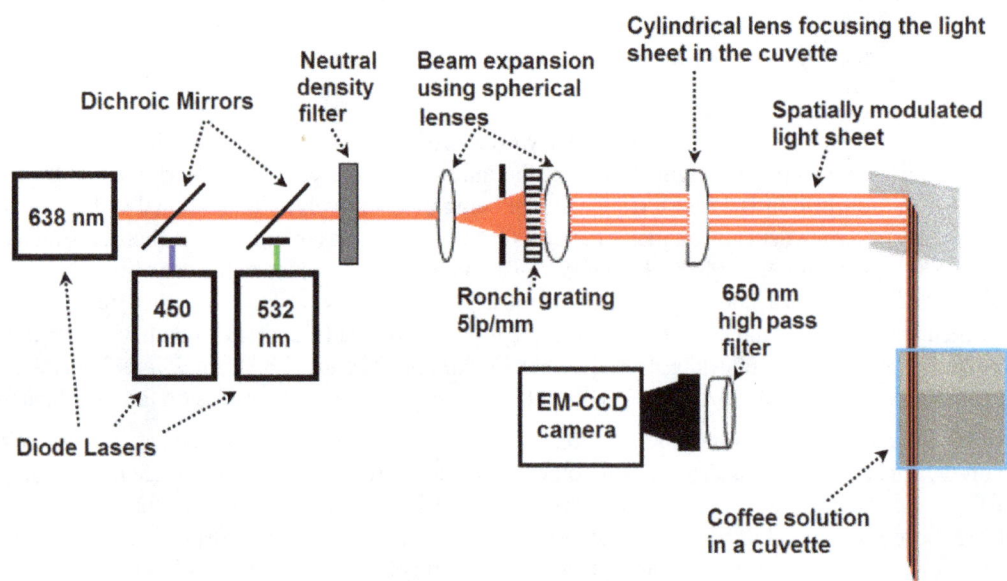

Figure 1. Description of the single-phase SLIPI optical arrangement: A light sheet with a vertically modulated light intensity profile is formed, illuminating the coffee solution. Images of the spatially modulated light sheet are recorded from the side using an EM-CCD camera. By then extracting the amplitude of the modulation, the exponential light extinction through the cuvette can be observed. The measurements are performed sequentially for three different illumination wavelengths corresponding to 450nm, 532 nm and 638 nm respectively

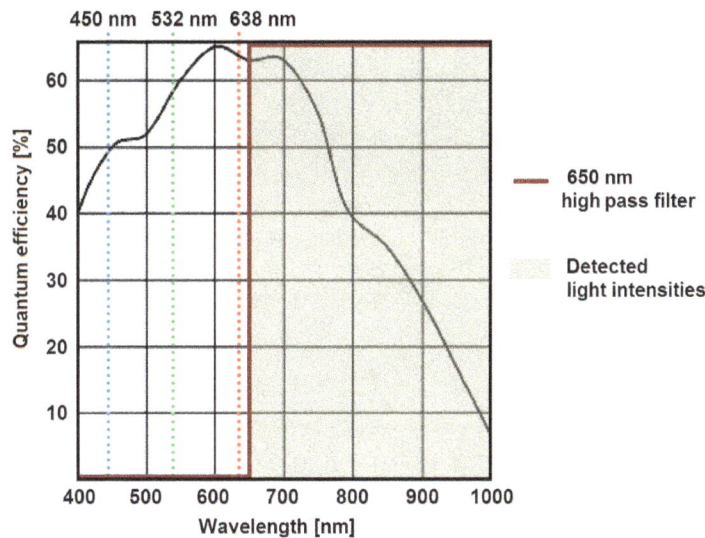

Figure 2. Quantum efficiency curve of the Andor Luca-R EM-CCD camera (data curve from Andor Technology). The wavelengths of corresponding to each illumination scheme and to the low pass filter fixed on the camera objective are also indicated

The images are acquired with an Andor Luca-R Electron Multiplying Charge-Coupled Device (EM-CCD) camera placed at an angle of $90°$ from the direction of the incident the light sheet. The quantum efficiency of the camera, cooled at $-20°C$, is shown in Figure 2.

2.3 Data Analysis

Our clustering algorithm uses the extinction coefficient and the optical depth of each sample for the different laser illumination. Therefore, each sample has six variables making this a multivariate statistical problem. In data sets containing many variables, groups of variables are often inter-related. This can be explained as one variable might be measuring the same underlying principle governing the behavior of the complete system. We can exploit this redundancy by replacing a group of variables with a single new variable. The best way to achieve this is to apply Principal Component Analysis (PCA). PCA generates a new set of variables called principal components. Each principal component is a linear combination of the original variables. All the principal components are orthogonal to each other, so there is no redundancy of information (Jolliffe, 2002; François Husson, 2014; Besse, 1992).

The first principal component accounts for the most variance and therefore has the most information; the second principal component has the second best variance, and so on.

With this information, one can reduce the original data to represent the significant contrast and trends with only a few variables rather than all contained in the original data by the removal of insignificant variables for the desired contrast. Adding more dimensions do not provide any additional contrast but only increases the noise and reduces the potential contrast of the outcome.

Hierarchical clustering and dendrogram representation (Hastie, Tibshirani, & Friedman, 2009) were applied to summarize the interdistance of the PCA scores to see if there were any discrete clusters of data points in the new coordinate system and how related these were.

3. Results

The coffee samples were identified as follows, based on the labels on their respective packages: Belier and Malaga are 100% Robusta while Brasil, Columbia and Familial are 100% Arabica.

3.1 SLIPI Results

The experimental results from SLIPI measurements are presented in Figure 3 (a,b,c,d,e) for the different ground coffee and extinction coefficient and optical density obtained with the different laser (450nm, 532nm, 638nm) grouped as shown in Tables 1, 2 and 3.

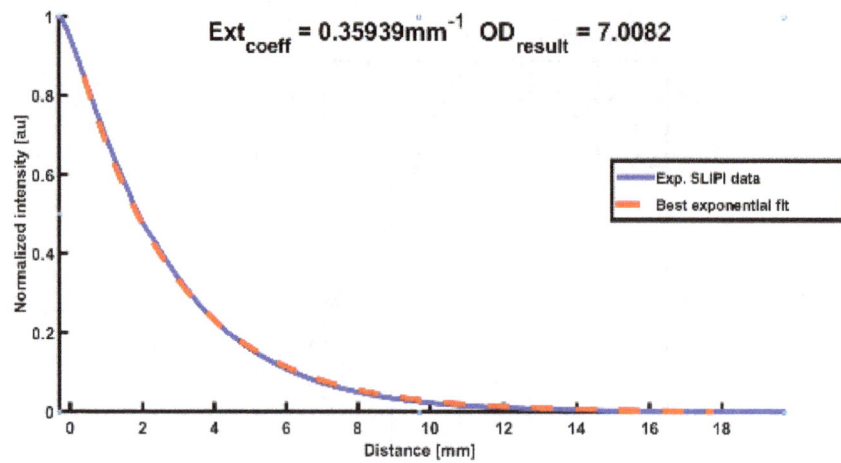

(a)

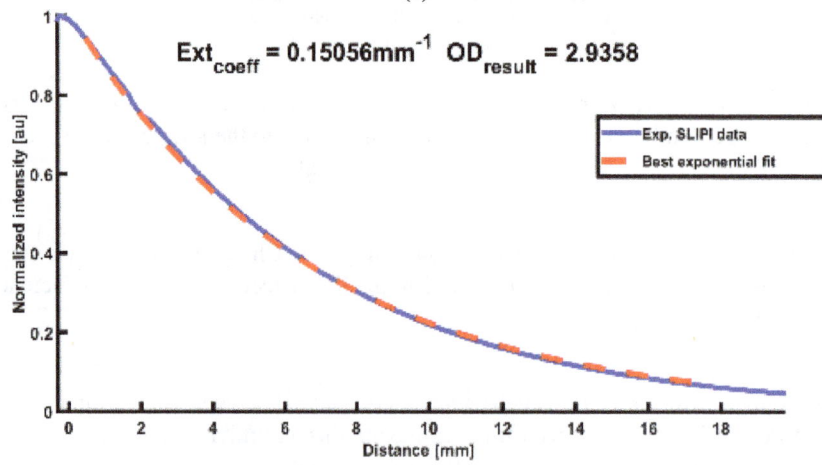

(b)

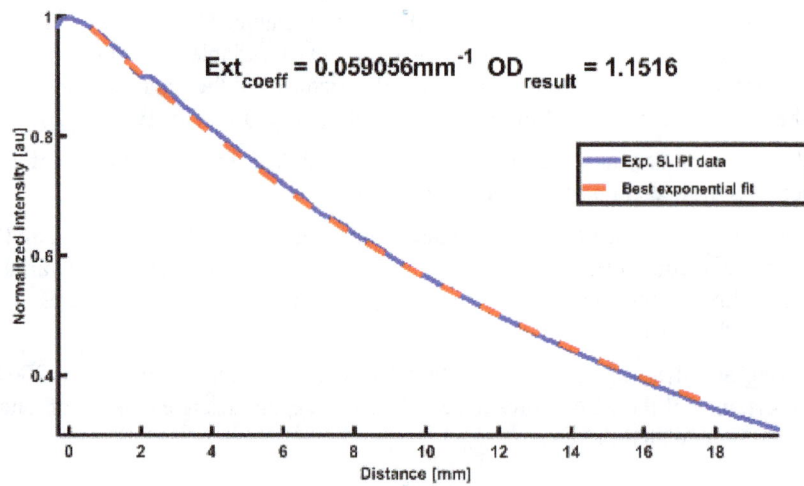

(c)

Figure 3. Experimental results of extinction coefficient and optical density from SLIPI measurements with Belier coffee at 450 nm (a), 532 nm (b) and 638 nm (c)

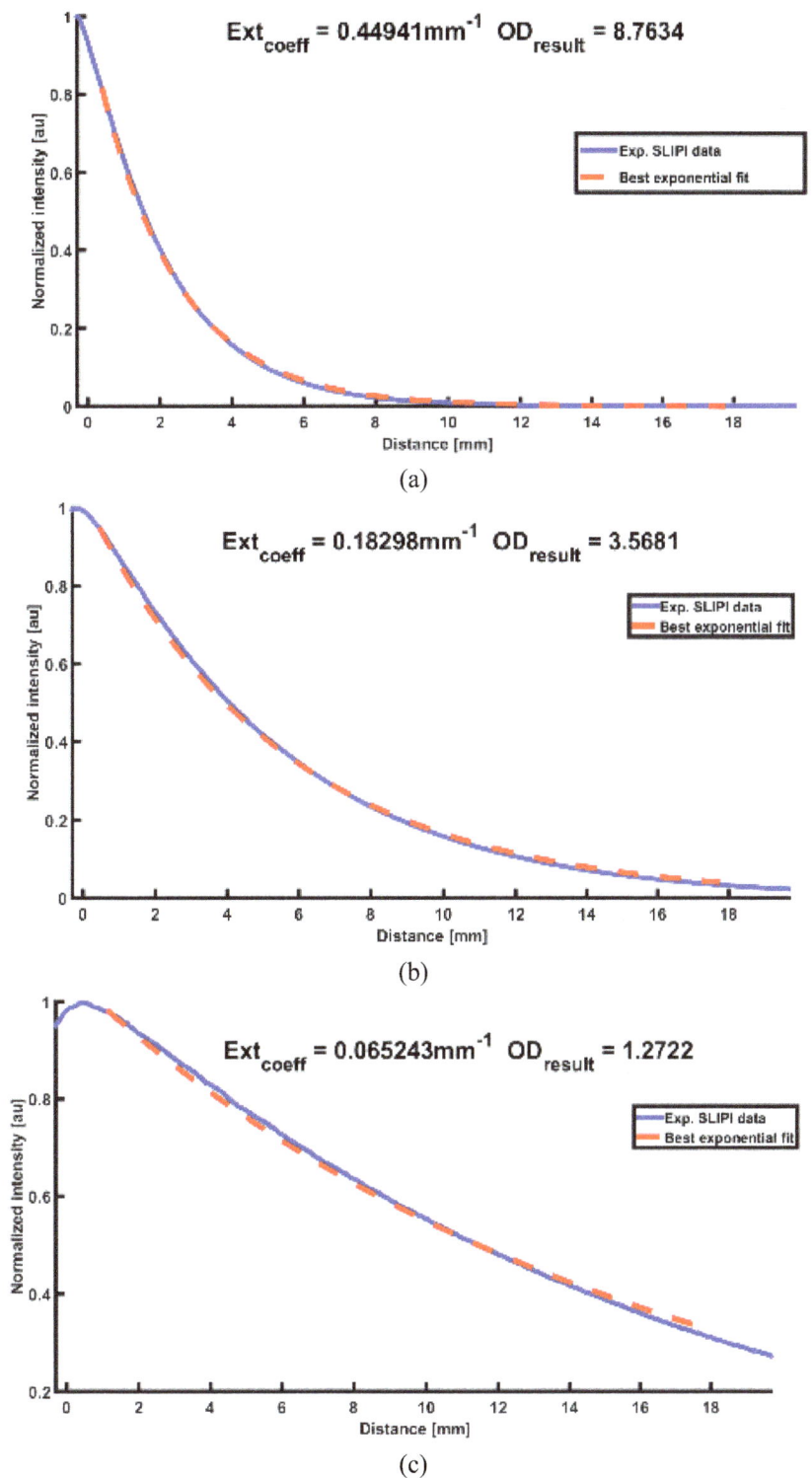

$Ext_{coeff} = 0.44941mm^{-1}$ $OD_{result} = 8.7634$

(a)

$Ext_{coeff} = 0.18298mm^{-1}$ $OD_{result} = 3.5681$

(b)

$Ext_{coeff} = 0.065243mm^{-1}$ $OD_{result} = 1.2722$

(c)

Figure 4. Experimental results of extinction coefficient and optical density from SLIPI measurements with Malaga coffee at 450 nm (a), 532 nm (b) and 638 nm (c)

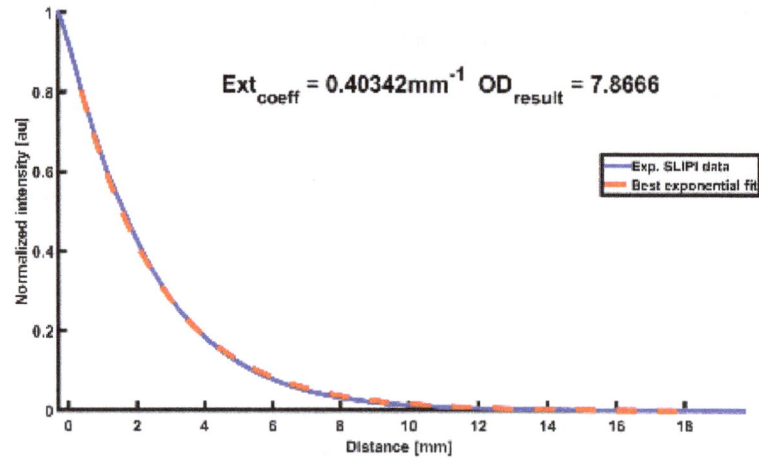

(a)

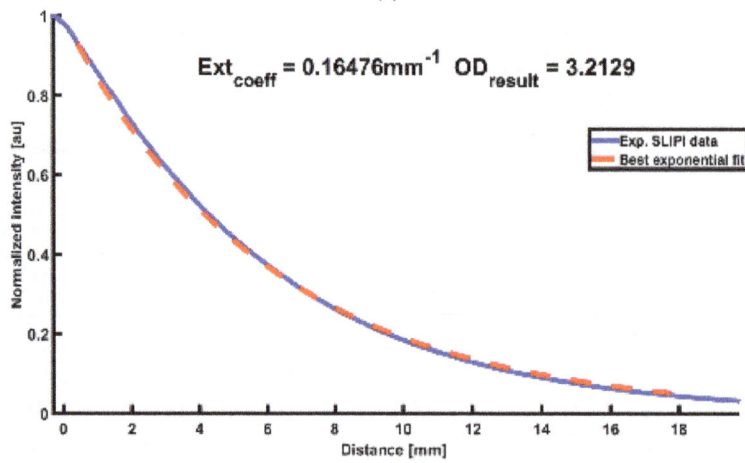

(b)

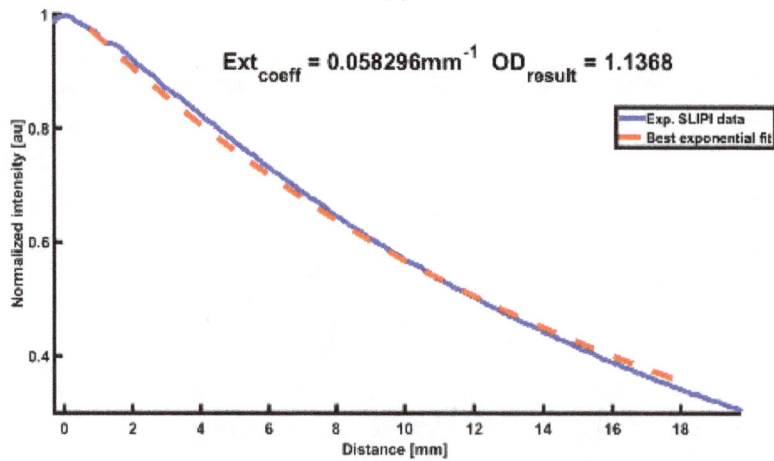

(c)

Figure 5. Experimental results of extinction coefficient and optical density from SLIPI measurements with Familial coffee at 450 nm (a), 532 nm (b) and 638 (c)

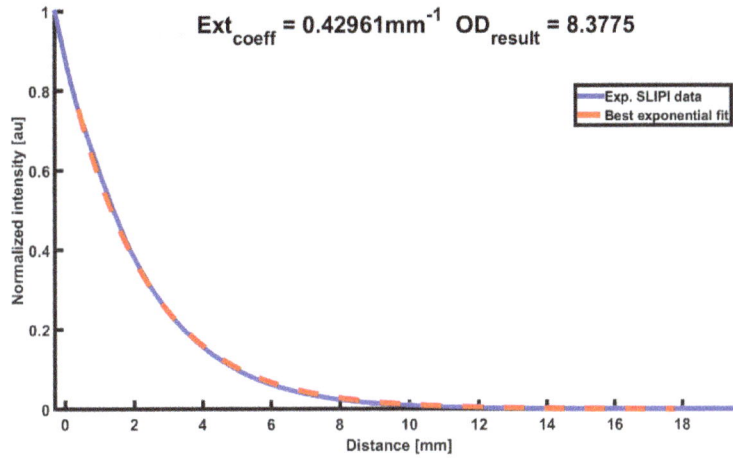

(a)

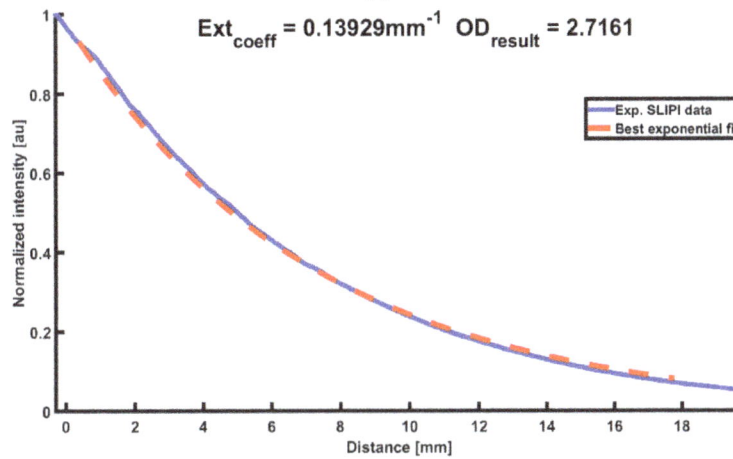

(b)

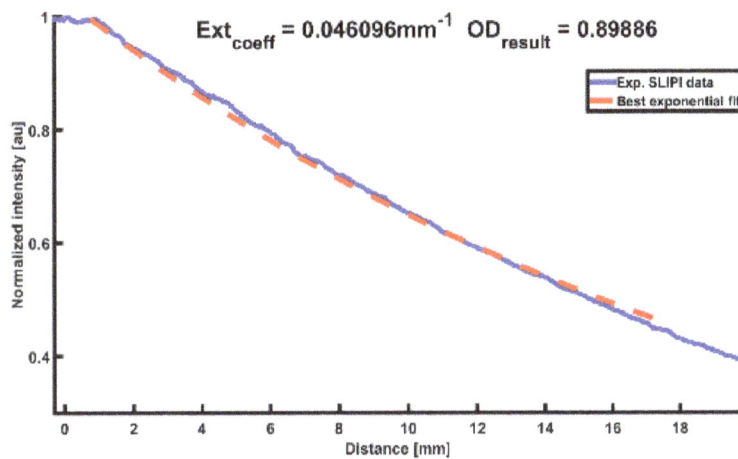

(c)

Figure 6. Experimental results of extinction coefficient and optical density from SLIPI measurements with Colombia coffee at 450 nm (a), 532 nm (b) and 638 nm (c)

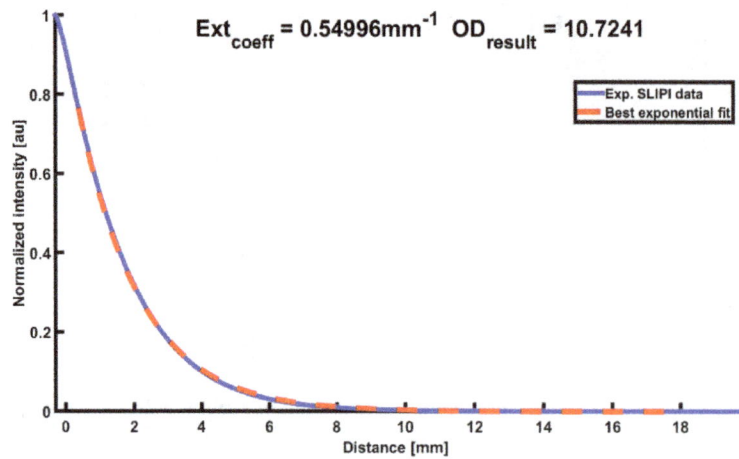

(a)

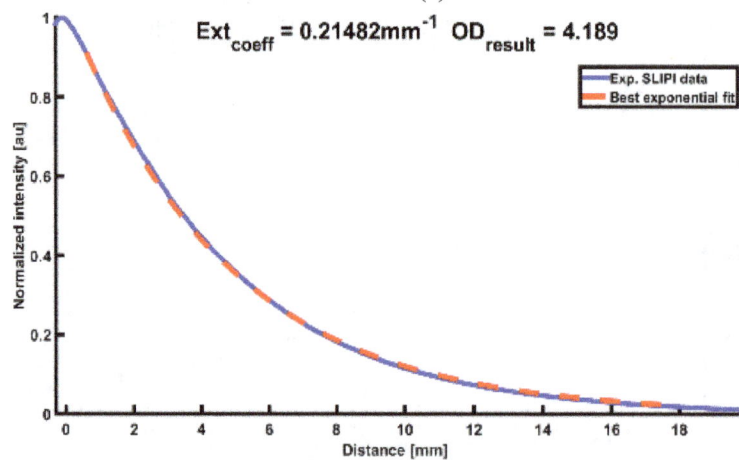

(b)

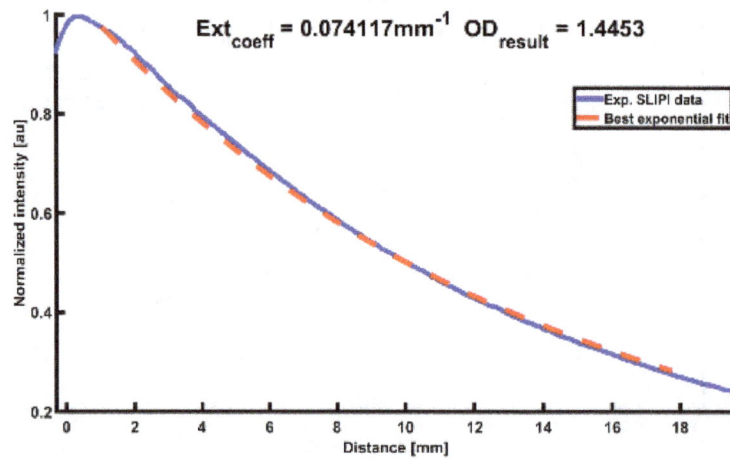

(c)

Figure 7. Experimental results of extinction coefficient and optical density from SLIPI measurements with Brasil coffee at 450 nm (a), 532 nm (b) and 638 nm (c)

Table 1. SLIPI results using laser illumination at 450 nm

	450 nm laser illumination	
	Extinction Coefficient (mm^{-1})	Optical Depth (-)
Belier	0.35939	7.0082
Malaga	0.44941	8.7634
Familial	0.40342	7.8666
Colombia	0.42961	2.7161
Brasil	0.54996	10.7241

Table 2. SLIPI results using laser illumination at 532 nm

	532 nm laser illumination	
	Extinction Coefficient (mm^{-1})	Optical Depth (-)
Belier	0.15056	2.9358
Malaga	0.18298	3.5681
Familial	0.16476	3.2129
Colombia	0.13929	0.046096
Brasil	0.21482	4.189

Table 3. SLIPI results using laser illumination at 638 nm

	638 nm laser illumination	
	Extinction Coefficient (mm^{-1})	Optical Depth (-)
Belier	0.059056	1.1516
Malaga	0.065243	1.2722
Familial	0.058296	1.1368
Colombia	0.046096	0.89886
Brasil	0.074117	1.4453

The results in all the tables show a difference between all coffees with regard to the extinction coefficient and optical density parameters. The corresponding plots of the extinction coefficients are given in Figure 8(a) together with the ratio of the extinction coefficients (for each illumination scheme) in Figure 8(b). However, it is very difficult to classify in which type of coffee they are belonging (whether Arabica or Robusta). In order to find a relevant feature to describe the classification it is very important that the variables (extinction coefficient and optical density) are independent with regard to coffee species and laser illumination. To do this we used chemometrics which gives many different ways to solve the discrimination problem in the analysis of data.

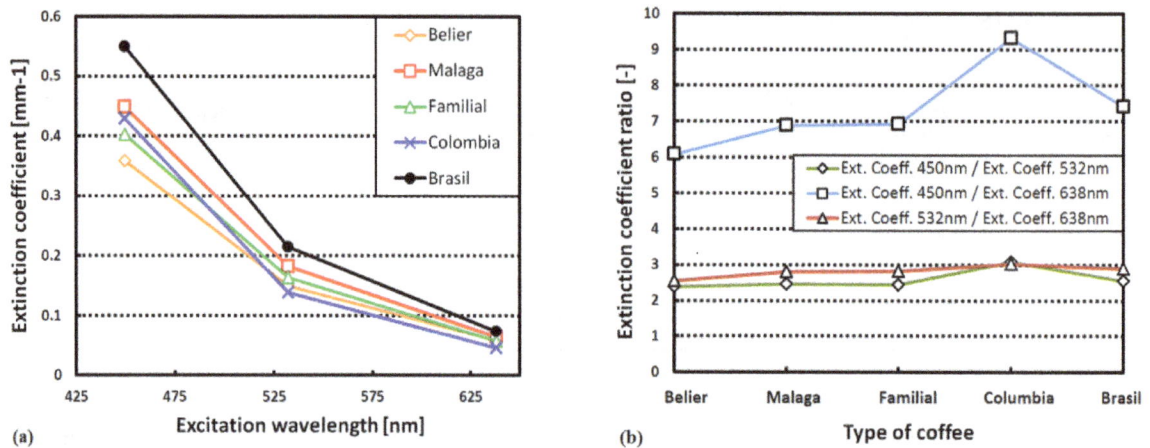

Figure 8. (a) Results of the measurement of the extinction coefficient for each type of coffees for the three illumination wavelengths. (b) Ratio of the extinction coefficients for each type of coffee

3.2 Chemometric Results

Before performing the analysis, we checked the correlation between all the variables. The correlation among some variables is as high as 60 % (Figure 8).

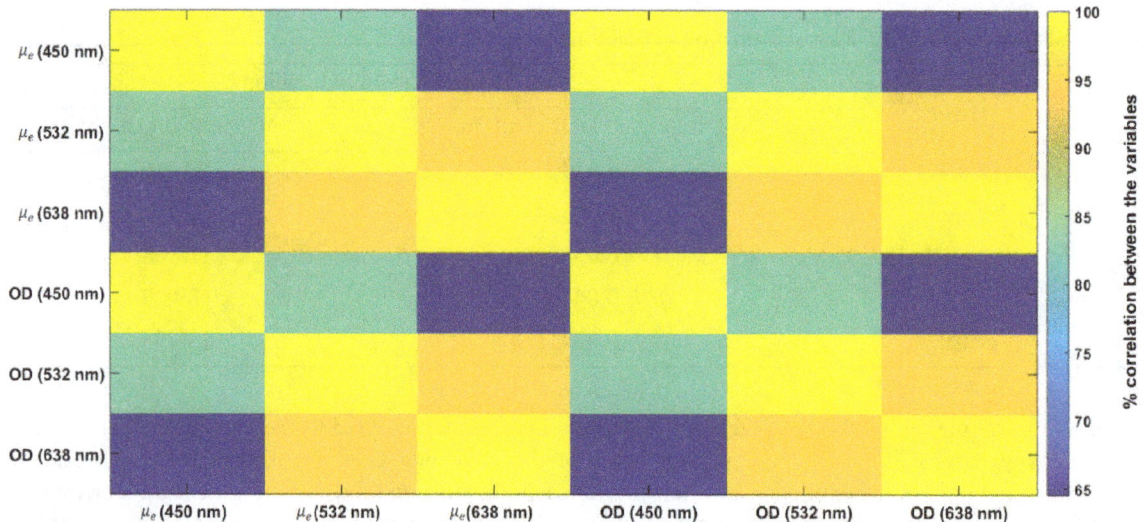

Figure 8. correlation between the variables

Note that there are a large correlation between extinction coefficient and optical density at the same wavelength. This high correlation can be explained by equation 1.

$$OD = l\mu_e \tag{1}$$

where μ_e is the extinction coefficient. The optical depth (OD) is an approximation of the mean number of scattering events occurring through a scattering medium of length l. The extinction coefficient is equal to the sum of the scattering coefficient and the absorption coefficient (Equation 2):

$$\mu_e = \mu_a + \mu_S \tag{2}$$

PCA was then applied to construct independent new variables which are linear combinations of the original variables. The variables do not have the same units so we apply PCA using the inverse variances of the data as weights. To determine which components have high variance and must be retained to describe the data, we made a scree plot of the percent variability explained by each principal component (Figure 9). The scree plot only shows the first two (instead of the total seven) components that explain 99.7% of the total variance. Thus only the first and the second principal component can be retained.

We then applied Hierarchical Clustering and Euclidean distance as a metric using these two new variables. Hierarchical Clustering groups data over a variety of scales by creating a cluster tree. We used the silhouette criteria to determine where to truncate the cluster. The silhouette value for each point is a measure of how similar that point is to points in its own cluster, when compared to points in other clusters. The silhouette value for the ith point (S_i) is defined as:

$$S_i = \frac{b_i - a_i}{\max(a_i, b_i)} \quad (3)$$

where a_i is the average distance from the ith point to the other points in the same cluster as i, and b_i is the minimum average distance from the ith point to points in a different cluster, minimized over clusters. A high silhouette value indicate that it is well-matched to its own cluster, and poorly matched to neighboring clusters (Kaufman, L., & Rousseeuw, P. J., 2009). Figure 6 show the silhouette criterion values for the number of clusters tested. The plot shows that the highest silhouette value occurs at five clusters, suggesting that the optimal number of clusters is five. After this we grouped data over a variety of scales by creating a cluster tree using HCA (Figure 10).

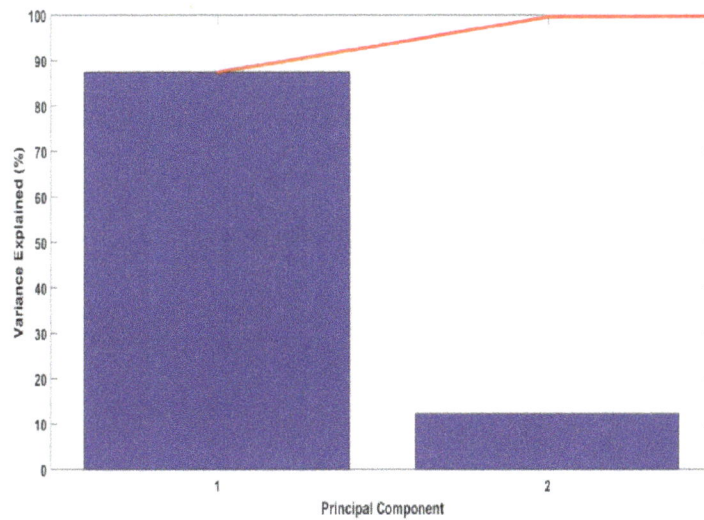

Figure 9. Scree plot of the percent variability explained by the first and second principal component

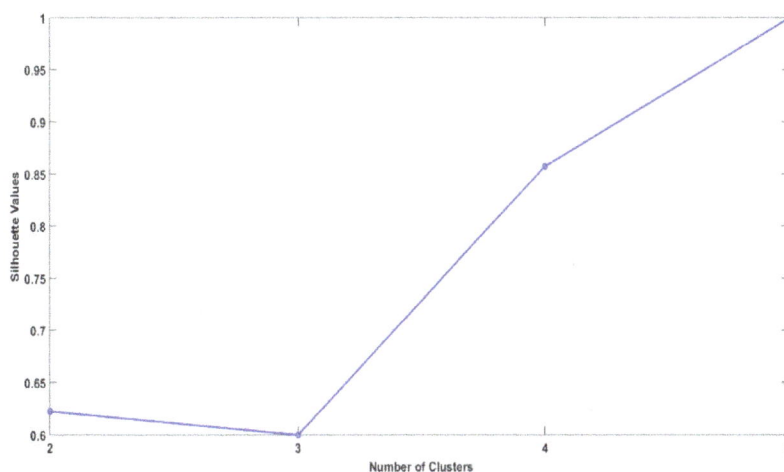

Figure 10. Silhouette values for each clusters tested

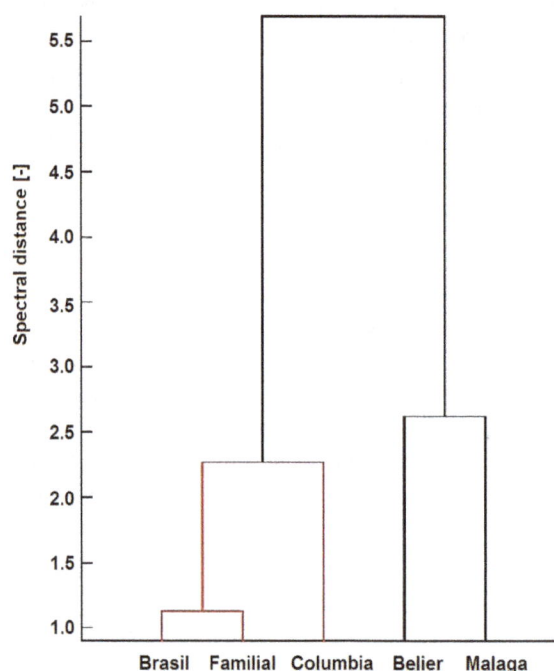

Figure 11. Representation of the hierarchical cluster tree

In the Figure 11, the numbers along the horizontal axis represent the name of the different ground coffee in the original data set. The links between grounds coffees are represented as upside-down U-shaped lines, with the height of the U indicates the distance between the objects.

Based on the silhouette criterion each coffee sample represents an independent group. The links in the dendrogram show that Brasil, Familial and Colombia are near and can form a group. In the same graph, Belier and Malaga are near and can be considered as another group. These results are in agreement with the identity of the coffee as indicated on the packaging labels. Brasil, Familial and Colombia are Arabica types and Belier and Malaga are from the Robusta variety.

4. Discussion

Chemometrics and SLIPI are both powerful techniques for spectroscopic studies; they have been used as complementary methods in this study.

The multivariate approach dealt with the following steps: pre-processing, PCA, variable selection and HCA classification. The data collected with SLIPI technique for each dataset (Belier, Malaga, Brasil, Colombia and Familial ground coffee) were used to show the suitability of this technique to detect similarity between the ground coffee samples.

For every pre-treated dataset, PCA was performed as an explanatory tool in order to get the overspread of data. PCA variance was used to retain the best components to use in describing the variability in the different coffee types and sample groupings.

The HCA plot shows a good grouping of the samples on the basis of the tow classes in the space defined by the two first components. This strategy shows that extinction coefficient and optical density measured with SLIPI technique could be useful in the discrimination of coffees species.

5. Conclusion

The strategy showed a clear coffee grouping on the basis of the tow classes (Arabica and Robusta). We can conclude that, the SLIPI technique combined with chemometric analysis of coffee samples offer complementary results for the discrimination of products and can be used to accurately classify and evaluate coffee samples.

Acknowledgement

The authors wish to thank the International Science Program (ISP) of Uppsala University for equipment and financial support as well as the Lund Laser Center (LLC).

References

Berrocal, E., Johnsson, J., Kristensson, E., & Aldén, M., (2012). Single scattering detection in turbid media using single-phase structured illumination filtering. *Journal of the European Optical Society-Rapid Publications, 7*. http://dx.doi.org/10.2971/jeos.2012.12015

Berrocal, E., Kristensson, E., Richter, M., Linne, M., & Alden, M. (2008). Application of structured illumination for multiple scattering suppression in planar laser imaging of dense sprays. *Optics express, 16*(22), 17870-17881. http://dx.doi.org/10.1364/OE.16.017870

Besse, P. (1992). PCA stability and choice of dimensionality. *Statistics & Probability Letters, 13*(5), 405-410. http://dx.doi.org/10.1016/0167-7152(92)90115-L

Carelli, M. L. C., Fahl, J. I., & Ramalho, J. D. C. (2006). Aspects of nitrogen metabolism in coffee plants. *Brazilian Journal of Plant Physiology, 18*(1), 921. http://dx.doi.org/10.4067/S0718-9516201400500 0018

Clifford, M. N., & Willson, K. C. (1985). *Coffee: Botany, Biochemistry, and Production of Beans and Beverage.* Westport, CT: AVI. http://dx.doi.org/10.1007/978-1-4615-6657-1

François HUSSON. (2014). Analyse en composantes principales (ACP) Théorie et pratique.

Hastie, T., Tibshirani, R., & Friedman, J. (2009). *Hierarchical clustering.* The elements of statistical learning, SPRINGER 2. 520-528.

International Coffee organization. (2015). *List of countries by coffee production. Wikipedia.* Retrieved January 22, 2016, from https://en.wikipedia.org/wiki/List_of_countries_by_coffee_production

Jolliffe, I. T. (2002). Principal component analysis, ser. *Springer Ser. Statist.* (2nd ed.). *New York: Springer.*

Kaufman, L., & Rousseeuw, P. J. (2009). Finding groups in data: an introduction to *cluster analysis* (Vol. 344). John Wiley & Sons.

Kristensson, E. (2012). *Structured Laser Illumination Planar Imaging, SLIPI: Applications for Spray Diagnostics.* Lund University, PhD Thesis.

Kristensson, E., Berrocal, E. & Aldén, M. (2012). Quantitative 3D imaging of scattering media using structured illumination and computed tomography. *Opt. Express, 20*, 14437-14450.

Kristensson, E., Berrocal, E., & Aldén, M. (2011). Extinction coefficient imaging of turbid media using dual structured laser illumination planar imaging. *Opt. Lett., 36*, 1656-1658.

National Coffee Association of USA. (2015). Retrieved October30, 2015, from http://www.ncausa.org/

Oder, T. (2015). How coffee changed the world. *Mother Nature Network.* Retrieved October 30, 2015, from http://www.mnn.com/food/beverages/stories/how-coffee-changed-the-world

2

Quantum Gravity as Higher Dimensional Perspective

Rob Langley[1]

[1] Norwich, UK

Correspondence: Rob Langley,   Norwich, UK. E-mail: rhlangley@hotmail.co.uk

Abstract

Although highly predictive in their respective macroscopic and microscopic domains of applicability, General Relativity and quantum mechanics are mathematically incompatible, perhaps most markedly in assumptions in their formalisms concerning the nature of space and time. In *perspective* we already have a conceptual structure that links the local, macroscopic frame and the remote, apparently microscopic frame. A mathematical principle is invoked as a natural limit on $D(n)$, so that effects which are clearly perspectival at $D = 3$ become 'more real' (*effectively* observer-independent) with each $D(n)$ increment. For instance, the apparently microscopic becomes the effectively microscopic and *scale extremes are juxtaposed*, so that black holes are local, macroscopic vanishing-points, in a similar way to that in which in projective geometry the point at infinity is incorporated into the foreground. (In other words, *a black hole is a blown-up 'Planck-scale' singularity*.) Characteristics of the earthbound frame are applied to $D > 3$, suggesting a physical basis for entanglement, and perspectival interpretations of quantum gravity, dimensional reduction and the information paradox. We claim that the familiar processes whereby multiple physical states become describable by a single state in which composition information appears to be lost (e.g., 'falling into a black hole', the state of quantum linearity, and the state of freefall) are all examples of effective convergence of a space or n-surface to a single point of perspective.

Keywords: black holes, cosmological coincidence problem, dimensional reduction, firewalls, information paradox, measurement paradox, observers, quantum gravity, time.

1. Introduction

1.1 Background

Whilst General Relativity (GR) and quantum mechanics (QM) are rigorously proven and highly predictive, their mathematical formalisms and respective macroscopic and microscopic domains of applicability have remained conceptually disparate. GR is typified by Einstein's field equation (shown with the cosmological parameter),

$$R_{ab} - \tfrac{1}{2} R g_{ab} + \lambda\, g_{\mathrm{ab}} = \tfrac{8\pi G}{c^4} T_{ab}\,, \tag{1}$$

which equates geometry and matter-energy degrees of freedom. Time is a covariantly transforming coordinate of a dynamic, arbitrarily differentiable, semi-Riemannian (asymptotically Minkowskian) spacetime manifold, and observers are incidental. The field equations are highly nonlinear, and unconstrained geodesic convergence leads to the formation of a black hole, with a central singularity at which GR breaks down. The information and firewall paradoxes also point to profound conceptual issues in our understanding of black holes.

In QM, states $|\psi\rangle$ are vectors in Hilbert space $(\mathcal{H})$ and unitary state development is typically described by the Schrödinger equation,

$$i\hbar \tfrac{\partial}{\partial t}(|\psi\rangle \boldsymbol{r}, t) = \widehat{H}(|\psi\rangle \boldsymbol{r}, t). \tag{2}$$

The linearity of $\mathcal{H}$ is such that any linear combination $\psi(\boldsymbol{r}, t) = a_1\psi_1(\boldsymbol{r}, t) + a_2\psi_2(\boldsymbol{r}, t)$ of $n \geq 2$ solutions, where $a_1\psi_1(\boldsymbol{r}, t)$ and $a_2\psi_2(\boldsymbol{r}, t)$ are complex weighted, linearly superposed alternatives, is also a solution. The Copenhagen Interpretation (CIQM) comprises two, conceptually conflicting dynamics; (i) the **U**-process, which describes the linear, unitary, continuous, deterministic, and reversible development of complex-weighted, superposed alternatives, and (ii) the **R**-process, which describes the nonlinear, nonunitary, discontinuous, nondeterministic and irreversible projection (in von Neumann's scheme) of $|\psi\rangle$ onto an eigenbasis of mutually orthogonal eigenvectors. Classical probabilities p_C are derived via the Born rule $p_C = |a|^2$ in $|\psi\rangle$-collapse to a single definite outcome, but there is no axiomatic prescription for when to invoke this second postulate. CIQM is irreducibly probabilistic, inspiring the Einstein-Bohr debates concerning EPR entangled states such as $1/\sqrt{2}\left(|0\rangle_A \otimes |1\rangle_B - |1\rangle_A \otimes |0\rangle_B\right)$, the measurement outcomes of which demonstrate statistical correlations that violate the Clauser, Horne, Shimony and Holt inequality, $\rho(a,b) + \rho(a,b') + \rho(a',b) - \rho(a',b') \leq 2$, indicating that the predictions of QM cannot be reproduced by any local hidden variables theory. In QM, time is an external, Newtonian parameter, and observers are implicated to the extent that their intervention appears to disturb deterministic state development.

1.2 Historical Background

Newton thought of gravitation as a universal force (Newton, 1687) but did not attempt to explain it. Einstein's GR (Einstein, 1915) therefore represented a significant scientific development in describing it as local curvature of the spacetime manifold. However, the subsequent, highly successful unification of the strong, weak and electromagnetic interactions as the gauge symmetries (e.g., Glashow, 1961, Weinberg, 1967, Salam, 1968) that together comprise the Standard Model (SM) of particles and interactions has motivated attempts to incorporate gravitation in terms of the exchange of gravitons, the hypothetical gauge boson of the gravity field. Since 'gravity gravitates' it is nonrenormalisable at high energies and is therefore in need of UV completion. Most approaches therefore advocate a minimum length; $\ell \sim \ell_P = \sqrt{\hbar G/c^3} = 1.62 \times 10^{-35}$m. For instance, string theory invokes a one-dimensional cut-off on a flat spacetime background, and loop quantum gravity proposes an intrinsically relativistic, granular spacetime structure. Amongst many other approaches are AdS/CFT (Maldacena, 1998), canonical quantum gravity, causal dynamical triangulation and twistor theory (see Kiefer, 2014, for a general review).

1.3 Perspective approach

The method adopted in the approach outlined in this paper is simply to examine the main characteristics of the familiar perspectival relationship between the earthbound observer and the horizon, and to note those that demonstrate conceptual similarities with mathematical issues in quantum gravity. Whilst we appear to inhabit a Universe with three spatial dimensions and one time dimension, it is noted that it also allows the freedoms to accelerate, to nonuniformly accelerate, and so on. These temporal derivatives are taken seriously as indications of higher dimensionality, so characteristics are extrapolated to $D > 3$ in order to assess conceptual fitness.

1.4 Observations

The notion of physical 'composition' (i.e., that larger things are made of smaller things) is reconsidered. In particular, it is noted;

(i) that when matter enters a *black hole*, composition information is effectively *missing*, being replaced by a black hole state defined by just electric charge (Q), mass (M) and angular momentum (J),

(ii) that since matter is subject to the universality of *freefall*, its composition information is effectively *missing*, being replaced by a single geodesic equation, $(d^2 x^i/d\lambda^2) + \sum_{j,k} \Gamma^i_{jk} (dx^j/d\lambda)(dx^k/d\lambda) = 0$ (in which λ is the affine parameter and Γ^i_{jk} are the connection coefficients), and

(iii) that when a quantum system enters a state of *linear superposition* or *entanglement*, information about the composition or configuration of the experimental set-up (e.g., which-path data) is effectively *missing*, being replaced by a single state $(|\psi\rangle)$ that seems able to describe system, measurement apparatus *and* environment.

In each case, a '**MANY**' system with multiple variables and interdependencies 'falls' into a simple '**ONE**' state defined by one or very few parameters, and possibly without composition at all.

1.5 Notation

The case in which a $D(n)$ surface is spherically closed and embedded in $\mathbb{R}^{D(n+1)}$ is denoted typically as $D(n)|D(n+1)$; for instance, the earthbound scenario is denoted D2|D3.

1.6 Structure

The paper is structured as follows. In Section 2, the main characteristics of perspective and the D2|D3 frame are explored and applied to $D > 3$, with a direct equivalence claimed at D4|D5 between relativity and quantisation. In answer to the question of whether or not application to $D > 3$ is infinitely extendable, a simple mathematical principle is invoked in Section 3 as a natural limit on $D(n)$, and the possible implications of this are discussed. In Section 4, a correspondence between perspective and gauge symmetry is proposed, including a perspectival interpretation of the Standard Model of particles and interactions and the QED sector. A definition of gravitation in terms of universal expansion is outlined in Section 5, and the notion of macroscopic charge is introduced. In Section 6 we propose perspectival interpretations of black holes, entropy and the information paradox. Time is considered as a perspectival effect in Section 7, and Section 8 looks briefly at observer degrees of freedom and includes a short summary of the perspective model.

2. Perspective and curvature

2.1 Main Characteristics

Whereas perspective in $\mathbb{R}^3$ is that illusory effect whereby the apparent height of a receding object appears to contract to a point at spatial infinity, perspective from a point of view above a surface that is closed with constant positive curvature means that a receding object will disappear over a horizon before the object's apparent height goes to zero. Using metres as units, we first consider the earthbound observer Alice, who stands with 3-height $h_3 = 2$ above the 2-surface, which is closed with constant radius of curvature, $r_2 = 6.371 \times 10^6$ (Figure 1b).

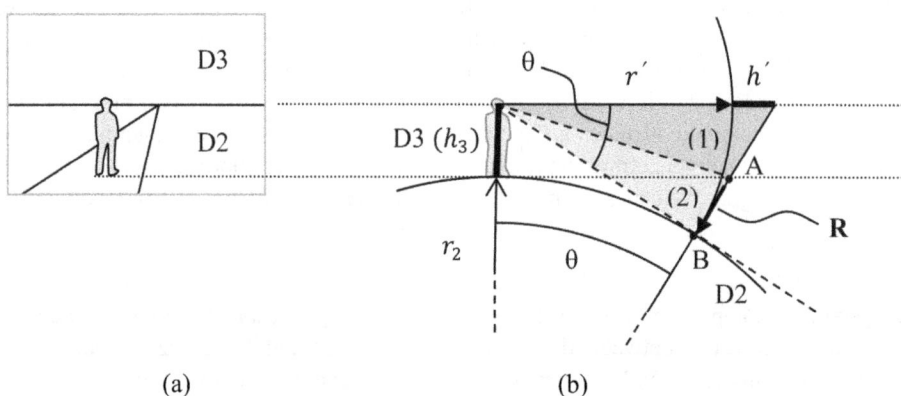

(a)   (b)

Figure 1. The familiar geometry of the earthbound (D2|D3) scenario.

The following are the basic properties of the relationship between the earthbound observer and the horizon.

1) Assuming $h_3 \lll r_2$, the 2-surface looks flat to good approximation (Figure 1a).
2) With increasing distance from Alice, there is an increasing divergence between apparent flatness of the 2-surface and actual global closure (Figure 1b).
3) Parallel lines on the 2-surface (e.g., railway lines) appear to converge to an apparent D0 point or *singularity* on the S^1 horizon (at a distance $\ell \approx 5.05 \times 10^3$).
4) With increasing distance from Alice, objects appear to **contract** and *space appears to **expand***, and Alice reasons that these rates must be exactly equivalent; **CON $\equiv$ EXP**.
5) There is a $D(n-3)$ **dimensional reduction** from Alice's 3-height to an apparent D0 point on S^1.
6) There are two such points; A, where Alice (naively) believes she is looking, and B, where she is actually looking. In 3-space, these are clearly the same point; i.e., A and B are identified (A $\equiv$ B).
7) If Alice gains sufficient 3-height, she will see the D0 *point* singularity transform or *open up* to become a D1 *line* singularity (i.e., there is now a non-zero separation, $\ell > 0$, between railway lines as they disappear over the horizon). This ***forced transition***, A $\rightarrow$ B, acts as a *gate* that reveals the embedding of $\mathcal{M}^2$ in $\mathbb{R}^3$ (i.e., the D2|D3 case).
8) The difference between apparent flatness and actual global closure is just Alice's local 3-height (h_3), ***projected*** (and rotated through $\theta = 0.045°$) to the remote, S^1 horizon. (Alice's apparent 3-height, as projected to the remote frame and seen from the local frame, is less than the thickness of the average human hair; $h_{3(LOC)} \rightarrow (h_{3(REM)} \approx 6.3 \times 10^{-5})$.)
9) The angle $\theta = 0.045°$ subtended at Alice's eye between the *true* horizontal line of sight (i.e., tangential at a radius of $r_2 + h_3$) and a line of sight to the actual horizon is equal to that subtended at the centre of the Earth between radial lines to the locations of Alice and the horizon, so *the relationship (and transition) between points A and B is proportional to and associated with those between the local observer and the remote horizon.*
10) The transition $\mathbf{R} = (A \rightarrow B)$ is also that from triangle (1) with $a^2 = b^2 + c^2 + 2\cos\theta$, to triangle (2), with $a^2 = b^2 + c^2$, so the transition A $\rightarrow$ B is associated with the loss of the $2\cos\theta$ term.

11) **Observer-dependency of the S^1 horizon can be expressed as the fact that 2-distance to the S^1 horizon is *a constant in every frame*.**

12) The h_3 of an object that disappears over S^1 does not cease to exist but is 'conserved in 3-space'.

2.2 Scenarios

2.2.1 The D2|D3 Scenario

Alice and Bob are standing on Earth. As Bob stands close to Alice, his 3-height (h_3) is at its maximum value, but as Bob recedes from Alice, his apparent height in Alice's frame is $h_{3(B)app.} \rightarrow \sigma_3$, where $\sigma_3 > 0$ is an effective minimum (as they stand at each other's curvature limit) before Bob disappears over the S^1 horizon in Alice's frame. As noted above, the rate at which Bob appears to *contract* with increasing distance from Alice's location must be exactly the same as that with which space appears to *expand*; **CON $\equiv$ EXP**. We also note that at D2|D3, observer-dependency of the S^1 planetary horizon is just the fact that 2-distance to the S^1 horizon is a constant in every frame.

2.2.2 The D3|D4 Scenario

We assume that the 3-dimensional hypersurface $\mathcal{M}^3$ is spherically closed (Einstein, 1923), and that 4-height (h_4) is *velocity* (ds/dt), rather than the Minkowski spacetime interval, $\Delta s^2 = \Delta x^2 + \Delta y^2 + \Delta z^2 - (c\Delta t)^2$. Also assuming $h_4 \lll r_3$, the 3-manifold $\mathcal{M}^3$ appears to be flat, but as Bob recedes from Alice, his apparent velocity in her frame is $h_{4(B)app.} \rightarrow \sigma_4$, where $\sigma_4 > 0$ is some irreducible constant (since $\sigma_4 = 0$ would represent a preferred rest frame). The respective local and remote curvature limits are taken to be a maximum velocity, $v_{max} = c = 3 \times 10^6 \text{ms}^{-1}$ and a minimum *energy*, $E_{min} = hv, h/2\pi = 1.055 \times 10^{-34}\text{m}^2$ kg/s; i.e., the characteristic constants of GR and QM. At D3|D4, apparent expansion of space with increasing 3-distance is interpreted as ***universal expansion with constant velocity*** (Lemaître, 1927), (Hubble, 1929), and observer-dependency of the S^2 cosmological horizon is just the fact that 3-distance to the S^2 horizon is a constant in every frame.

2.2.3 The D4|D5 Scenario

Again, we take the 4-manifold $\mathcal{M}^4$ to possess constant positive curvature, and 5-height (h_5) to be *acceleration* (d^2s/dt^2). Assuming $h_5 \lll r_4$, the 4-surface *feels flat*; in other words, Alice does not detect higher curvature (i.e., she never experiences her own mass). As Bob recedes from Alice, freefalling towards the curvature limit (the S^3 black hole event horizon), his apparent acceleration $(h_{5(B)app.})$ decreases until he appears to hover at $r = 2GM$, as the photons comprising his image redshift to infinity. However, $\sigma_5 = 0$ would represent a preferred *velocity* frame, which is precluded in GR since it is impossible to say that a body is not being gravitationally accelerated. Therefore, $h_{5(B)app.} \rightarrow (\sigma_5 > 0)$, providing an IR cut-off. The respective local and remote curvature limits are taken to be a *maximum* acceleration, $a_{max} = \alpha$ (where $\omega_\alpha \approx \ell_P$), and a *minimum* acceleration, $a_{min} = \lambda$ (where $\lambda = 1 \times 10^{-52}\text{m}^{-2}$ is the *cosmological constant*) (Note 1.). At D4|D5, apparent expansion with increasing 4-distance from Alice is interpreted as ***accelerating universal expansion*** (Reiss et al,1998), (Perlmutter et al, 1999). We also note that at D4|D5, ***observer-dependency of the S^3 black hole event horizon is just the fact that 4-distance (i.e., $v = c$) to S^3 is a constant in every frame***. In other words, we have derived Lorentz invariance, in which frames differ by the factor, $\gamma = 1/\sqrt{1 - (v^2/c^2)}$. At the microscopic scale, *constant 4-distance to the S^3 horizon is interpreted as the quantum*; the irreducible unit that prevents the electron spiralling into the atomic nucleus. Therefore *relativity at the macroscopic level is equivalent to quantisation at the microscopic level*, so we get;

$$\textbf{RELATIVITY} = \textbf{QUANTISATION}. \qquad (3)$$

We next explore whether this sequence is infinitely extendable.

3. Limit Principle

We first specify two references, s_1 and s_2. At step 1, we specify a new reference representing the *difference* between them, $s_1 - s_2$, and at step 2, using the new reference, we then specify a further reference representing the difference, $s_1 - (s_1 - s_2)$. At the third iteration, the difference becomes *detached* from the original references; $(s_1 - (s_1 - s_2)) - (s_1 - s_2)$; i.e., s_1 only appears in brackets after this level (Figure 2).

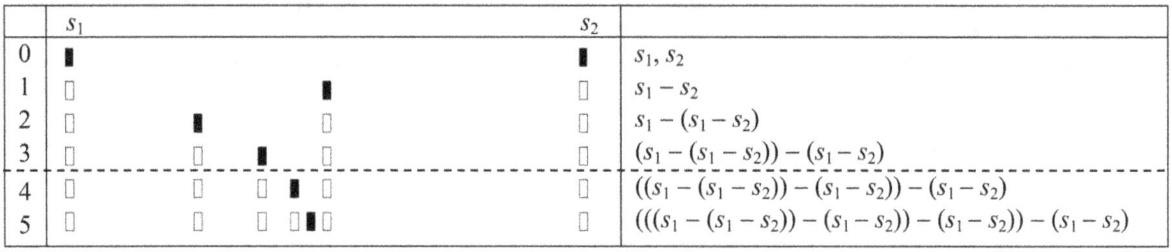

	s_1						s_2	
0	■						■	s_1, s_2
1	▢				■		▢	$s_1 - s_2$
2	▢		■		▢		▢	$s_1 - (s_1 - s_2)$
3	▢		▢	■	▢		▢	$(s_1 - (s_1 - s_2)) - (s_1 - s_2)$
4	▢		▢	▢ ■ ▢			▢	$((s_1 - (s_1 - s_2)) - (s_1 - s_2)) - (s_1 - s_2)$
5	▢		▢	▢ ▢■▢			▢	$(((s_1 - (s_1 - s_2)) - (s_1 - s_2)) - (s_1 - s_2)) - (s_1 - s_2)$

Figure 2. At step 3, the new reference is detached from the original references

We now interpret references s_1 and s_2 as *locations*, so that the difference $s_1 - s_2$ is then a *change* of location, or *velocity* (ds/dt), the difference $s_1 - (s_1 - s_2)$ represents a *change* of velocity, or *acceleration* (d^2t/dt^2), and the difference $(s_1 - (s_1 - s_2)) - (s_1 - s_2)$ corresponds to a *change* of acceleration, or nonuniform acceleration (d^3t/dt^3). Of critical noteworthiness is the fact that from this level onwards, there is no further reference to the original locations; i.e., *there is no longer a physical reference or datum against which to distinguish higher derivatives* (Figure 3).

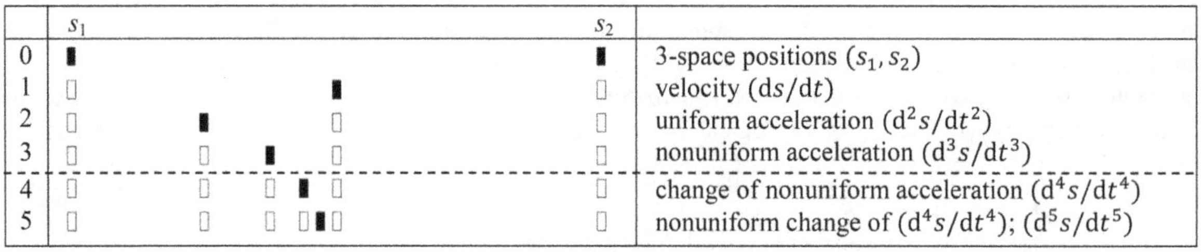

	s_1						s_2	
0	■						■	3-space positions (s_1, s_2)
1	▢				■		▢	velocity (ds/dt)
2	▢		■		▢		▢	uniform acceleration (d^2s/dt^2)
3	▢		▢	■	▢		▢	nonuniform acceleration (d^3s/dt^3)
4	▢		▢	▢ ■ ▢			▢	change of nonuniform acceleration (d^4s/dt^4)
5	▢		▢	▢ ▢■▢			▢	nonuniform change of (d^4s/dt^4); (d^5s/dt^5)

Figure 3. Beyond step 3, detachment from original references implies indistinguishability of derivatives

3.1 Diminishing Freedoms

The limit principle implies the existence of a maximum physical reality, and therefore that effects which are clearly perspectival at D3 become 'more real' (effectively observer-independent) with each $D(n)$ increment. Alice therefore possesses diminishing freedoms with respect to 'effective physical reality'. For this reason we define dimensionality in terms of freedoms. Euclidean 3-space ($\mathbb{R}^3$) may be defined in terms of Alice's freedom to orientate according to the set of axes $\{x, y, z\}$. In $\mathbb{M}^{3,1}$ Alice's freedom is curtailed; whilst she can orientate in $\{x, y, z, t\}$ to the extent that relativistic velocity can be thought of as rotation in spacetime (Penrose, 1959), (Terrell, 1959), she is now subject to time. A five-dimensional continuum ($\mathbb{M}^{3,1,1}$) may be defined in term of Alice's freedom to *accelerate*, but when she does she elicits an inertial reaction and is subject to gravitation. A similar diminishing of observer freedom can be seen from a brief study of horizon characteristics. At D2|D3, the S^1 planetary horizon is completely observer-dependent, moving with Alice's change of location. At D3|D4, the S^2 cosmological horizon is semi-observer-dependent, moving with Alice's change of velocity, but implying loss of causal contact, and therefore being semi-permeable. At D4|D5, the S^3 black hole event horizon appears to Alice in Minkowski coordinates to be a completely observer-independent, strictly one-way membrane. To further illustrate this general process, we note that the separation of points A and B (which is clearly illusory in $\mathbb{R}^3$) becomes increasingly (effectively) 'physically real' with each $D(n)$ increment, until at the highest level, the identification $A \equiv B$ may be interpreted as the physical underpinning of entanglement.

3.2 Observations and Claims

Based on our consideration of the limit principle, we make the following observations and claims regarding gravitation, mass and time.

(i) Given a massive central body and gravitationally bound satellite on various orbits, gravitation *cancels* the pseudoforces $(\vec{F}_{pseudo}^{(d^n s/dt^n)})$ that would otherwise be experienced by that satellite if it was forced to follow identical trajectories in free (sourceless) space, with no known or expected limit to $d^n s/dt^n$ as

$n \to \infty$. For instance, a body forced to trace a circular path would experience constant centrifugal acceleration (d^2s/dt^2), a body forced to trace an elliptical path would experience nonuniform acceleration (d^3s/dt^3), a body forced to trace an elliptical path with perihelion advance would experience a change of nonuniform acceleration (d^4s/dt^4), and so on.

(i) It follows that Newton's second law, $\vec{F} = m\boldsymbol{a}$, generalised as $\vec{F}(d^ns/dt^n) = m(d^ns/dt^n)$, also has no known or expected limit as $n \to \infty$. In accordance with his third law (the action-reaction principle), the reaction force must be equal to any derivative of force applied; $\vec{F}(d^ns/dt^n)_{ACT.} \equiv \vec{F}(d^ns/dt^n)_{REACT.}$

(ii) Derivatives of 3-space position become indistinguishable at defined thresholds, as demonstrated by the following simple, testable thought-experiments. Firstly, a passenger in a very powerful car is prevented from detecting motion except by the experience of pseudoforces felt against the back of the seat. S/he will be able to distinguish between constant velocity (ds/dt) and constant acceleration (d^2s/dt^2) (i.e., increased but constant effective mass, or m_E), and between constant acceleration (d^2s/dt^2) and nonuniform acceleration (d^3s/dt^3) (i.e., between increased but constant m_E and uniformly increasing m_E, or $(d/dt)m_E$), but will *not* be not able to distinguish between nonuniform acceleration (d^3s/dt^3) and a change of nonuniform acceleration (d^4s/dt^4) (i.e., between $(d/dt)m_E$ and $(d^2s/dt^2)\,m_E$). (Defining dimensionality in terms of freedoms, the boundary D6|D7 is therefore undetectable.) Secondly, we note that on a graph plotting time $(x = t)$ against velocity $(y = v)$ (Figure 4), (ds/dt) is a horizontal line at $y > 0$, (d^2s/dt^2) is diagonal, and (d^3s/dt^3) is a parabola, the gradient of which is of course nowhere vertical. No higher derivative of motion can therefore be represented on this graph; since a parabola is an infinitely extended ellipse, which is an extended circle, any tighter curve is effectively a spiral. A computer with input readings of pressure (necessarily compared to functions held in memory) would be able to detect the difference between nonuniform acceleration (d^3t/dt^3) and a change of nonuniform acceleration (d^4t/dt^4) (i.e., between a circle and spiral, or equivalently between D6 and D7), but *not* between (d^4t/dt^4) and (d^5t/dt^5), since all tighter curves are just varying classes of spiral. In other words, *detection of the D7|D8 boundary is noncomputable*.

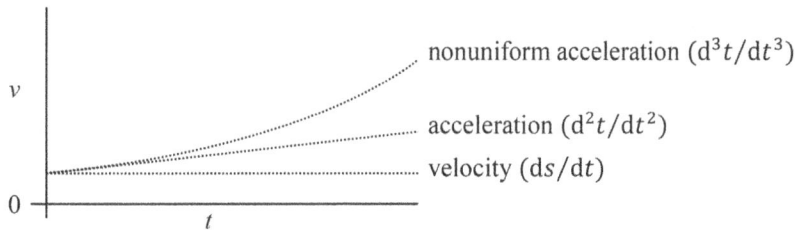

Figure 4.

3.3 Implications of the Limit Principle

The conclusions that may be drawn from the limit principle are;

(i) that effects that are clearly perspectival at D3 become 'more real' (*effectively* observer-independent) with each D(n) increment,

(ii) that *the Universe only differentiates twice with respect to time*,

(iii) that therefore there exists a *maximum physical reality* at D6,

(iv) that the Universe runs out of physical references with which to distinguish derivatives; i.e., it does not distinguish beyond this level, and so *the notion of physicality is therefore not valid at* D($n > 6$),

(v) that D7 is *non-physical*, and corresponds to the *mathematical* realm, and

(vi) that D8 is associated with *observer degrees of freedom* (i.e., the D7|D8 boundary is noncomputable).

4. Perspective-Gauge Equivalence

4.1 Perspective as Gauge

We imagine astronauts Alice and Bob, who have no prior understanding of space or perspective, moving apart in free 3-space. They disagree on who is contracting, and theorise the existence of an embedding 3-space $\mathbb{R}^3$ to account for this. Now on the Earth's surface, the difference between apparent flatness and actual global closure in

Alice's frame is just her 3-height (h_3) projected (and rotated through θ) to the remote S^1 horizon, where it is the collapse vector $\mathbf{R}$ (Figure 1b), and if Alice and Bob stand at each other's curvature limit (horizons) they will now disagree on contraction *and* phase. Taking the 2-surface as the base space $\mathcal{M}^2$, then the 3-height h_3 is the fibre, $\mathcal{V}$, representing an internal space at each point on $\mathcal{M}^2$, generating a fibre bundle, $\mathcal{B}$, with $\mathcal{B} = \mathcal{M} \times \mathcal{V}$. The *perspective* field comprises gauge connections that correspond to the group $\mathcal{G}$ of continuous symmetries of $\mathcal{B}$, and in that there is a canonical projection, submersion or surjective map, $\pi \colon \mathcal{B} \to \mathcal{M}$, then the phase ($\theta$) is a continuous image of $\mathcal{M}$ in $\mathcal{B}$. Alice and Bob then theorise the existence of $\mathbb{R}^3$ plus curvature of $\mathcal{M}^2$, in which case, the collapse vector $\mathbf{R}$ is the *gauge boson* or *field quantum* of the theory; i.e., the coupling constant or interaction strength (curvature) that translates between frames and restores *local* gauge invariance.

4.2 Gauge as Perspective: the Standard Model

The Standard Model (SM) of particle fields and interactions is described mathematically as the non-Abelian gauge group $SU(3) \times SU(2) \times U(1)/Z_6$, and physically as a sequence of early-epoch phase transitions in which the fundamental interactions split off from a primordial, unified force following a hot Big Bang at $t = 0$, some 13.82 billion years ago. In the perspective approach, this structure is thought of as a series of global, spherical closures, each generating a characteristic fermion; e.g., closure of D0 generates closed 0-forns (1-dimensional strings) that live on D1, closure of D2 generates closed 1-forms (2-dimensional quarks) that live on D2, etc. This forms a set of concentric, nested *n*-spheres, or 'onion' layers, in which a nucleus is composed of hadrons, which are composed of quarks, etc (Figure 5). In this approach, each composite form is therefore a '**MANY**' system, in which 'larger' *n*-forms are thought of as being composed of 'smaller' *n*-forms.

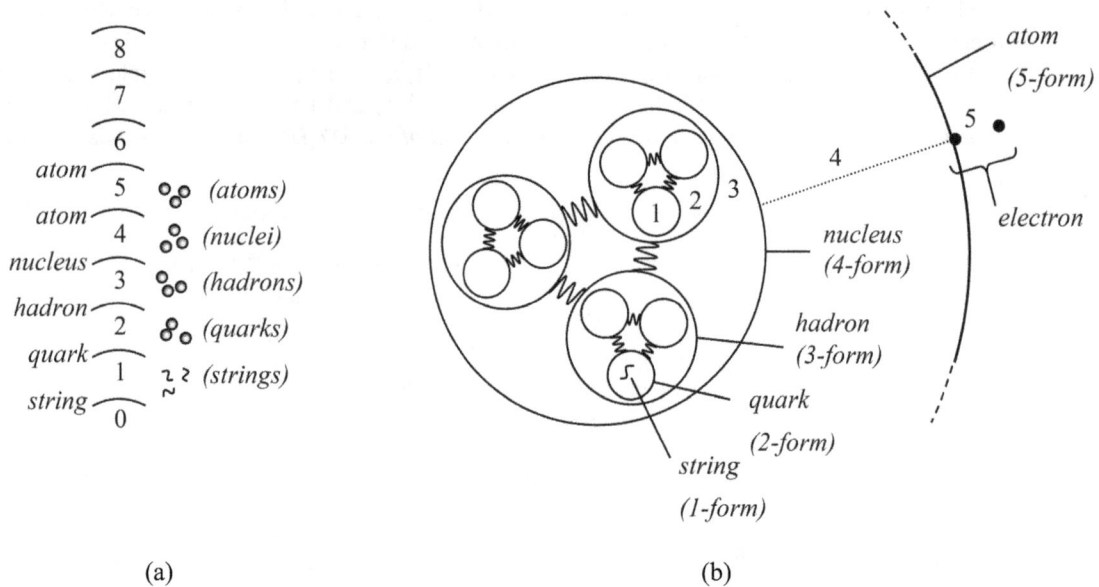

Figure 5. Onion-skin interpretation of composite particles' lower internal spaces

The structure can also be seen as two sets of D($n-3$) reductions; (i) $D(4 \leq n \leq 8) \xrightarrow{D(n-3)} D(1 \leq n \leq 5)$, from each *n*-height to a Planck-dimension limit point of D($n-3$) perspectival convergence (or fermion), and (ii) $D5 \xrightarrow{D(n-3)} D2$, which juxtaposes leptons with quarks. The *n*-height, projected into the remote frame, corresponds to the coupling constant (the curvature of the gauge connection). This is the associated boson, which in the perspective model is just the collapse vector $\mathbf{R}_{D(n)}$ (see Figure 1b), as depicted in Figure 6. In this approach, larger *n*-forms are *not composite*, but are instead, single '**ONE**' states (i.e., ℓ_P^n perspectival convergence limits).

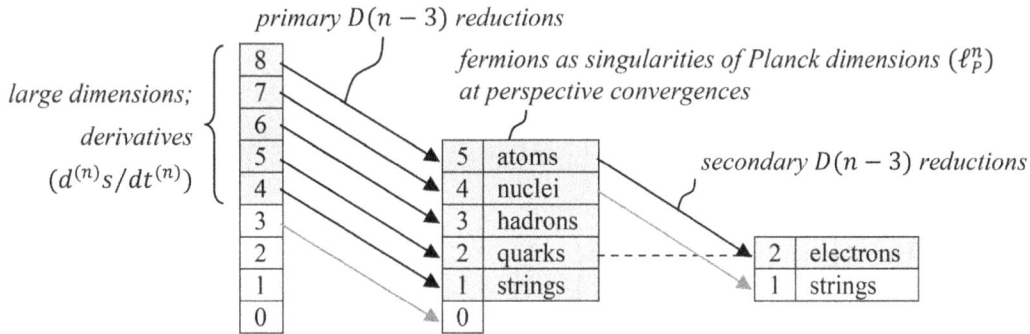

Figure 6. Two groups of $D(n-3)$ reductions form particles as perspectival convergence limit points, juxtaposing quarks and leptons (dashed line)

4.2.1 Quarkworld

Each particle splits the $D(0 \leq n \leq 8)$ spectrum into lower and higher internal spaces. For instance, a quark is a closed 1-form (i.e., a 2-form) that lives on $\mathbb{R}^2$, the spectrum being $[0)1)\mathbf{2}(3(4(5(6(7(8]$, where brackets denote spherical closures. We may then define quark orientation on $\mathbb{R}^2$ as a superposition of generators labelled $\{R, G\}$. Although the quark possesses 3-height (h_3) above the closed 2-surface, since $h_3 \lll r_2$, the 2-surface 'looks flat' (i.e., appears to be $\mathbb{R}^2$) and labelling is therefore arbitrary (*globally* gauge invariant) in the quark's frame. However, since D2 is spherically closed, the third generator $\{B\}$ corresponds to the *absolute* $\mathbb{R}^3$ reference frame; i.e., the SU(3)-colour embedding space. The gauge field or Ehresmann connection is mediated by eight gluons, the basis states of the Lie algebra that transform in the adjoint representation of SU(3), which is 8-dimensional, thereby excluding the outlawed colour singlet state, $(r\bar{r}+g\bar{g}+b\bar{b})/\sqrt{3}$. The exchange of a gluon represents the *absolute* $\{R, G, B\}$ orientation or rotation of quarks with respect to $\mathbb{R}^3$, restoring symmetry via *local* gauge invariance.

The 2-surface that quarks inhabit is the interior of a hadron, the closed 2-form (i.e., 3-form) that splits the spectrum as $[0)1)2)\mathbf{3}(4(5(6(7(8]$. There is a $D(n-3)$ projection from the quark's 3-height to an apparent D0 point on the S^1 boundary of 'Quarkworld' (in effect, point A in Figure 1b). A cyclic oscillation between points A and B (D0 and D1) corresponds to an 'inside-out hadron', which we interpret as a Quarkworld 'black hole'. As quarks fall towards it, the SU(3)-colour charge they carry (corresponding to 3-height, h_3) appears to decrease as $r \to max$ at S^1 and finally disappears. Thus SU(3)-colour bleaches out of Quarkworld with the formation of colour-neutral hadrons.

4.2.2 Mixing-space

Assuming $h_5 \lll r_4$, the 4-surface is defined by four directions or generators that appear to be gauge-equivalent (i.e., labelling is arbitrary). However, since $h_5 > 0$, the 4-surface is spherically closed and orientation is no longer gauge-invariant. In the SM this is the spontaneous electroweak symmetry breaking (EWSB) mechanism from $SU(2) \times U(1)_Y$ to $U(1)_{em}$, whereby the $W^\pm$ and Z^0 IVBs acquire mass. As part of the same process, charge associated with $U(1) = \{e^{i\theta}\}$ is given by the Gell-Mann-Nishijima formula, $Q = Y/2 + I_3$ (where Y is weak hypercharge relating baryon number and strangeness, and I_3 is the third component of weak isospin, in which $SU(2)$ is defined by the Pauli matrices, σ_j). In the perspective model, each (non-identical) generator then produces its own copy or *family* of 3-space, the directions of which correspond to *generations*. The superposition of directions (generators) at higher $D(n)$ are rotations that transform lower $D(n)$ particles (see Figure 7). In terms of the SM, this is to say that the Cabibbo-Kobayashi-Maskawa (CKM) matrix describes expectation-values for transitions between families and generations of quarks (with the Cabibbo angle corresponding to probabilities of decay modes), and the Pontecorvo-Maki-Nakagawa-Sakata (PMNS) matrix summarises weightings for lepton-mixing.

D4 (4)		W^+			W^-			Z°			γ	

D3 (12)	I	II	III	I	II	III	I	II	III	I	II	III
	u	c	t	d	s	b	ν_{e-}	$\nu_{\mu-}$	$\nu_{\tau-}$	e^-	μ^-	τ^-

D2 (24)												
	$u,\bar{u}$	$c,\bar{c}$	$t,\bar{t}$	$d,\bar{d}$	$s,\bar{s}$	$b,\bar{b}$	$\nu_e,\bar{\nu}_e$	$\nu_\mu,\bar{\nu}_\mu$	$\nu_\tau,\bar{\nu}_\tau$	e^-,e^+	μ^-,μ^+	τ^-,τ^+

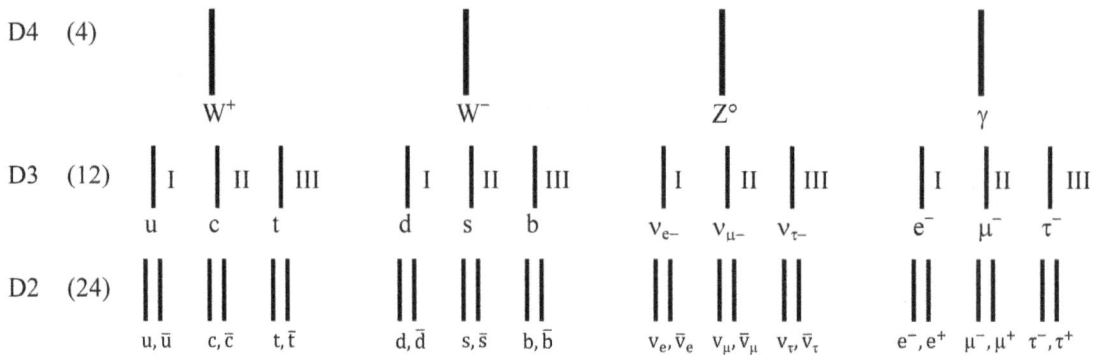

Figure 7. $D(n)$ directions are *particles* at $D > n$, and IVBs at $D > n$ are *fermionic* generations at $D < n$. In this progression, *space* at $D(n)$ is a lower internal space of a *fermion* at $D(n + 1)$

4.3 QED

In the perspective model, the atomic nucleus is a closed 3-form (a 4-form) on $\mathcal{M}^4$ that possesses 5-height (h_5) above the 4-surface. From h_5 there is a $D(n-3)$ projection to an apparent D2 (*area*) singularity (ℓ_P^2) on the S^3 boundary separating D4|D5. This is point A in Figure 1b, and the *oscillation* between points A and B corresponds in the perspective model to the coupling between the unprimed and primed Dirac spinors of the Penrose zigzag electron, $|\psi\rangle_{e-} = (\alpha_A, \beta_{A'})$ (Penrose, 2004). In this picture, each component acts as a source for the other, with the 'coupling constant' $2^{-1/2}M$ representing the interaction strength between the spinors, $\nabla_B^A{}'\alpha_A = 2^{-1/2}M\beta_{B'}$, $\nabla_A^B{}'\beta_{B'} = 2^{-1/2}M\alpha_{A'}$, in which $\nabla_B^A{}'$ and $\nabla_A^B{}'$ are 2-spinor translations of the gradient operator ∇. Each component has $m = 0$ and $v = c$, but the electron as a whole has $m > 0$ and a drift velocity, $v < c$. In the perspective model, the oscillation between spinors is just the continual cycle of divergence of apparent flatness from actual global closure and subsequent collapse.

Maximum infilling of electron shells is defined by the Aufbau Principle, with $N_{(e^-)} = 2, 8, 18, 32, 50 \ldots 2n^2$, repeating the periodicity of blocks 2, 6, 10, 14. However, no known element has more than 32 electrons in any one shell, and there is no stable element beyond period E_7 in the Periodic Table. In the perspective model this is due to the limit principle, in which, by the third iteration, the new reference becomes detached from the original physical references, introducing partial indistinguishability beyond this level. Spectral lines (energy levels) become completely indistinguishable at the ionisation level (calculation of orbitals reaches a threshold of prohibitive computational complexity), beyond which lies the continuum, as $n_E \to \infty$. We thus have a picture of the atom as nucleus surrounded by multiple S^3 horizons, which, due to $h_5 \lll r_4$, is a quantum superposition, $|\psi\rangle_{atom} = c_1\psi_1 + c_2\psi_2 + c_3\psi_3 + \ldots + c_n\psi_n$, in which $\psi_1, \psi_2, \psi_3, \ldots \psi_n$ are the eigenstates of an observable (e.g., the principle quantum number, n) corresponding to eigenvalues $a_1, a_2, a_3, \ldots a_n$. The positive charge of the nucleus is perfectly balanced by the collective negative charges of electrons; $N_{(p^+)} = N_{(e^-)}$.

The earliest visualisation of the atom was Ernest Rutherford's model of a central nucleus surrounded by electrons on planet-like orbits. However, being based on classical electrodynamics and the equipartition theorem (the assumption that the energy of a standing wave is continuous), it elicited the UV catastrophe, in which the expectation that a black body should radiate equally across all available frequency modes caused the blackbody spectrum to blow up at short wavelengths. One consequence of this was that an electron would be accelerated and therefore radiate, spiralling rapidly into the atomic nucleus. The crisis in theory was averted by application of Planck's constant and the quantisation of energy, $E = h\nu$, meaning that energy could only be emitted or absorbed in quantised values that exactly matched atomic energy level spacings. The Rutherford-Bohr model was the first to incorporate Planck's radiation law into a model of energy levels as stationary harmonic states, with electrons still on circular orbits; i.e., at constant 3-distances from the central nucleus.

In the modern picture, electrons populate s, p, d, and f orbitals, or standing probability waves, the densities of which define the classical probability $|\psi(x)|^2$ of finding an electron at any given position, in accordance with Heisenberg's relation, $\Delta x \cdot \Delta p \geq \hbar$. In the perspective model, the nucleus is bounded by S^2 and possesses 5-height (h_5), with electron shells being multiple S^3 horizons. Therefore, observer-dependency of the S^3 horizon in the nucleus frame is just the fact that 4-distance ($E = h\nu$) is a constant in every frame. Whilst s-orbitals are concentric (i.e., are defined by fixed 3-distances from the nucleus), since 4-distance is the conserved quantity (rather than 3-distance), an electron probability wave (e.g., the lobe-shaped p-orbitals) can pass straight through the atomic nucleus, since the probability of finding the electron at the nucleus is the theoretical minimum.

At D2|D3, to gain extra 3-height is to see the apparent D0 (point) at perspectival convergence transform or 'open up' to become an actual D1 (line) that reveals the embedding, D2|D3. This is the *forced transition* D0 → D1 (i.e., revealing the identification A ≡ B in Figure 1b). Likewise, at D4|D5, to gain extra 5-height is to *non-gravitationally accelerate*, and to see the apparent D2 (area) singularity open up to become an actual D3 (volume) singularity, revealing the embedding, D4|D5. This is to force the transition D2 → D3. The unit of D4|D5 (i.e., of the $\mathbb{R}^5$ absolute embedding space) is interpreted as a photon, which is absorbed (emitted) by a decelerated (accelerated) electron. (Whilst in the conventional view it would be reasonable to say that the forced transition D2 → D3 is associated with the emission of a photon, in the perspective model the picture is that of *projection* of the local to the remote frame. The boson is then a 'unit of projection', in which case, strictly speaking, ***photons do not travel*.**)

5. Gravitation

5.1 Macroscopic Charge

We next ask if there is a precedent for a notion of 'charge' at the macroscopic scale. In this regard, we note that matter tends towards horizons; either *expanding* as part of the Hubble flow towards the S^2 cosmological horizon, or *contracting* towards S^3 black hole event horizons. We interpret these tendencies, **EXP** and **CON**, as *charge* at the macroscopic level (see Figure 8).

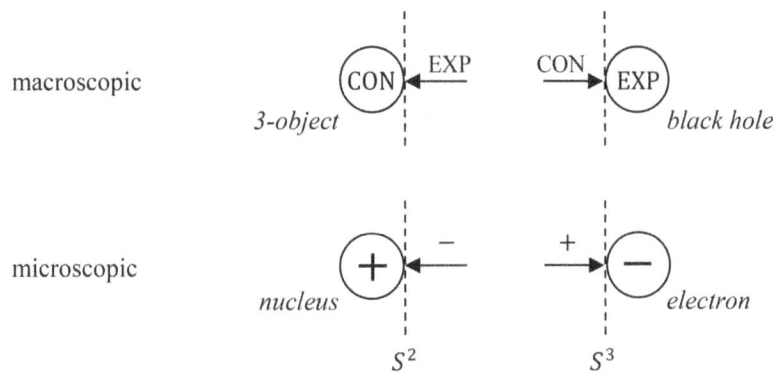

Figure 8. Macroscopic charge. Labelling of **EXP** and **CON** is of course arbitrary, demonstrating our inability to define macroscopic charge in any absolute sense

This allows us to consider the action of the Weyl (conformal) tensor C_{abcd} in terms of **EXP** and **CON** components, so that static tidal influence (falling off as the inverse cube of the separation distance) would be thought of as a standing wave, and the traceless component of the Riemann tensor as (transverse) gravitational waves (Figure 9) that correspond to the *transport* of curvature; i.e., bodies warped by gravity-waves react (with *varying* tidal forces) as though the massive source of the waves was in the vicinity.

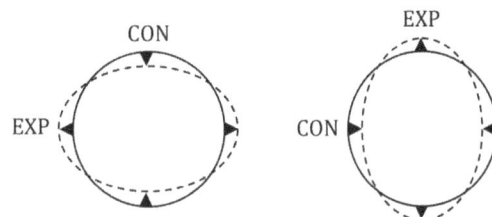

Figure 9. Plus-polarised (transverse) gravitational waves depicted as

an oscillation of **EXP** and **CON** components

5.2 Gravity As Perspective

Whilst at D2|D3, expansion of space with increasing 2-distance from the observer is a purely apparent effect, at D4|D5 it is an 'effectively actual' accelerating universal expansion. Gravity is a perspectival effect in the following sense. The cosmological models of Copernicus and Brahe with circular orbits were generalised by Kepler's elliptical orbits. In Kepler's second law, a radial line linking the central body at one focus of the ellipse to the orbiting body sweeps out equal areas in equal times, which is redolent of perspective. In the frame of an observer with 3-height (h_3) above the Earth's 2-surface, equal calibrations of the 2-surface appear to become more dense as $r \to max$ at S^1. A receding object moving at constant velocity therefore appears to slow down (decelerate), whilst continuing to cross equal calibrations at equal times in its own frame, just as in Kepler's model, a body actually accelerates (Note 2) and decelerates as calibration spacing varies.

At D4|D5 (i.e., in a Universe with accelerating expansion), while the spacing of calibrations of the 4-surface from the viewpoint of the 'stationary' local observer with 5-height (h_5) decreases exponentially as $r \to max$ at S^3 (i.e., a receding body's apparent acceleration appears to decrease), the spacing of calibrations relative to the local observer increases exponentially. The receding body still sweeps out calibrations in equal times in its own frame (i.e., in GR, proper time, τ, is maximised). The 'point at infinity' of projective geometry is incorporated *everywhere* in the finite foreground. In the perspective model this is due to the fact that the local frame is 'effectively actually' projected into the remote frame, and thus a linear, perspectival effect at $D(n)$ is incorporated into a closed, 'gravitationally bound' foreground at $D(n + 1)$. Gravity is therefore the 'effectively actual' accelerated expansion of space towards the remote frame; i.e., gravity is 'accelerated expansion (**EXP**) towards points (**CON**)'. In terms of the SM, gravity is the bleaching of **EXP/CON** charge from the D4 Universe.

5.3 Quantum Gravitational Description of Freefall

At the microscopic level, the perspective model picture is of a positively-charged nucleus, a closed 3-form (i.e., a 4-form) with 5-height (h_5) projecting $D(n - 3)$ to (and perfectly balanced by) negatively-charged electrons (actually to ℓ_P^2 or a 'zag' spinor) populating the electron shells (multiple S^3 horizons) that surround it. Likewise, the notion of macroscopic charge suggests the picture of a massive 4-object (i.e., possessing an h_5 potential) surrounded by multiple 'derivative shells' $(d^n s/dt^n)$ populated (in an appropriate sense, to be defined below) by black holes (or more correctly, to an ℓ_P^2 singularity on S^3), as depicted in Figure 10.

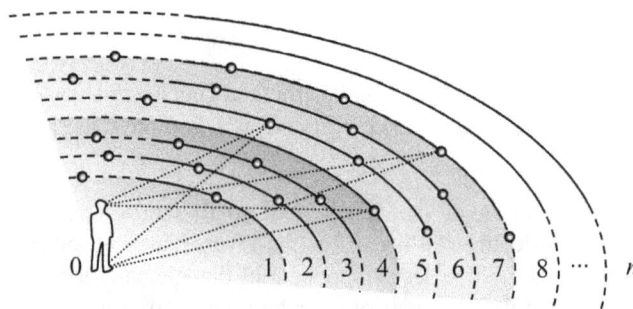

Figure 10. The generalised model for perspective at $D > 3$. The $D(n)$-height (depicted here as observer height) generates a spectrum of multiple horizons (H) with an effective limit of $N_H = 7$. Perspectival reduction $D(n - 3)$ is to apparent singularities on horizons

Just as there is a quantum description of the atom as a superposition of 'nucleus-plus-electron shells', so it would be plausible to describe freefall as a 'quantum-gravitational' superposition of 'massive-4-object-plus-derivative-shells'. In the same way that the limit principle imposes a threshold (at E_7) beyond which energy levels are no longer distinguishable, so in this approach to QG, derivative shells are no longer distinguishable beyond D7. (In this context, it is reasonable to interpret the spectrum of derivatives as a temporal quantisation.)

6. Black holes

6.1 Black Holes as Slits

Likewise, since electrons demonstrate wave-like behaviour, we would expect the same from black holes. According to the Correspondence Principle (CP), classical dynamics (typified by Newton's $\vec{F} = ma$) is recovered in the statistical limit and for well-defined wavepackets from quantum dynamics (typified by the

Schrödinger equation, $i\hbar(\partial/\partial t)(|\psi\rangle r, t) = \hat{H}(|\psi\rangle r, t))$, and so the CP must therefore also relate the respective GR and QM 'charges', **mass** and **information**, and the associated dynamics (**acceleration** and **measurement**) by which each is registered. This therefore validates a comparison between black holes and the slits in the Young double-slit experiment. Assuming no which-path detection, then in the frame of an emitted particle, *the slits themselves must become superposed or wavelike*. Therefore, if interference fringes are to appear, then just as the particle cannot 'know' by which slit it is entering the set-up, so *Bob cannot know towards which black hole he is falling*. Bob is therefore *falling into N black holes* and the tidal forces (fluctuations and oscillations) he experiences are in effect a 'Weyl interference field' that corresponds to a quantum gravitational linear superposition of an eigenbasis of distinct (mutually orthogonal) black hole spacetime geometries; $|\psi\rangle = \sum_{p=1}^{n} p = p_1 g_{\mu\nu}^{(1)} + p_2 g_{\mu\nu}^{(2)} + p_3 g_{\mu\nu}^{(3)} \ldots + p_n g_{\mu\nu}^{(n)}, n \leq 8$.

In QED, the location of a bound electron is subject to a minimum amount of uncertainty. In the perspective model, position and velocity variables ($\Delta x \cdot \Delta v \geq \hbar$) do not commute (and therefore produce probabilistic outcomes) because they are dimensionally distinct; $\Delta x \Rightarrow \mathbb{R}^3: \{x, y, z\}$ and $\Delta v \Rightarrow \mathbb{R}^{1,3}: \{x, y, z, t\}$. Whilst the Bohr model implies a fixed 3-distance separating nucleus and electron shells, in the perspective model we define a fixed 4-distance ($E = h\nu$) to the S^3 horizon that is a constant in every frame. This means that the 'trajectory' of an electron (described by orbital configurations) is probabilistic, thereby allowing *p*-orbital lobes to cross the nucleus. Likewise in QG, falling 'into' a black hole is also probabilistic, and it seems reasonable to suggest that ***black hole entropy*** ($S_{BH} = k_B A / 4\ell_P^2 = c^3 A / 4G\hbar$) ***measures the probability of finding a black hole at any given spacetime location***. Thus, $N(\ell_P^2)_{BH}$ measures the black hole's 'connectedness' with the rest of the Universe, in the sense that m_{BH} is dispersed (*via* $E = mc^2$) by $N(\gamma_{HR})$, where HR is Hawking radiation (i.e., *projection-units*).

6.2 Falling towards a Black Hole

We will now explore these ideas using Alice and Bob, who are in orbit around a black hole. They synchronise clocks before Bob begins a freefall descent towards the black hole. As Bob approaches what is $r = 2GM$ in Alice's frame, she sees him begin to decelerate. This is to say that his apparent acceleration (in Alice's frame) decreases to some irreducible minimum ($\sigma_5 > 0$) as he approaches the S^3 horizon, providing an IR cut-off for the apparently arbitrary redshift of the photons that comprise his rapidly fading image in her frame. However, since $h_5 \lll r_4$, the 4-surface *feels* flat (i.e., Bob never experiences his own mass) and the ℓ_P^2 limit of effective perspectival convergence towards which he is heading in his frame *never leaves the S^3 horizon*. Therefore, Bob sees an illusory event horizon recede in front of him (Hamilton, 2012); what Alice thinks of as $r = 2GM$ does not exist in his frame. As Alice sees Bob decelerate asymptotically to $r = 2GM$, she reasons that he must be burned up by what is in effect a *firewall* (Almheiri et al, 2013), which is a requirement of the Equivalence Principle (EP) when applied to Hawking radiation. In the perspective model, as Bob approaches what is $r = 2GM$ in Alice's frame, she and Bob are located at each other's curvature limits. That is to say that Bob has a relative recessional acceleration with respect to Alice (due to the accelerating expansion of space) that goes to some maximum, which we interpret as that threshold in free space at which the frequency of Unruh radiation becomes transplanckian ; $a_{max} \Rightarrow (\omega_{HR} < \ell_P)$, where ω is wavelength.

Depending on the size of the black hole, at some value of r, Bob begins to experience increasingly extreme tidal deformations. (In Alice's frame, if these begin outside $r = 2GM$, they will be *exactly* obscured by the redshift generated by the huge Lorentz and gravitational shifts.) These tidal fluctuations are interpreted as the Weyl interference field corresponding to superposition of distinct black hole spacetime geometries. Bob's body is subject to an effectively real higher perspective convergence and is therefore spaghettified. The (D2) quarks and leptons that comprise the remains of his body reach (in finite time) what in the conventional picture is $r = 0$, where they merge with the illusory horizon. Had Bob's clock been intact, it would have recorded *exactly* the time at which Alice, allowing for dilation caused by the Lorentz and gravitational shifts, calculated that Bob would have crossed $r = 2GM$. Therefore, a black hole, defined by the Schwarzchild radius (r_{Sch}) is in effect the difference in spacetime path lengths as measured in the frames of asymptotic Alice and infalling Bob. What are $r = 0$ and $r = 2GM$ in the conventional picture are the entangled (identified) points A and B in Figure 1b.

6.3 Bootstrap Space

Bob's infall is really his projection, $\text{D5} \xrightarrow{D(n-3)} \text{D2}$, from the local, macroscopic frame to the remote, effectively microscopic frame. Like Alice's fall into a rabbit-hole (Carroll, 1865), Bob (had he remained intact) would see what is microscopic in the conventional picture (e.g., the ℓ_P^2 point of effective perspectival convergence) as *macroscopic*. In the perspective model, therefore, black holes are not merely *dual* to electrons, but, somewhat preposterously, *black holes **are** electrons*. This means that the macroscopic and microscopic scales are juxtaposed and thoroughly mixed. The conventional '3-ness of space' is thus replaced by the notion of the $D(n-3)$ *absence* of 3-dimensionality between a given n-height (h_n) and the apparent $D(n-3)$ limit of effective perspective convergence. As seen from the highest level at D7|D8, Bob is *effectively* actually projected from the local, macroscopic frame to the (previously *apparently*, but now, *effectively*) microscopic frame, where it then appears to him to be macroscopic. In the perspective model this is what we would call '**bootstrap space**'.

6.4 Dark Energy, Dark Matter

Black hole-electron correspondence also suggests that, just as polarisation of the QED vacuum is the dressing of the bare electron charge by a slight displacement of virtual $e^- + e^+$ pairs (respectively, away from and towards the charge), so polarisation of the GR vacuum would correspond to the dressing of the bare black hole '**EXP**' charge by a displacement of 'virtual' **EXP** (dark energy) and **CON** (dark matter) charges (respectively, away from and towards the charge); i.e., by a halo of dark matter. (Baryonic matter would then be the residue resulting from the mutual cancellation of displacement, or lattice distortion between the two virtual states.)

6.5 The Graviton Frame

In Alice's asymptotic frame, the black hole emits Hawking photons, but as Bob's remains hit the so-called curvature singularity at $r = 0$ (point A in Figure 1b) they encounter a *graviton*; i.e., they enter the *graviton* frame. Since points A and B are entangled (identified), Bob's remains are transported instantaneously (in Alice's frame) to $r = 2GM$, where, in Alice's frame, Bob is burned up by a firewall. (In quantum gravitational terms, this is the *teleportation* of curvature.) Bob's spacetime path length to the so-called $r = 0$ is clearly greater than that measured in Alice's frame to the so-called $r = 2GM$, but there is no backwards-in-time violation of causality since a graviton is by definition a minimum unit of spacetime uncertainty; $\Delta t \cdot \Delta E \geq \gamma_G$. In QG models involving a minimum length-scale, there are no gravitational radiation modes at scales $\ell < \ell_P$ because spacetime ceases to exist at that scale. Since projection effectively blows up the Planck scale to the macroscopic level, the graviton frame – in the conventional picture, the 'black hole interior' – is gravity-free. That is to say that the tidal force is gauged away in perspective coordinates. This transformation from photon (rank 1) to graviton (rank 2) frame is interpreted in the perspective model as the maximal extension of the Schwarzschild solution. Covered by Kruskal-Szekeres coordinates, the solution describes a non-traversible Einstein-Rosen (ER) bridge linking spacelike-separated black hole throats in two (asymptotically flat) exterior regions.

6.6 The Information Paradox

In the conventional picture, if Hawking radiation is completely thermal (Hawking, 1971), the black hole scattering matrix is nonunitary, clearly violating QM. Complementarity, based on the holographic paradigm, allows information to exist both in the exterior and interior regions as long as observers cannot communicate (Susskind et al, 1993), (Stephens et al, 1994), but Almheiri, Marolf, Polchinski and Sully (AMPS) (Almheiri et al, 2013) demonstrated inconsistencies in the theory's postulates (i.e., that evaporation is unitary, that QFT adequately describes $r > 2GM$, that the black hole exhibits a discrete spectrum and that there is no drama at $r = 2GM$). They reasoned (i) that if evaporation is unitary and maximal entanglement is monogamous, early radiation (C) and late radiation (B) must be entangled, generating highly energetic quanta (i.e., a firewall) at $r = 2GM$ where entanglement between B and interior modes (A) is broken, or, if (ii) the EP is to be preserved, then B must be only entangled with A, resulting in nonunitarity. The ER = EPR conjecture of Maldacena and Susskind (Maldacena & Susskind, 2013), in which ER bridges are dual to Einstein-Podolsky-Rosen (EPR) entanglement (with non-traversibility and non-signalling therefore also being necessarily dual), allows entanglement of A and B (ensuring no drama), and entanglement of B and C via ER bridges (ensuring unitarity). Given certain assumptions about computational complexity (Susskind, 2014), the asymptotic observer Alice may measure Hawking radiation and thereby generate a Schenker shockwave in the ER bridge (Stanford & Susskind, 2014) and thus send a firewall to the entangled black hole throat.

In the perspective model, the maximal extension of the Schwarzschild solution is invoked when infalling mass-energy reaches the conventional $r = 0$. Therefore, the information content of infalling mass-energy, characterised in terms of *photons* or qubits that define inter-electron spacings, when it reaches $r = 0$, then becomes characterised in terms of *gravitons* that define inter-black hole spacings. In other words, the photonic 'essence' of infalling mass-energy is replicated as gravitonic essence on the cosmic scale. Bob's remains fall into $N_{(BH)}$, where $N_{(BH)} \propto N_{(e^-)}$ of the infalling matter. The information paradox is therefore a perspectival effect; the information content of infalling matter defined as inter-electron connectivity at D4|D5 is conserved 'on the other side of the vanishing-point' as inter-black hole connectivity at D7|D8. The catch is that it is now in-principle inaccessible since it exists in the 'non-computable realm'. In terms of the SM, this process appears in the h_5 frame as *the bleaching of information from the D4 universe*.

Just as at the microscopic level, an atom is a quantum superposition of nucleus-plus-multiple-energy-levels-populated-by-electrons, so at the macroscopic level, a 'cosmic atom' is a quantum gravitational superposition of massive-4-object-plus-derivative-shells-populated-by-black-holes. In QED, the UV crisis (in which an electron would radiate and spiral into the nucleus) was averted because an atom can only absorb or emit energy as photons with frequencies that match inter-electron spacing. Likewise, in QG, the 'gravitational crisis' (the notion that 'mass-energy can fall into a black hole') is averted because a 'cosmic atom' can only absorb or emit energy as gravitons with frequencies that match inter-black hole spacing. The rate of state development is defined as the number of distinct (mutually orthogonal) states it transitions in a given time (Note 3). In QED these are electrons at energy levels and in QG they are black holes on derivative shells. The strong suggestion is therefore that 'cosmic atoms' and microscopic atoms *are the same thing*, just seen from two scale extremes.

6.7 Antimatter

According to the CPT theorem, the present overwhelming preponderance of matter over antimatter is associated with a violation of CP-symmetry (a combined charge-parity transformation) that has been experimentally observed in weak decay of neutral kaons (Christenson et al, 1964), and which was presumably also present in the early Universe. The theorem also states that CP-violation implies T-violation (direction reversal). A complex phase is observed in the PMNS matrix, but not in the CKM matrix, and the strong CP problem suggests that the apparent invariance of QCD under combined CP transformations is due to fine-tuning. However, such phases are not enough anyway to account for the matter-antimatter imbalance. Antimatter occurs inside mesons, in beta-decay and other particle interactions, and as virtual particle-antiparticle pairs in vacuum polarisation and vacuum fluctuations (e.g., as implicated in the heuristic model of black hole evaporation).

In the perspective model, local $D(n)$ height ($h_{D(n)}$) is effectively projected to the remote horizon where it represents the difference (much reduced, when seen from the local frame) between apparent flatness and actual global closure. Where $h_{D(n+1)} \lll r_{D(n)}$, the $D(n)$ surface appears to be flat to very good approximation. The limit point of effective perspectival convergence is the 'bare charge' that is 'dressed' by virtual displacements of the charge associated with that n-surface. In other words, vacuum polarisation or the generation of virtual particle-antiparticle pairs (and more generally, matter-antimatter imbalance) is thought of as *entanglement* between point A (associated with divergence of apparent flatness of any given n-surface from actual global closure) and point B (associated with subsequent collapse to actual global closure), as shown in Figure 1b. The collapse vector **R** then corresponds to the (near) mutual annihilation of displaced charges, leaving a residue proportional to $h_{D(n+1)}/r_{D(n)}$, where $h_{D(n+1)} \lll r_{D(n)}$. In the D4|D5 frame, points A and B correspond respectively to the zag (primed Dirac spinor) and zig (unprimed Dirac spinor) components that together comprise a fermion, At D7|D8, points A and B correspond respectively to an on- or off-shell antiparticle ($\bar{p}$) and particle (p) pair. This is to say that $\bar{p}p$ pairs are linked by ER bridges (Jensen et al, 2014). In that case, the weakness of Hawking radiation must be proportional to the smallness of the amplitude for the scattering matrix of a lepton-nucleon interaction to enable a $W^{\pm}$ or Z^0 boson to, in effect, reach back into the quark-gluon epoch to rotate a quark between flavours (e.g., in beta-decay), where 'small' means 'fine-tuned'.

Matter-antimatter symmetry imbalance (allowed by $\Delta t \cdot \Delta E \geq \hbar$) at $r = 2GM$ is generally depicted as outgoing (positive frequency) and infalling (negative frequency) modes, denoted as $\leftarrow(\rightarrow$, where the left-pointing arrow indicates the mode that escapes to infinity, the bracket is the horizon and the right-pointing arrow is the infalling mode. However, the Feynman-Stueckelburg time-reversal interpretation implies $\leftarrow(\leftarrow$. In the perspective model, as infalling Bob reaches the so-called $r = 0$ he enters the graviton frame that transports him instantaneously to the so-called $r = 2GM$ in Alice's frame. In the perspective model, temporal asymmetry is just the gradient of $D(n) \rightarrow D(n-3)$, so the reversal corresponds to the fact that the graviton is the minimum unit of spacetime uncertainty ($\Delta t \cdot \Delta E \geq \gamma_G$), the gauge boson that restores *local* invariance under T-transformations.

6.8 Experimentally Testable Claim

Just as there is a microwave background resulting from the surface of last scattering (recombination) at D4|D5, so we argue that there should be a *mathematical* background at D7|D8, corresponding to a distribution of *primes*. In other words, variation in homogeneity that in conventional terms seeded galaxy formation (and by inference, determines the distribution of black holes) is due to eigenvalue asymptotics (Berry, 1986), (Berry & Keating, 1999). Therefore, given also that the act of quantum measurement is *not* a process of amplification, but is rather the perspectival projection from the local frame to the remote (or *effectively* microscopic) frame, then *the pattern of impacts on the second screen is a quantum distribution reflecting the cosmic distribution of black holes*. This is to say that black holes can be thought of as *paths*, and that individual impacts on the second screen are faithfully replicating (modulo the wavelike distortion of the interference fringes) the distribution of black holes in the direction that the experiment is oriented, as depicted in Figure 11. In other words, each impact will indicate the direction of the next nearest and most massive black hole as the superposed state transitions through distinguishable states. In effect, the double-slit experiment is a form of *camera* (Note 4). Therefore, the pattern of impacts on the second screen (together with the implied particle trajectories) is *deterministic but noncomputable* (i.e., *effectively* or *sufficiently* random).

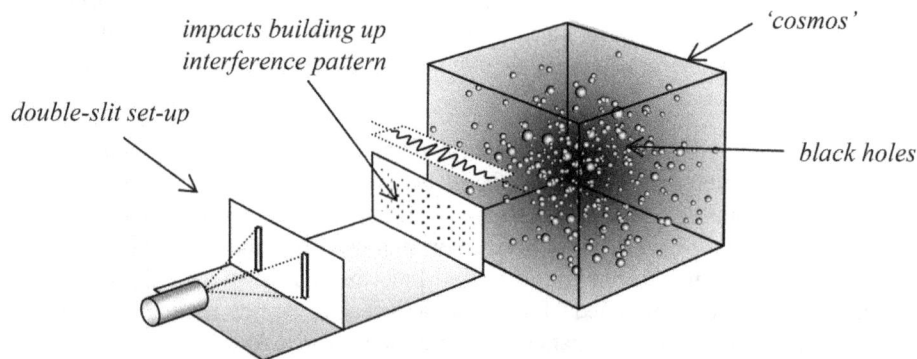

Figure 11. (Not to scale.) The Young double-slit set-up should replicate the distribution of black holes in the direction in which the experiment is orientated, subject to the waveform due to the interference pattern. Particle impact positions should reflect black hole proximity and mass, so the first impact would correspond to the nearest and most massive black hole, and so on

7. Time

7.1 Cosmological Time

In GR, time is just another covariantly transforming coordinate of an arbitrarily differentiable, dynamic spacetime manifold, and in QM it is an external (absolute), Newtonian parameter. In the canonical approach to QG, the Hamiltonian that generates state development is a constraint, leading to the Wheeler-de Witt equation, $\hat{H}(x)|\psi\rangle_U = 0$, which implies that time is frozen. Time is also implicit in the SM, in which the physical Universe is conceived of in terms of early-epoch phase transitions associated with the collective gauge group $SU(3) \times SU(2) \times U(1)/Z_6$. However, the cosmological coincidence problem (and variations of it) ask why we appear to find ourselves at a certain 'distance' in time (1.382×10^9yr from the initial singularity, $t = 0$) at which the evolving densities of dark energy (e.g., the cosmological constant) and matter are comparable. In (Avelino & Kirshner, 2016) the question is posed in terms of the dimensionless age of the Universe, $H_0 t_0$. As the authors explain, the claim that $H_0 t_0 = 1$ is in effect the assertion that the age of the Universe is exactly equal to the *Hubble time*, implying a *constant* rate of expansion since $t = 0$; i.e., as required in the perspective model.

In asking if QG has a bearing on the issue, we could note that in the perspective model, just as in QED the photon mediates between + and – charges and restores local gauge invariance (i.e., compensates for perspectivality at the microscopic D4|D5 level), so in QG, the graviton mediates between **CON** and **EXP** charges and restores local gauge invariance (i.e., compensates for perspectivality at the macroscopic D4|D5 level). It is interesting to note that in the Conformal Cyclic Cosmology (CCC) conjecture of Penrose (Penrose, 2006), (Penrose, 2008), there is a conformally smooth transition between the state of maximum universal **expansion** of the previous phase and the maximally **contracted** state of the present phase. In the perspective model, the geometry of CCC is seen as being analogous to the cycle of **U**- and **R**-processes of QM (Penrose, 1989), with a single graviton mediating between the **EXP** and **CON** phases (Figure 12). In this approach, black holes are local 'points at infinity' that act as perspectival, 'macro-micro flip-over' points.

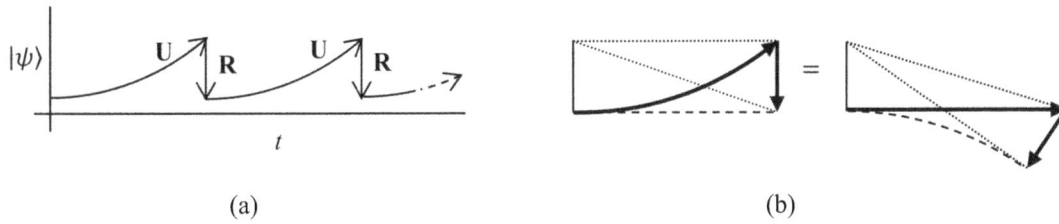

(a) (b)

Figure 12. The cycle of **U**- and **R**-processes of QM (a) are interpreted as divergence of the apparently flat surface from actual global (spherical) closure and subsequent collapse (b). In the perspective model, the collapse vector **R** is the gauge boson that restores local ('pre-perspectival') symmetry. The same geometry may be applied to the transition between the maximal universal expansion $(\mathfrak{I}^+)$ and maximal contraction $(t = 0)$ phases in the Conformal Cyclic Cosmology conjecture of Penrose (i.e., 'the collapse of the universal wavefunction')

7.2 Gravitons and Time

In the perspective model, the projection $\text{D5} \xrightarrow{D(n-3)} \text{D2}$ maps the local, macroscopic (black hole) frame in the present era to the primordial photon epoch of the SM after nucleosynthesis but prior to electron acquisition by ions, when the densities of relativistic radiation (photons) and non-relativistic matter (atomic nuclei) were equal. (The subsequent decoupling of matter and radiation and the generation of charge-neutral atoms corresponds in the perspective model to the global closure of D4.) At recombination, or the surface of last scattering, the photon mean path length became effectively infinite, and the Universe thus became transparent to electromagnetic radiation. At the present, macroscopic scale, the coincidence (or 'why now?') problem questions why the energy densities of dark energy (i.e., the cosmological constant, λ) and matter should be comparable. The implication in is that there is a perspective map $f(P): (\rho_\lambda \approx \rho_M) \to (\rho_\gamma \approx \rho_M)$, from the coincidence problem at the present, macroscopic era (scale) to the pre-decoupling radiation-matter densities of the SM photon epoch. Therefore we claim that *gravitons at the macroscopic scale are photons at the microscopic scale*, in which case, the experimental absence of gravitons would indicate that ***the Universe is not 'yet' transparent to gravitons***. In other words, black holes exchange gravitons in a higher or *future* space.

7.3 Bootstrap Time

The perspective model therefore links events that in the conventional picture are separated by billions of years. The thermodynamic specialness of the Big Bang that most strongly characterises time-symmetry violation (i.e., the basis of the second law) is defined by Penrose in terms of the accuracy with which the Creator would have had to specify the initial singularity $(t = 0)$, given as one part in $10^{10^{123}}$ (Penrose, 1989), (Penrose, 2004). In the perspective model, the transition phases of the SM comprise a coherent superposition or spectrum of S^6 horizons in the D7|D8 frame, and the specialness of the initial singularity is just the projection of that coherence from h_8 to the effectively temporally remote frame at $t = 0$. In this sense, the present era h_8 (observer) frame is 'effectively actually' projected into the effectively primordial epoch, and so we would expect, for instance, that quark masses would be ranged *perspectivally*; i.e., with the 'biggest' (top and bottom) quarks being 'closest' to 'the observer at $t = 0$'. Therefore, in the 'current era', we see a perspectival projection of local, 'human-scale' time to the temporally remote frame, where it is seen from the local frame as highly contracted, and juxtaposed with expanded, 'cosmic scale' time. The initial singularity, $t = 0$, is therefore (however *effectively* real) still a perspectival illusion, and so time in the perspective model is in effect '**bootstrap time**'.

7.4 Inversion from Microscopic to Macroscopic Scales

To illustrate further, we imagine a 2-sphere in 3-space, orbited at a fixed distance by an object with a trajectory such that, *in time*, the object will map the entire surface of the 2-sphere. Ignoring physical constraints, we arbitrarily increase the orbital velocity of the object until, in the frame of the 2-sphere, the 2-sphere is surrounded by a 'blurred' S^2 horizon. At some higher, critical (infinite) velocity, an abrupt inversion will occur, such that the object will become suddenly stationary (and massive), and surrounded at the fixed 3-distance by the *inverted* 2-sphere surface; i.e., the entire 2-surface is now visible to the object *all at the same time*. This is of course suggestive of an atomic nucleus on spherically closed D4 spacetime with an electron (a Dirac spinor pair) orbiting at a fixed 4-distance $(v = c)$ on the S^3 horizon. In the nucleus frame, the electron appears as an S^3 surface, smeared-out due to $\Delta x \cdot \Delta p \geq \hbar$. The critical inversion velocity is $v = c$, which in the Penrose zigzag picture of the electron is the fundamental velocity of each Dirac spinor. At the point of inversion, the orbiting object is suddenly stationary and massive, and is orbited by the 2-sphere. In this analogy, the point of inversion represents a transition from the microscopic level (nucleus orbited by electron) to the macroscopic level (massive object orbiting a black hole). In effect, the inversion from the 'nucleus-orbited-by-electron' picture to the 'massive-object-orbiting-a-black-hole' picture means that we are seeing a 'freeze-frame' image of the SM epoch; i.e., we surmise but do not *observe* galactic rotation, etc.

8. Discussion and Summary

8.1 Quantum Mechanics and Observers

Observer degrees of freedom are necessarily associated with the highest frame, D7|D8, and (from the limit principle, D8 being 'beyond the mathematical realm'), are therefore to be associated with *noncomputability*. We take the view that this has a direct bearing on QM, as follows. In order to ensure peaceful coexistence of relativity and EPR-entanglement, all that is needed is conservation of no-signalling. Outcomes of measurements performed on entangled qubits by spacelike separated observers Alice and Bob are irreducibly probabilistic, or *stochastic*. However, since 'true randomness' can by definition never be proven, all that is required to conserve no-signalling is 'sufficient randomness', which we interpret as *noncomputability*.

Now, since the CP relates the 'charges' **mass** and **information**, it must also relate the dynamics by which those charges are registered, namely **acceleration** and **measurement**, respectively. In the perspective model, information and mass are related by a $D(n-3)$ projection. At D4|D5, assuming $h_5 \lll r_4$ (i.e., minimum 5-height) we have a superposition comprising an h_5 **mass** potential plus eigenbasis of multiple (orthonormal) S^3 horizons, and at D7|D8 assuming $h_8 \lll r_7$ (i.e., minimum 8-height) we have a superposition comprising an h_8 **information** potential plus eigenbasis of multiple (orthonormal) S^6 horizons. (In the perspective model, the ontological status of the wavefunction $|\psi\rangle$ is therefore that of a '**ONE**' state; i.e., an ℓ_P^5 perspectival convergence limit). Just as at D4|D5, to gain extra 5-height is to *accelerate*, so at D7|D8, to gain extra 8-height is to *measure*. Therefore the projection $D8 \xrightarrow{D(n-3)} D5$ applies noncomputability to measurement outcomes at D4|D5; probability enters the process with von Neumann projection of the state vector onto the eigenbasis. To the extent that observer choice is independent, the same independence is therefore applied to quantum systems. In other words, particles must demonstrate as much freedom as experimenters (Conway and Kochen, 2008).

8.2 What does it Mean to Quantise Gravity?

In the perspective model, gravitons are exchanged in a higher, 'future' space (D7|D8) between black holes. The preparation of a pure state at D7|D8 (an h_8 information potential) and the interpretation of a quantum measurement outcome requires *understanding*, which Penrose believes is essentially non-mathematical and therefore the observer's highest faculty (Penrose, 1989), (Penrose, 1994). In the perspective model, the mathematical limit on $D(n)$ implies the existence of a 'maximum physical reality', and that effects which are clearly perspectival at D3 become 'more real' (effectively observer-independent) with each $D(n)$ increment, as the observer loses degrees of freedom with respect to 'physical reality' (i.e., becomes subject to external forces). From the highest frame (D7|D8), I observe 'maximum physical reality'. I am therefore both maximally objective (of 'physical reality') and maximally subject (to physical phenomena), represented by the following equivalence of statements that apply to D4|D5;

(1a)		(1b)	
'I am inside the Universe'		**'The Universe is inside me'.**	(4)
('Physical reality is [necessarily] maximally convincing'.)	=	('I can report internal representations of physical reality'; i.e., sensory perceptions.)	

There then follows from this a further equivalence of statements that apply to D7|D8;

(2a)		(2b)	
'I am inside my sensory perceptions'		**'My sensory perceptions are inside me'.**	(5)
('My sensory perceptions are maximally convincing'; i.e., qualia.)	=	('I can report internal representations of my sensory perceptions'; i.e., consciousness.)	

In terms of fibre bundles, a particle's (nested) internal spaces are isometric to its (nested) embedding spaces; e.g., quark interactions that appear to quarks to be free actually indicate their higher (*future*) confinement inside hadrons. In the same way, it should be possible to define a *fibre bundle* (or extended SM gauge) structure of consciousness, in which brain states represent higher or future embedding spaces. Although a detailed consideration is beyond the scope of this paper, the basic idea would be that $N_{BH} \approx N_{NEU}$ (where NEU denotes neurons). Thus an observer corresponds to a *centre* or point of perspective with 8-height (h_8) at D7|D8 (in effect, the celestial sphere) from which frame the observer sees a linear quantum gravitational superposition of multiple S^6 horizons populated by black hole (ERB) and quantum EPR (or BCS) pairs. In other words, this is the way in which it appears to us that we are 'in the Universe'; i.e., in our present era, to quantise gravity (to switch off tidal forces) is to *measure* or *think*.

8.3 Summary

We have considered macroscopic and microscopic versions of two $D(n-3)$ reductions; (i) the D5 → D2 reduction, which describes gravitational collapse and represents the ***splitting*** of mass and information (i.e., in which *properties become indefinite*), and (ii) the D8 → D5 reduction, which describes wavefunction collapse and represents the ***joining*** of mass and information (i.e., in which *properties become definite*). The D8 → D5 and D5 → D2 reductions are themselves linked by a $D(n-3)$ reduction, and therefore one horizon **splits** or **joins** these two process;

$$\text{'SOMETHING ENTERS A BLACK HOLE'} \equiv \text{'SOMETHING HAPPENS'}. \tag{6}$$

In other words, the processes 'falling into a black hole' (D5 $\xrightarrow{D(n-3)}$ D2) and '$|\psi\rangle$-collapse' (D8 $\xrightarrow{D(n-3)}$ D5) both generate 'macroscopic properties'.

In the D4|D5 frame, 4-distance was interpreted at the macroscopic level as $v = c$ and at the microscopic level as $E = hv$. Since observer-dependency of the S^3 horizon is just the fact that 4-distance to the S^3 horizon is a constant in every frame, this therefore suggested a direct equivalence, RELATIVITY = QUANTISATION. The information paradox was interpreted as a perspectival effect in which information is not lost, but does become noncomputable. The approach is falsifiable on the basis of experiments proposed.

From the D4|D5 frame at the macroscopic level, massive 4-objects are orbited by black holes (entangled $r = 0$ plus $r = 2GM$ pairs) on S^3 horizons, and from the D4|D5 frame at the microscopic level, atomic nuclei (4-forms) are orbited by electrons (Dirac spinor pairs) on S^3 horizons. From the D7|D8 frame at the macroscopic level, information potentials are orbited by ERBs (black hole pairs) on S^6 horizons, and at the microscopic level, information potentials are orbited by EPR-entangled particle pairs (e.g., $\bar{p}p$ or BCS pairs) on S^6 horizons. The microscopic and macroscopic versions of the two principle reductions therefore generate four sectors, as summarised in Figure 13.

	D8 → D5	D5 → D2
macro	QG	GR
micro	BCS	QED

Figure 13. The macroscopic and microscopic versions of the $D(n-3)$ perspective reductions, D8 → D5 and D5 → D2

References

Almheiri, A., Marolf, D., Polchinski, J., & Sully, J (2013). Black Holes: Complementarity or Firewalls? *Journal of High Energy Physics, 2.* http://dx.doi.org/10.1007/JHEP02(2013)062

Avelino, A., & Kirshner, R.P. (2016). *The Dimensionless Age of the Universe: a Riddle for Our Time.* arXiv:1607.00002v1 [astro-ph.CO]

Berry, M. V. (1986). Riemann's zeta function: a model for quantum chaos?. In *Quantum chaos and statistical nuclear physics* (pp. 1-17). Springer Berlin Heidelberg. http://dx.doi.org/10.1007/3-540-17171-1_1

Berry, M. V., & Keating, J. P. (1999). The Riemann zeros and eigenvalue asymptotics. *SIAM review, 41*(2), 236-266. http://dx.doi.org/10.1137/S0036144598347497

Carroll, L. (1865). *Alice's Adventures in Wonderland.*

Christenson, J. H., Cronin, J. W., Fitch, V. L., & Turlay, R. (1964). Evidence for the 2π Decay of the K_2^0 Meson System. *Physical Review Letters, 13*, 138. http://dx.doi.org/10.1103/PhysRevLett.13.138

Conway, J., & Kochen, S. (2008). *The Strong Free Will Theorem.* http://dx.doi.org/10.1017/CBO9780511976 971.014

Einstein, A. (1913). Entwurf einer verallgemeinerten Relativitätstheorie und eine Theorie der Gravitation. I. Physikalischer Teil von A. Einstein II. Mathematischer Teil von M. Grossmann (Outline of a Generalized Theory of Relativity and of a Theory of Gravitation. I. Physical Part by A. Einstein II. Mathematical Part by M. Grossmann). *Zeitschrift für Mathematik und Physik, 62*, 225–244, 245–261.

Einstein, A. (1915). *Erklärung der Perihelbewegung des Merkur aus der allgemeinen Relativitätstheorie* (Explanation of the Perihelion Motion of Mercury from the General Theory of Relativity). Preussische Akademie der Wissenschaften, Sitzungsberichte, 1915 (part 2), 831–839.

Einstein, A. (1915). *Feldgleichungen der Gravitation* (The Field Equations of Gravitation). Preussische Akademie der Wissenschaften, Sitzungsberichte, (part 2, pp. 844–847).

Einstein, A. (1923). *The Meaning of Relativity: Including the Relativistic Theory of the Non-Symmetric Field.* Fifth Edition. Princeton University Press. http://dx.doi.org/10.1007/978-94-011-6022-3

Einstein, A., Podolsky, B., & Rosen, N. (1935). Can quantum-mechanical description of physical reality be considered complete? *Phys. Rev., 47*, 777. http://dx.doi.org/10.1103/PhysRev.47.777

Gingerich, O., & Kenneth Lang, K. (1979). *A Source Book in Astronomy and Astrophysics*, 1900–1975. Harvard University Press. http://dx.doi.org/10.4159/harvard.9780674366688.c114

Glashow, S. L. (1961). Partial Symmetries of Weak Interactions. *Nuclear Physics, 22*(4), 579-588. http://dx.doi.org/10.1016/0029-5582(61)90469-2

Hamilton, A. J. (2014). Illusory horizons, thermodynamics, and holography inside black holes. In *Relativity and Gravitation* (pp. 99-110). Springer International Publishing. ttp://dx.doi.org/10.1007/978-3-31906761-2_13

Hawking, S. W. (1971). *Gravitational radiation from colliding black holes. Phys. Rev. Lett., 26*, 1344-1346. http://dx.doi.org/10.1103/PhysRevLett.26.1344

Hubble, E. (1929). A relation between distance and radial velocity among extra-galactic nebulae. *Proceedings of the National Academy of Sciences, 15*(3), 168–73. http://dx.doi.org/10.1073/pnas.15.3.168

Jensen, K., Karch, A., & Robinson, B. (2014). The holographic dual of a Hawking pair has a wormhole. *Physical Review D, 90*. http://dx.doi.org/10.1103/PhysRevD.90.064019

Kiefer, C. (2014). *Conceptual Problems in Quantum Gravity and Quantum Cosmology.* http://dx.doi.org/10.1155/2013/509316

Lemaître, G. (1927). *Un Univers homogène de masse constante et de rayon croissant rendant compte de la vitesse radiale des nébuleuses extra-galactiques.* Annales de la Société Scientifique de Bruxelles (in French) 47: 49.

Maldacena, J. (1998). The Large N Limit of Superconformal field theories and supergravity. *Advances in Theoretical and Mathematical Physics, 2*, 231-252. http://dx.doi.org/10.4310/ATMP.1998.v2.n2.a1

Maldacena, J., & Susskind, L. (2013). Cool horizons for entangled black holes. *Fortschritte der Physik, 61*, 781. http://dx.doi.org/10.1002/prop.201300020

Newton, I. (1687). *Philosophiæ Naturalis Principia Mathematica.*

Penrose, R. (1959). The Apparent Shape of a Relativistically Moving Sphere. *Proceedings of the Cambridge Philosophical Society, 55*(01), 137–139. http://dx.doi.org/10.1017/S0305004100033776

Penrose, R. (1989). *The Emperor's New Mind: Concerning Computers, Minds and the Laws of Physics.* Oxford University Press.

Penrose, R. (1994). *Shadows of the Mind: A Search for the Missing Science of Consciousness.* Oxford University Press.

Penrose, R. (2004). *The Road to Reality: A Complete Guide to the Laws of the Universe.* Jonathan Cape, London.

Penrose, R. (2006). *Before the big bang: An outrageous new perspective and its implications for particle physics* (pp. 2759–2767). Edinburgh, Scotland: Proceedings of EPAC 2006.

Penrose, R. (2008). Causality, quantum theory and cosmology. In S. Majid (Ed), *On Space and Time* (pp141-195. p184). Cambridge University Press. http://dx.doi.org/10.1017/CBO9781139644259.004 and http://dx.doi.org/10.1017/CBO9781139197069.004

Perlmutter, S. et al. (The Supernova Cosmology Project). (1999). *Measurements of Omega and Lambda from 42 high redshift supernovae. Astrophysical J., 517*, 565-86.

Reiss, A. G. et al. (Supernova Search Team). (1998). *Observational evidence from supernovae for an accelerating universe and a cosmological constant. Astrophysical J., 116*, 1009-38. http://dx.doi.org/10.1086/300499

Salam, A. (1968). Elementary Particle Physics: Relativistic Groups and Analyticity. In N. Svartholm (Ed.), *Eighth Nobel Symposium*. Almquvist and Wicksell, Stockholm: p367.

Stanford, D. & Susskind, L. (2014). Complexity and Shock Wave Geometries. *Physical Review D.. 90*. http://dx.doi.org/10.1103/PhysRevD.90.126007

Stephens, C.R., 't Hooft, G., & Whiting, B. F. (1994). Black hole evaporation without information loss. *Class. Quant. Grav., 11*, 621. http://dx.doi.org/10.1088/0264-9381/11/3/014

Susskind, L. (2014). *Computational Complexity and Black Hole Horizons.* http://dx.doi.org/10.1002/prop.201500092

Susskind, L., Thorlacius, L., & Uglum, J. (1993). *The Stretched Horizon and Black Hole Complementarity. Phys. Rev. D, 48*, 3743. http://dx.doi.org/10.1103/PhysRevD.48.3743

Terrell, J. (1959). Invisibility of the Lorentz Contraction. *Physical Review, 116*(4), 1041–1045, http://dx.doi.org/10.1103/PhysRev.116.1041

Weinberg, S. (1967). *A Model of Leptons. Physical review Letters, 19*(21), 1264-1266. http://dx.doi.org/10.1103/PhysRevLett.19.1264

Notes

Note 1. Indeed, it is expected in the perspective model that all apparently fine-tuned values may be determined, and hierarchical problems resolved, in terms of the effective projection of values associated with the local, macroscopic frame into the remote or effectively microscopic frame.

Note 2. In a general relativistic setting, of course, there is strictly no such thing as gravitational acceleration.

Note 3. In QM this is absolute, Newtonian time, but in a metric treatment of QM it would be (quantum) proper time.

Note 4. This departure from CIQM would be lab-testable.

Statistical Study of Solar Forcing of Total Column Ozone Variation Over Three Cities in Kenya

Carolyne M. M. Songa[1,2], Jared H. O. Ndeda[2] & Gilbert Ouma[3]

[1] Natural Science Department (Physics), Faculty of Science, The Catholic University of Eastern Africa, Nairobi, Kenya

[2] Physics Department, Jomo Kenyatta University of Agriculture and Technology, Nairobi, Kenya

[3] Meteorological Department, University of Nairobi, Kenya

Correspondence: Carolyne M. M. Songa, The Catholic University of Eastern Africa, P.O. Box 62157 - 00200, Nairobi, Kenya. E-mail: cmutambi@cuea.edu

Abstract

In this study, a statistical analysis between three solar activity indices (SAI) namely; sunspot number (ssn), F10.7 index (sf) and Mg II index (mg) and total column ozone (TCO) time series over three cities in Kenya namely; Nairobi (1.17° S; 36.46° E), Kisumu (0.03° S; 34.45° E) and Mombasa (4.02° S; 39.43° E) for the period 1985 - 2011 are considered. Pearson and cross correlations, linear and multiple regression analyses are performed. All the statistical analyses are based on 95% confidence level. SAI show decreasing trend at significant levels with highest decrease in international sunspot number and least in Mg II index. TCO are highly correlated with each other at ($0.936 < r < 0.955$, $p < 0.001$). SAI are also highly correlated with each other at ($0.941 < r < 0.976$, $p < 0.001$) and are significantly positively correlated with TCO over the study period except Mg II index at Kisumu. TCO and SAI have correlations at both long and short lags. At all the cities, F10.7 index has an immediate impact and Mg II index has a delayed impact on TCO. A linear relationship exists between the two variables in all the cities. An increase in TCO of about 2 – 3 % (Nairobi), 1 – 2% (Kisumu) and 3 – 4 % (Mombasa) is attributed to solar activity indices. The multiple correlation coefficients and significant levels obtained show that 3 – 5% of the TCO at Nairobi, Kisumu and Mombasa can be predicted by the SAI.

Keywords: correlation, F10.7 index, Mg II index, regression, solar activity indices, sunspot number, time series, total column ozone

1. Introduction

The Sun, undergoes changes characterized by variations in its output ranging from minutes to months and decades due to the different degrees of activity (Lockwood, 2004) and these variations reflect the inhomogeneous emission of radiation due to the presence and absence of active regions on the solar disk. The variations occur because of the changing impacts of the solar features whose opposing influences depend on the wavelengths. The three manifestations of solar variability are variations of solar structures, electromagnetic radiations and energetic solar particles. Variations of the solar electromagnetic radiation are the cause of the radiated solar output variability both total and in various wavelengths (Spruit, 2000).

The magnetic fields generated by a self-exciting magneto hydrodynamics (MHD) dynamo in the interior of the sun controls the solar activity (Charbonneau, 2005). The distinctive features perpetrated by these magnetic fields are displayed on different solar spheres, namely, the photosphere, chromosphere, and corona. Even though different solar layers are coupled, the variability is strongly wavelength dependent.

Solar cycle variations affect stratospheric ozone through changes in the UV fluxes that affects the dissociation of chemical species. In particular, the solar Ultraviolet (UV) radiation plays a major role in the temperature, dynamics and photochemistry of the stratosphere (Newman, 2004; Gray, Rumbold, & Shine, 2009; Gray, 2010). UV radiation between 120 and 300 nm is absorbed in the earth's outer atmosphere by ozone and molecular oxygen. Increased levels of UV radiation heat the stratosphere and mesosphere due to absorption by ozone and oxygen. Therefore variations in the different spectral regions can have a major impact on the Earth's atmosphere

(Geller., 1988; Marsh et al., 2007) since the origins and fate of life on Earth are intimately connected to the way different layers of the Earth's atmosphere respond to these variations (Tsiropouli, 2003).

Many different indicators have been employed to measure solar activity (Lean, 2000b). The sunspots, visible on the solar disk at any given time, has been the most commonly used parameter and the basic indicator of solar activity hence, a proxy for the general state of solar activity. Since no single proxy can produce the solar variability over the whole UV spectrum various other indicators and proxies for various UV spectral bands that encompass emissions coming from the solar corona down to the photosphere do exist (Dudok, Kretzschmar, Lilensten, & Woods, 2009) and have been tested (Dudok & Watermann, 2007).

Relative sunspot numbers are the collection of sunspot numbers that provide spot counts averaged over different time intervals. It is given by the sum of the number of individual sunspots and ten times the number of groups. On average, most sunspot groups have about ten spots; hence this formula for counting sunspots gives reliable numbers even when the observing conditions are less than ideal and small spots difficult to observe (Jana, Bhattacharyya, & Midya, 2013)

The 10.7-cm solar radio flux, F10.7 was first observed by Covington (1948). F10.7 index is the solar output originating in the high solar atmospheric layers of the chromosphere and in lower corona. The solar flux, measured in solar flux units (Solar Flux Unit [SFU], Tobiska, Bouwer & Bowman, 2008), is the amount of radio noise or flux emitted at a frequency of 2800 MHz (10.7 cm). Representing a measure of diffuse, non-radiative heating of the coronal plasma trapped by magnetic fields over the active regions, the F10.7 index is an excellent indicator of overall solar activity levels in the UV region. Each value of F10.7 index is a measurement of the total emission at a wavelength of 10.7 cm from all sources present on the solar disk, made over a 1 h period centered on the epoch given for the value (Tapping, 2013)

The Mg II core-to-wing index has been shown to be a good measure of solar chromospheric activity for solar features and wavelengths that have strong chromospheric components (Viereck et al., 2001). It is a ratio of the Mg II chromospheric emission at 280 nm to the photospheric radiation in the line wings (de Toma, White, Knapp, Rottman, & Woods, 1997). Introduced by Heath and Schlesinger (1986), the Mg II core-to-wing ratio is one of the most widely used indices of solar activity in the UV region. The Mg II core to wing ratio is calculated by taking the ratio between the highly variable chromospheric Mg II h and k lines at 279.56 and 280.27 nm respectively and weakly varying wings or nearby continuum which are photospheric in origin. The result is mostly a measure of chromospheric solar active region emission that is theoretically independent of instrument sensitivity change through the time (Bruevich, Bruevich, & Yakunina, 2013)

Several studies, for example, (Angell, 1989; Labitzke & Van Loon, 1997; Isikwue, Agada, & Okeke, 2010) have shown increase in solar ultraviolet radiation with increase in sunspot number and also a more or less in phase variation of TCO with sunspot number. The relationship between ozone concentration and sunshine in Nigeria carried out by Obiekezie (2009) for a period 1997 – 2005 using linear regression analysis gave a significant negative correlation between the two parameters. The linear regression analysis carried out by Rabiu and Omotosho (2003) to determine the relationship between solar activity and TCO variation in Lagos again in Nigeria, showed a significant negative correlation between mean total column ozone and solar activity both at monthly level and annual level, showing that the total column ozone decreases with increasing solar activity.

Isikwue, Agada and Okeke (2010) performed a multiple regression analysis in their study on the contribution of solar activity indices on the stratospheric ozone variations in some cities in Nigeria from 1998 – 2005 and obtained a positive significant correlation between sunspot numbers and ozone concentration and negative non-significant correlation between Mg II core to wing ratio and ozone concentration.

A statistical relationship (Pearson product moment correlation coefficient) between surface ozone and solar activity in a tropical rural site on the east coast of south India (1996 - 2004) investigated by Selvaraj, Gopinath, & Jayalakshmi (2010b), found a high positive rank correlation coefficient of 0.76 and 0.62 obtained for the years 2000 and 2002 respectively indicating the influence of higher solar activity over the surface ozone levels.

A study on the relationship between ozone and solar activity has not been carried out in cities and towns in Kenya. Hence to give an idea of solar activity effect in this low latitude region, the contributions or inputs of the solar activity indices namely; the international sunspot number (index of the sun's photosphere), the 10.7 cm solar radio flux from the sun's corona and the Mg II core-to-wing ratio (index of the sun's chromospheres), are considered for the sun's influence on the stratospheric ozone in Kenyan cities namely Nairobi, Kisumu and Mombasa using statistical analysis.

2. Materials and Methods

2.1 Data

The dataset used in this study consists of the daily values of TCO (at Nairobi, Kisumu and Mombasa) and SAI namely, international sunspot number (ssn), F10.7 cm solar radio flux (sf) and Mg II core to wing ratio (mg).

2.1.1 Total Column Ozone Dataset

A near-global TCO database at 1.25° longitude by 1° latitude, combining measurements from multiple satellite-based instruments (Bodeker, Hassler, Young, & Portmann, 2013) has been used to interpolate data to the three locations of interest in Kenya. The Bodeker scientific database extends from 1st November 1978 to 2013 and can be obtained from data site: http://www.bodekerscientific.com/data/total-column-ozone. Comparisons against TCO measurements from the ground-based Dobson and Brewer spectrophotometer network are used to remove offsets and drifts between a subset of the satellite-based instruments. The corrected subset is then used as a basis for homogenizing the remaining data sets.

2.1.2 Solar Activity Indices Dataset

These datasets are selected for the period January1985 to December 2011from existing data sources; The independent variable of solar indices are obtained from data sites as follows; The daily international sunspot number which is the photospheric index are taken from data site http://www.ngdc.noaa.gov/stp/space-weather/solar-data/solar-indices/sunspot-numbers/international/listings/listing_international-sunspot-numb ers_daily.txt the F10.7 cm solar radio flux which is the coronal index are obtained from data site: ftp://ftp.ngdc.noaa.gov/STP/space-weather/solar-data/solar-features/solar-radio/noontime-flux/penticton/penticto n_observed/ flux-observed_daily.txt and Mg II core-to-wing ratio, which is the chromospheric index are taken from data site: ftp://ftp.ngdc.noaa.gov/STP/SOLAR_DATA/SOLAR_UV/NOAAMgII.dat.

2.2 Study Area Descriptions

Kenya is located in the Eastern part of Africa (5°N, 4.40°S and 33.53°E, 41.56°E) and has a total area of 582,646 km². Table 1 shows the codes, positions and elevations of the three Kenyan cities.

Table 1. Names, codes, position and elevation of the three Kenyan cities

Station	Code	Latitude (°S)	Longitude (°E)	Altitude (m, asl)
Nairobi	nrb	1.17	36.46	1661
Kisumu	ksm	0.03	34.45	1131
Mombasa	msa	4.02	39.43	17

asl = above sea level.

Nairobi is centrally located in the country and covers an area of 684 km². It is the country's capital and largest city and holds a population of 3.1 million people (Kenya National Bureau of Statistics [KNBS], 2010). Mombasa is Kenya's second largest city, located on the South Eastern coast of the country, along the Indian Ocean. The city has an area of 295 km² and a population of 939 370 people (KNBS, 2010). Kisumu is the third largest city in Kenya. It is located in Western Kenya and covers an area of approximately 417 km², with a population of 968 909 people (KNBS, 2010).

2.3 Methodology

Data reduction is done to get the mean monthly values of both TCO at each location and the SAI variables. For data to represent the monthly and yearly scales, means and the respective standard deviation for the concerned time periods are calculated. The monthly means of the daily values of the SAI and TCO is evaluated by taking the averages of all the days in every month for each year of the data. The statistical relationship between SAI and TCO over the period 1985 to 2011 is examined in Nairobi, Kisumu and Mombasa. The R software version 3.2.3 is employed to obtain the results.

Time series plots of mean monthly of both TCO at Nairobi, Kisumu and Mombasa and the SAI variables are shown in Figure 1. A descriptive statistical summary including TCO in the various locations and SAI variables (ssn, sf and mg) along with their attributes are produced in Table 2 and Table 3 respectively

Analysis is carried out in four aspects namely; Pearson correlation analysis, cross correlation analysis, linear and multiple regression analyses. All statistical analyses are based on 95% confidence level where p value or significant value is less than 0.05 for the result to be significant.

2.3.1 Pearson Correlation Analysis

Pearson correlation analysis is conducted to examine the relationship between monthly mean SAI variables ssn, sf and mg and mean monthly TCO at each city over the study period. Correlations among SAI variables and among TCO are also performed.

2.3.2 Cross Correlation Analysis

The cross correlation function between the time series of both mean monthly TCO at each city and SAI variables is analyzed to determine the time lag(s) of SAI preceding TCO at which the series showed strongest correlations.

2.3.3 Linear Regression Analysis

The time lags with maximum correlation coefficient values are used in the linear regression of SAI and TCO to obtain linear relationships between them. The linear dependence significance is justified by the t – test and further confirmed by the analysis of variance (ANOVA) results.

2.3.4 Solar Forcing Effect

SAI variables are employed to model TCO at Nairobi, Kisumu and Mombasa by using a linear multivariate model and applying the least square fittings. Table 10 presents the values of multiple correlation coefficients obtained and their level of significance. The models are subjected to test using a 20 month data SAI values from January 2012 – August 2013 The correlation between the predicted and observed values of the models at $\alpha = 0.05$ are given in Table 11.

3. Results and Discussions

3.1 Descriptive Statistics Analysis

The various statistical attributes of both mean monthly TCO and SAI variables during the study periods are shown in Table 2 for TCO and Table 3 for SAI variables.

Table 2. Descriptive Statistics of mean monthly TCO at Nairobi (nrb), Kisumu (ksm) and Mombasa (msa) between January 1985 and December 2011

Variable	mean	SD	Range	Maximum	Minimum
nrb	254.75	9.76	47.39	277.93	230.55
ksm	253.92	9.54	45.38	275.85	230.47
msa	257.07	8.84	41.77	277.86	236.09

Table 3. Descriptive Statistics of mean monthly SAI variables between January 1985 and December 2011

Variable	mean	SD	Range	Maximum	Minimum
ssn	57.77	50.81	200.00	200.00	0.00
sf	117.82	49.02	182.35	247.20	64.86
mg	0.271	0.006	0.025	0.2880	0.263

3.2 Trend Analysis

The monthly mean TCO time series from 1985 – 2011 are shown in Figures 1 (a) – (c) and the mean monthly SAI time series for the same period are shown in Figures 1(d) – (f). Songa, Ndeda and Ouma (2015) discussed in details the TCO trends and variability in Nairobi, Kisumu and Mombasa.

A decrease in ssn, sf and mg with time is observed. The decrease could be due to declining solar activity since all the indices of solar activity are in most cases closely related as the main source of all their variations is variable magnetic field. The trends are different but at significant levels for each SAI variable as shown by the linear fits in Table 4. The linear trends suggest that, over the 27 year period of study, a decrease of 94% of the mean for ssn,

34 % of the mean for sf and ≈ 2 % of the mean for mg. Hence there is a decreasing trend of SAI variables showing that during the study period, SAI is decreasing with highest decrease in ssn and least in mg.

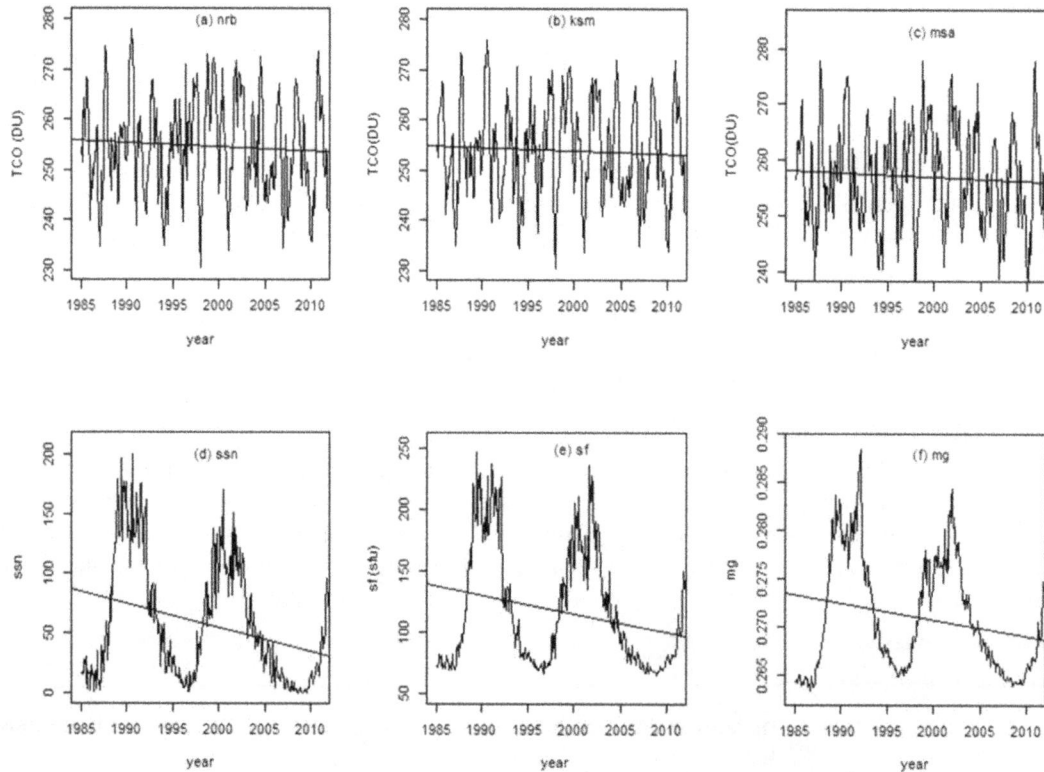

Figure 1. Mean monthly time series for TCO at (a) Nairobi (nrb), (b) Kisumu (ksm), (c) Mombasa and SAI variables (d) international sunspot number (ssn), F10.7cm solar radio flux (sf) and Mg II core to wing ratio (mg) from 1985 – 2011 with their linear fits

Table 4. Trends derived from linear fits to the mean monthly TCO and SAI time series from 1985 – 2011

Variables	Trend/yr.	r^2	p values
TCO (nrb)	- 0.0799 ± 0.06953 DU	0.004084	0.2514
TCO (ksm)	- 0.0651± 0.06800 DU	0.002852	0.3379
TCO (msa)	- 0.08016±0.06297 DU	0.005014	0.2037
ssn	- 2.0154±0.3449	0.09587	< 0.001
sf	- 1.4968±0.3399 sfu	0.05681	< 0.001
mg	- 0.0001726±0.0004.183	0.05084	< 0.001

3.3 Pearson Correlation Analysis

Pearson correlation analyses are performed relating mean monthly TCO at the three cities, monthly SAI variables and mean monthly TCO at the three cities to monthly SAI variables. Figure 2 shows the distribution of each variable on the diagonal. On the bottom of the diagonal, the bivariate scatter plots with a fitted line are displayed while on the top of the diagonal, the value of the correlation and the significance level as stars are displayed.

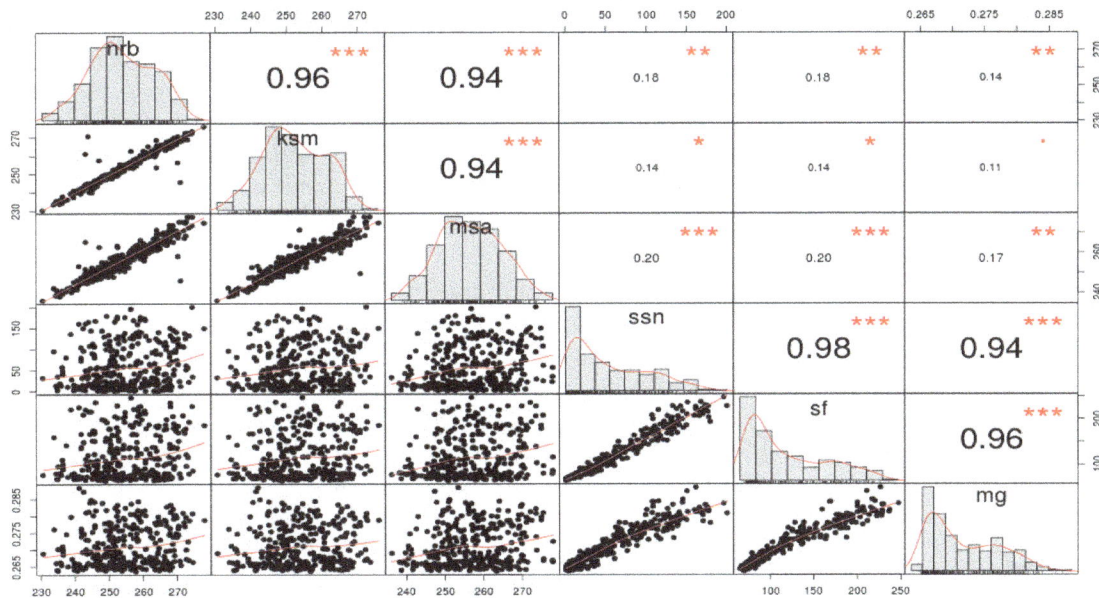

Figure 2. Correlation matrix of TCO (nrb, ksm and msa) and the SAI variables (ssn, sf and mg)

P-values (0, 0.001, 0.01, 0.05, 0.1, 1) <=> symbols ("***", "**", "*", ".", " ") ,2 – tailed.

Pearson's correlation analysis demonstrate that TCO at Nairobi, Kisumu and Mombasa are highly correlated with each other at p < 0.001 (2- tailed) significant level. The highest correlation of 0.955 is found to exist between TCO at Nairobi and Kisumu (Table 5). A lowest correlation coefficient of 0.936 is observed between TCO at Nairobi and Mombasa among the cities.

Table 5. Association among mean monthly TCO at the three cities for the period 1985 – 2011 and significant level

	TCO (nrb)	p	TCO (ksm)	p	TCO (msa)
TCO (nrb)	1				
TCO (ksm)	0.955	< 0.001	1		
TCO (msa)	0.936	< 0.001	0.938	< 0.001	1

The association among the SAI variables is shown in Table 6. SAI variables are also highly correlated with each other at p < 0.001 (2- tailed) significant level with highest correlation coefficient for the association between ssn and sf (r = 0.976, p< 0.001). The association between ssn and mg (r = 0.941, p< 0.001) is the least. sf and mg has a correlation of (r = 0.960 , p < 0.001).

Table 6. Association among mean monthly SAI variables (ssn, sf and mg) for the period 1985 – 2011 and the level of significance

	ssn	p	sf	p	mg
ssn	1				
sf	0.976	< 0.001	1		
mg	0.941	< 0.001	0.960	< 0.001	1

These results agree well with Pearson's correlation coefficient between F10.7 cm solar radio flux and international sunspot number values range between 0.94 and 0.98 (Tharshini & Shanthi, 2015). Hathaway et al., (2003) reported that a correlation between F10.7 cm solar radio flux and relative sunspot number exceeds 0.98.

The high correlation coefficient between F10.7 cm solar radio flux and international sunspot number and also between F10.7 cm solar radio flux and Mg II core-to-wing ratio could be due to the fact that F10.7 cm solar radio flux comes from high part of chromosphere and lower part of corona of the sun where it tracks other important emissions that form in the same region of the solar atmosphere (Tapping, 2013). High degree of correlation of F10.7 cm solar radio flux and Mg II core-to-wing ratio suggests dependency upon common plasma parameters and that their sources are spatially close. High correlation between international sunspot number and Mg II core-to-wing ratio could be due to the 280 nm Mg II solar spectral band containing photospheric continuum and chromospheric line emission hence chromospheric in origin while weakly varying wings or nearby continuum are photospheric in origin.

The monthly averaged ssn, monthly averaged sf and monthly averaged mg (Table 7), correlations though low, are significantly positively correlated with monthly mean TCO over the study period except for mean monthly TCO and averaged mg at Kisumu where $p > 0.05$ hence insignificant. Among the SAI variables, sf is most closely correlated to monthly TCO at nrb ($r = 0.1753$, $p < 0.05$) and Mombasa($r = 0.2039$, $p<0.001$) while ssn is most correlated to monthly TCO at Kisumu ($r = 0.1372$, $p< 0.05$. Mg was least correlated to monthly TCO at nrb ($r = 0.1440$, $p< 0.05$), at ksm ($r = 0.1089$, $p > 0.05$) and at msa ($r = 0.1740$, $p <0.05$). Correlation of SAI and ozone is due to the fact that most of the solar EUV radiation is absorbed in the upper terrestrial atmosphere some of which include international sunspot number, F10.7 cm solar radio flux and Mg II core-to-wing ratio. Correlation in the equatorial region is low where ozone is produced and in the sub Polar Regions where the larger amounts of ozone is found. The effect could be, according to Labitzke and Van Loon (1997) due to solar induced changes on the pole ward transport of ozone rather than of due to radiative interaction between sun and ozone (Tsiropoula, 2003)

Table 7. Association between mean monthly TCO and mean monthly SAI variables for the period 1985 – 2011 and their level of significance

	TCO (nrb)	p	TCO (ksm)	p	TCO (msa)	p
ssn	0.1750	0.001	0.1372	0.05	0.1974	< 0.001
sf	0.1753	0.001	0.1368	0.05	0.2039	< 0.001
mg	0.1440	0.001	0.1089	0.10	0.1740	0.001

Hence Pearson analysis demonstrates that TCO at all the cities are highly correlated among each other at $p < 0.001$ (2- tailed) significant level. All correlations are positive and highest correlation is found to exist between TCO at Nairobi and TCO at Kisumu. SAI variables are also positive and highly correlated among each other with the highest correlation coefficient (0.976) between ssn and sf. SAI variables are found to bear a significant correlation with mean monthly TCO at Nairobi and Mombasa at $p < 0.05$ except at Kisumu.

3.3 Cross Correlation

Cross correlation function between the time series of TCO at each city and SAI variables is performed and the time lags of SAI preceding TCO at which the series showed strongest (maximum) and minimum correlation is then determined at different cities as shown in Table 8.

ssn exhibited the same maximum lag month 0 at Nairobi ($r = 0.175$) and Mombasa ($r = 0.197$) but a different maximum correlation coefficient ($r = 0.140$) at lag month 11 at Kisumu. A minimum correlation coefficient of 0.112 and 0.139 are observed between ssn and TCO at Nairobi and Mombasa respectively at lag month 6. A minimum correlation of 0.0870 at lag month 5 is observed at Kisumu. sf exhibited the same maximum at lag month 0 at all the cities while a local minimum varied between lags 4 to 6. mg increases gradually from lag 0 to a maximum value at lag month 9 and then drops to its minimum value at lag month 12 (Figure 3 - 4). mg showed a constant local maximum and minimum cross correlation at lag months 9 and 12 respectively at all the cities.

The lag months for ssn and sf (Table 8) are shorter ($\leq$ - 6) than lag months for mg (- 9 and - 12), except in Kisumu where a large lag month of - 11 is observed to give a maximum correlation of 0.140 between TCO and ssn. Hence sf has an immediate impact on TCO at all the cities while ssn has immediate impact on TCO at Nairobi and Mombasa. But a delayed impact at Kisumu. mg has a delayed impact on TCO at all the cities. This could also have an effect on the ozone variability.

Table 8. Cross correlation coefficients and lags (in months) between SAI and TCO

TCO		ssn	lag/months	sf	lag/months	mg	lag/months
nrb	Max	0.175	0	0.175	0	0.167	- 9
	Min	0.112	- 6	0.123	- 6	0.123	- 12
ksm	Max	0.140	- 11	0.137	0	0.150	- 9
	Min	0.0870	- 5	0.0980	- 4, - 5	0.108	- 12
msa	Max	0.197	0	0.204	0	0.190	- 9
	Min	0.139	- 6	0.140	- 5	0.142	- 12

Figures 3, 4 and 5 shows the local maximum and minimum of the lagged cross correlation monthly mean TCO at respectively Nairobi, Kisumu and Mombasa and the SAI over the study period. ssn and sf patterns are similar. Both start at a higher coefficient value at zero month then drop gradually up to lag months; - 6 at Nairobi, - 4 and - 5 at Kisumu and 5 at Mombasa. The correlation starts to increase up to lag month - 10 and again decreases. The difference in the patterns could be due to decreased accuracy during solar minimum which is exhibited by ssn and sf, not mg. Also, mg is a ratio and hence less sensitive to artifacts and instrumental degradation than a non-ratio measurement (Suess, Snow, Viereck & Machol, 2016)

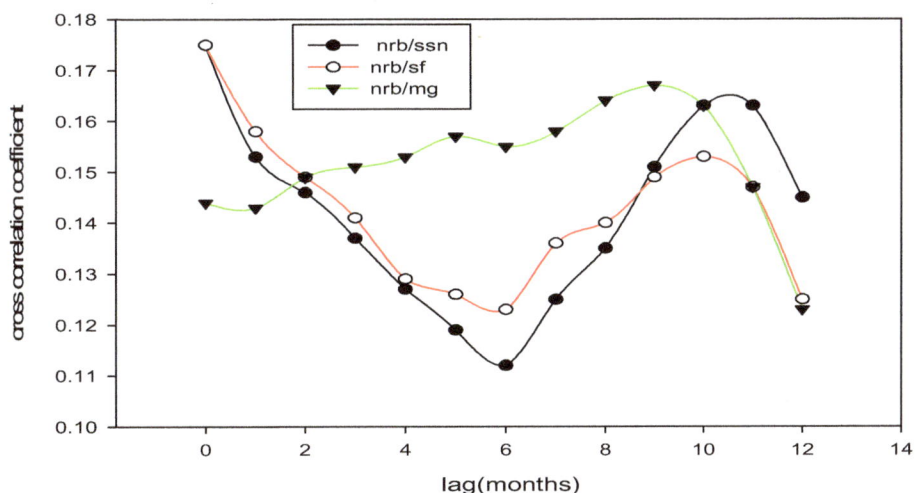

Figure 3. Plots of lagged cross correlation of monthly mean TCO at Nairobi and averaged SAI variables over the study period 1985 - 2011

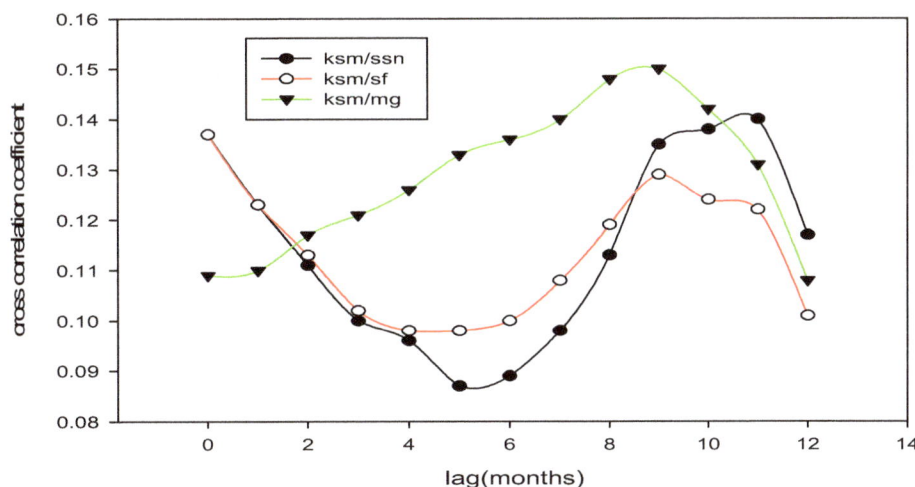

Figure 4. Plots of lagged cross correlation of monthly mean TCO at Kisumu and averaged SAI variables over the study period 1985 - 2011

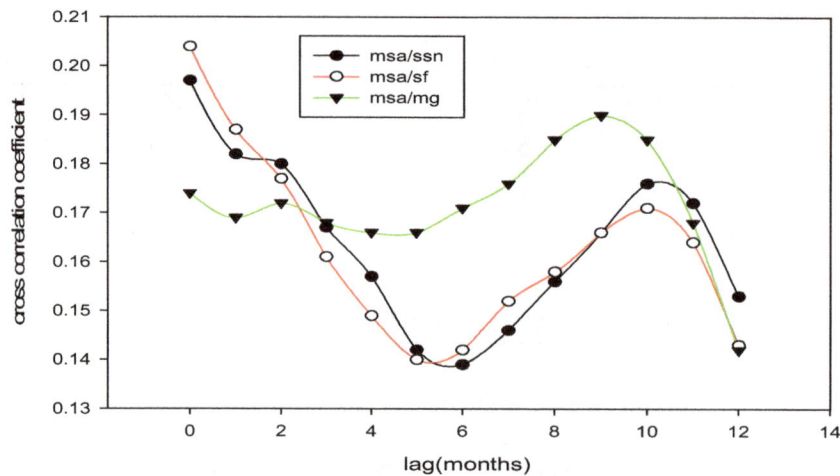

Figure 5. Plots of lagged cross correlation of monthly mean TCO at Mombasa and averaged SAI variables over the study period 1985 – 2011

Hence the mean monthly TCO and mean monthly SAI variables have correlations at both long and short lags. Positive maximum correlation is again observed between TCO at all the cities and SAI variables. A local maximum cross correlation between mean monthly TCO and mean monthly SAI variables is observed when ssn and sf variables are at month zero lag with TCO at all the cities as shown in Figure 3 for Nairobi, Figure 4 for Kisumu and Figure 5 for Mombasa. A lag month of - 9 is observed between mg and TCO at all the cities to give a maximum correlation.

3.4 Linear Regression Analysis

Tables 9(a) – (c) shows the results attained after subjecting both the mean monthly TCO at the three cities and SAI variables to linear regression. Linear regression analysis gave the correlation coefficient between TCO and SAI variables as 0.1750, 0.1753 and 0.1671 for ssn, sf and mg respectively at Nairobi, as 0.1372, 0.1368 and 0.1499 for ssn, sf and mg at Kisumu respectively and as 0.1974, 0.2043 and 0.1894 for ssn, sf and mg respectively at respectively at Mombasa.

Table 9(a). Results of linear regression and t-test for mean monthly TCO and mean monthly SAI at Nairobi

TCO	intercept	slope	r	r^2	t value	Pr (> |t|)
ssn	252.81051	0.03362	0.1750	0.03063	3.190	0.00156[**]
sf	250.64016	0.03490	0.1753	0.03073	3.195	0.00153[**]
mg	181.760	268.620	0.1671	0.02793	2.999	0.00293[**]

Table 9(b). Results of linear regression and t-test for mean monthly TCO and mean monthly SAI at Kisumu

TCO	intercept	slope	r	r^2	t value	Pr (> |t|)
ssn	252.42988	0.02576	0.1372	0.01882	2.486	0.0134[*]
sf	250.78070	0.02663	0.1368	0.01872	2.479	0.0137[*]
mg	189.960	235.170	0.1499	0.02248	2.683	0.00768[**]

Table 9(c). Results of linear regression and t-test for mean monthly TCO and mean monthly SAI at Mombasa

TCO	intercept	slope	r	r^2	t value	Pr (> |t|)
ssn	255.08534	0.03435	0.1974	0.03895	3.613	0.000351[***]
sf	252.73619	0.03678	0.2043	0.04175	3.737	0.000220[***]
mg	182.044	276.413	0.1894	0.03586	3.412	0.000729[***]

Signif. Codes: 0.001 '**', 0.01 '*', 0 '***'.

The straight line probabilistic model has slopes, intercepts and coefficient of determination values given in Table 9(a) – (c) for the respective cities. The least square prediction equation is given as;

Nairobi;
$$TCO = 252.81051 + 0.03362 \text{ ssn} \tag{1a}$$
$$TCO = 250.64016 + 0.03490 \text{ sf} \tag{1b}$$
$$TCO = 181.760 + 268.620 \text{ mg} \tag{1c}$$

Kisumu;
$$TCO = 252.42988 + 0.02576 \text{ ssn} \tag{2a}$$
$$TCO = 250.78070 + 0.02663 \text{ sf} \tag{2b}$$
$$TCO = 189.960 + 235.170 \text{ mg} \tag{2c}$$

Mombasa;
$$TCO = 255.08534 + 0.03435 \text{ ssn} \tag{3a}$$
$$TCO = 252.73619 + 0.03678 \text{ sf} \tag{3b}$$
$$TCO = 182.040 + 276.413 \text{ mg} \tag{3c}$$

The significance of this linear dependence are justified by the t – test where the t – statistic value 3.190, 3.195, 2.999 for ssn, sf and mg at Nairobi, 2.486, 2.479 and 2.683 for ssn, sf and mg at Kisumu and 3.613, 3.737, 3.412 for ssn, sf and mg at Mombasa are greater than the t critical Table 9(a) – (c) at $\alpha = 0.05$.

The linear dependence is further confirmed by the ANOVA results as the calculated F – values at the three cities and their corresponding SAI variables Table 10(a) – (c) are greater than the critical F at $\alpha = 0.05$.

Table 10(a). ANOVA table for Nairobi

TCO	F value	Pr $(>F)$
ssn	10.175	0.001564[**]
sf	10.210	0.001535[**]
mg	9.017	0.002926[**]

Table 10(b). ANOVA table for Kisumu

TCO	F value	Pr $(>F)$
ssn	6.1778	0.01344[*]
sf	6.1442	0.01370[*]
mg	3.861	0.007684[**]

Table 10(c). ANOVA table for Mombasa

TCO	F value	Pr $(>F)$
ssn	13.051	0.0003515[***]
sf	13.966	0.0002201[***]
mg	10.050	0.0007293[***]

Signif. Codes: 0.001[**], 0.01[*], 0 [***].

Hence these results reveal that the mean monthly TCO at Nairobi, Kisumu and Mombasa have positive correlation with SAI variables. A linear relationship exists between the two variables in all the cities. The best fit equations 1(a) – (1c) for Nairobi, equations 2(a) – 2(c) for Kisumu and equations 3(a) – 3(c) for Mombasa suggest an increase in TCO of about 2 – 3 % (Nairobi), 1 – 2% (Kisumu) and 3 – 4 % (Mombasa) is attributed to solar activity indices; ssn, sf and mg. The contribution is not the same for all the SAI variables. The F10.7 cm solar radio flux seems to have the best input while the Mg II core to wing ratio has the least input in the variation of TCO at Nairobi, Kisumu and Mombasa.

3.5 Linear Multivariate Models

A linear multivariate model of the form

$$y = b_0 + b_1 ssn + b_2 sf + b_3 mg$$

Where y is the mean monthly TCO, b_0, b_1, b_2 and b_3 are the coefficients determined by the least square fittings. The coefficients are presented in the multivariate models 4 to 6.

$$TCO_{nrb} = 382.58345 + 0.02036ssn + 0.07412sf - 508.24278mg \tag{4}$$

$$TCO_{ksm} = 370.28900 + 0.01859ssn + 0.06200sf - 460.30874mg \tag{5}$$

$$TCO_{msa} = 356.70976 + 0.00279ssn + 0.08728sf - 405.00741mg \tag{6}$$

Models in equations 4 to 6 shows solar forcing of TCO at the cities under study. Models in equations 4 and 5 show direct forcing of sf and ssn, but a negative forcing of mg at Nairobi and Kisumu. All the models indicate direct forcing of sf and inverse forcing of mg at Nairobi, Kisumu and Mombasa. A direct forcing of ssn is indicated only at Nairobi and Kisumu. The high coefficients of mg in the models indicate that forcing due to the chromosphere and hence the solar ultraviolet radiation is more prominent on TCO during this study period compared to the photosphere and corona (Ndeda, Rabiu, Ngoo & Ouma, 2010).

Table 11. Multiple correction coefficients results

Model equation	R	R^2	p
4	0.196825	0.03874	0.005425[***]
5	0.160000	0.025600	0.03995 [*]
6	0.218128	0.04758	0.001354[**]

Signif. Codes: 0.001'***' 0.01 '**' 0.05'*' (2 – tailed).

The multiple correlation coefficients, R and the significant levels for the three cities are shown in Table 11. At Nairobi (R = 0.196825, p = 0.01), Kisumu (R = 0.1600, p < 0.05) and Mombasa (R = 0.218128, p = 0.001) are obtained. Hence 4%, 3% and 5% of the TCO at Nairobi, Kisumu and Mombasa can be predicted by the SAI variables.

4. Conclusion

A decreasing trend at significant levels of SAI variables is observed showing that, during the study period, SAI is decreasing with highest decrease in ssn (94% of the mean), followed by sf (34%) and least in mg ($\approx$ 2%). Pearson correlation analysis demonstrated that both TCO and the SAI variables time series are highly correlated amongst each other. TCO are highly correlated with each other at (0.936< r < 0.955, p < 0.001, 2- tailed). SAI variables are also highly correlated with each other at (0.941< r < 0.976, p < 0.001, 2- tailed). Monthly mean SAI variables are significantly positively correlated with monthly mean TCO over the study period except for mean monthly TCO and averaged mg at Kisumu is not significant (p > 0.05)

All correlations are positive hence TCO and SAI are more or else in phase. The highest correlation is found to exist between SAI variables and mean monthly TCO at Mombasa with ssn (0.1974), sf (0.2039) and mg (0.1740). Among the SAI variables, sf and then ssn, seems to have more contribution (input) in ozone production in Nairobi and Mombasa while ssn and then sf, has more input in ozone production in Kisumu. Mg II core to wing ratio has the least contribution to ozone in Nairobi and Mombasa but insignificant in Kisumu.

The mean monthly TCO and mean monthly SAI variables have correlations at both long and short lags. Positive maximum correlation is again observed between TCO at all the cities and SAI variables. A local maximum cross correlation between mean monthly TCO and mean monthly SAI variables is observed when ssn and sf variables are at month zero lag with TCO at all the cities. Therefore in all the cities, sf has an immediate impact and mg has a delayed impact on TCO while ssn has immediate impact on TCO at Nairobi and Mombasa but a delayed impact at Kisumu.

A weak linear relationship exists between the two variables in all the cities. TCO and SAI bore a significant linear relationship at 5%. An increase in TCO of about 2 – 3 % (Nairobi), 1 – 2% (Kisumu) and 3 – 4 % (Mombasa) is attributed to solar activity indices. The contribution is not the same for all the SAI variables. The F10.7 cm solar radio flux seems to have the best input while the Mg II core to wing ratio has the least input in the variation of TCO at Nairobi, Kisumu and Mombasa.

All cities indicate positive forcing of the coronal index on TCO. The solar activity in the chromosphere through the mg, has negative forcing of TCO in all the cities. The index of the photosphere, the ssn, has positive forcing of TCO at Nairobi and Kisumu and a negative forcing at Mombasa. The multiple correlation coefficients and significant levels of ($R = 0.196825$, $p = 0.01$) at Nairobi, ($R = 0.1600$, $p < 0.05$) at Kisumu and ($R = 0.218128$, $p = 0.001$) at Mombasa are obtained showing 4%, 3% and 5% of the TCO at Nairobi, Kisumu and Mombasa can be predicted by the SAI variables.

Acknowledgements

We would like to thank Greg Bodeker and Jan Markus Diezel for providing the combined total column ozone database for the three Kenyan cities. The authors are thankful to the NOAA and NASA team for the solar indices data (international sunspot number, the F10.7 cm solar radio flux and the Mg II core to wing ratio. We are also grateful to the National Commission for Science, Technology and Innovation (NACOSTI) – the research authorizing body in Kenya, for provision of funds that enabled us to undertake this research. Finally, the authors wish to express their sincere thanks to the anonymous reviewers for providing constructive comments and suggestions to enhance the quality of the article.

References

Angell, J. K. (1989). On the relation between atmospheric ozone and sunspot number. *J Clim (USA), 2*, 1404-1416. http://dx.doi.org/10.1175/1520-0442(1989)002<1404:OTRBAO>2.0.CO;2

Bodeker, G. E., Hassler, B., Young, P. J., & Portmann, R. W. (2013). A vertically resolved, global, gap-free ozone database for assessing or constraining global climate model simulations. *Earth System Science Data, 5*, 31-43 http://dx.doi.org/10.5194/essd-5-31-2013

Bruevich, E. A., Bruevich, V. V., & Yakunina, G. V. (2013). Correlational study of some solar activity indices in the cycles 21 – 23. Retrieved from arXiv preprint arXiv: 1304. 4545, 2013 – arxiv.org

Charbonneau, P. (2005). Dynamo models of the solar cycle. *Living Reviews in Solar Physics, 7*(2), 20. http://dx.doi.org/10.12942/lrsp-2005-2

Covington, A. E. (1948). Solar noise observations on 10.7 centimeters. *Royal Astronomical Society of Canada Journal, 48*(4), 136. http://dx.doi.org/10.1109/jrproc.1948.234598

de Toma, G., White, O. R., Knapp, B. G., Rottman, G. J., & Woods, T. N. (1997). Mg II core-to-wing index: Comparison of SBUV2 and SOLSTICE time series. *J. Geophysical. Res., 102*(A2), 2597–2610. http://dx.doi.org/10.1029/96JA03342

Dudok de Wit, T., & Watermann, J. (2009). Solar forcing of the terrestrial atmosphere; Space Physics (physics. pace- ph). *Atmospheric and Oceanic Physics (physics.ao-ph).*

Dudok de Wit, T., Kretzschmar, M., Lilensten, J., & Woods T. N. (2009). Finding the best proxies for the solar UV irradiance. *Geophysical Research Letters 36, L10107.* http://dx.doi.org/10.1029/2009gl037825

Floyd, L., Tobiska, W. K., & Cebula, R. P. (2002). Solar UV irradiance, its variation, and its relevance to the Earth. *Advances in Space Research, 29*, 1427–1440. http://dx.doi.org/10.1016/S0273-1177(02)00202-8

Geller, M. A. (1988). Solar cycles and the atmosphere. *Nature, 332*, 584-585.

Gray, L. J. (2010). *Stratospheric equatorial dynamics.* In L. Polvani, A. H. Sobel, & D. W. Waugh (Eds.), *The stratosphere: Dynamics, transport, and chemistry,* volume 190 (pp. 93-108). Washington DC: American Geophysical Union. http://dx.doi.org/10.1029/2009GM000868

Gray, L. J., Rumbold, S. T., & Shine, K. P. (2009). Stratospheric temperature and radiative forcing response to 11-year solar cycle changes in irradiance and ozone. *Journal of the Atmospheric Sciences, 66*, 2402-2417. http://dx.doi.org/10.1175/2009JAS2866.1

Hathaway, H. D., Dibyendu, N., Wilson Robert, M., & Reichmann, E. J. (2003). Evidence that a deep meridional flow sets the sunspot cycle period, *The Astrophysical Journal, 589*, 665- 670. http://dx.doi.org/10.1086/374393

Heath, D. F., & Schlesinger, B. M. (1986). The 280 nm doublet as a monitor of changes in solar ultraviolet irradiance. *J. Geophys. Res., 91*, 8672-8682. http://dx.doi.org/10.1029/JD091iD08p08672

Jana, P. K.., Bhattacharyya, S., & Midya S. K. (2013). Equatorial, tropical and Antarctic ozone depletion and heir correlation with relative sunspot numbers. *Int. J. Res. Chem. Environ., 3*, 48-61.

Labitzke K., & Van Loon, H. (1997). Total ozone and the 11-yr sunspot cycle. *J. Atmos Sol-Terr Phys (UK), 59*(1) 9-19. http://dx.doi.org/10.1016/S1364-6826(96)00005-3

Lean, J. (2000b). Short term, direct indices of solar variability. *Space Science Reviews, 94*, 39-51. http://dx.doi.org/10.1023/A:1026726029831

Lilensten, J., Dudok de Wit, T., Kretzschmar, M., Amblard, P.-O., Moussaoui, S., Aboudarham, J., & Auch`ere, F. (2008). Review on the solar spectral variability in the EUV for space weather purposes. *Annales Geophysicae, 26*, 269-279. http://dx.doi.org/10.5194/angeo-26-269-2008

Lockwood, M. (2004). Solar outputs, their variations and their effects of Earth. In I. Redi, M. G¨udel, & W. Schmutz (Eds.), *The sun, solar analogs and the climate* (pp. 107–304). Berlin, Heidelberg, New York: Springer Proceedings of Saas-Fee Advanced Course, 34.

Marsh, D. R., Garcia R., Kinnison, D., Boville, B., Sassi, F., Solomon, S. C., & Matthes, K. (2007). Modeling the whole atmosphere response to solar cycle changes in radiative and geomagnetic forcing. *Journal of Geophysical Research, 112*, D23306. http://dx.doi.org/10.1029/2006JD008306

Ndeda, J. H. O., Rabiu, A. B., Ngoo, L. H. M., & Ouma, G. O. (2010). Estimation of climatic parameters from solar indices using ground based data from Kenya, East Africa. *Journal of Science and Technology, 31*(1), 131.

Newman, P. A. (2004). *Stratospheric ozone.* Retrieved fom: http://www. ccpo.odu.edu/SEES/ozone/class/ hapter_1/index.htm

Obiekezie, T. N. (2009). Sunshine activity and total column ozone variation in Lagos, Nigeria. Moldavian Journal of the Physical Sciences, 8(2), 169-172.

Rabiu, B., & Omotosho, T. V. (2003). Solar activity and TCO variation in Lagos, Nigeria. *Nigeria Journal of Pure and Appl. Physics, 2*(1), 17-20. http://dx.doi.org/10.4314/njpap.v2i1.21433

Selvaraj, R. S., Gopinath, T., & Jayalakshmi, K. (2010b). Statistical relationship between surface ozone and solar activity in a tropical rural coastal site, India. *Indian Journal of Science and Technology, 3*(7), 792- 794.

Songa, C. M. M., Ndeda, J., & Ouma, G. (2015). Total column ozone variability and trends over Kenya using combined multiple satellite – based instruments, *Applied Physics Research, 7*(5), 87-101. http://dx.doi.org/10.5539/apr.v7n5p87

Spruit, H. C. (2000). Theory of solar irradiance variations. *Space Sci. Rev., 94*, 113–126. http://dx.doi.org/10. 1023/A:1026742519353

Suess, K., Snow, M., Viereck, R., & Machol, J. (2016). Solar spectral proxy irradiance from GOES (SSPRING): a model for solar EUV irradiance. *J. Space weather climate*, 6, A10. http://dx.doi.org/10.1051/swsc/2016003

Tapping, K. F. (2013). The 10.7 cm solar radio flux (F10.7). *Space weather, 11*, 394 – 406. http://dx.doi.org/10. 1002/swe.20064

Tobiska W., Bouwer S., & Bowman B. (2008). The development of new solar indices for use in thermospheric density modeling. *Journal of Atmospheric and Solar-Terrestrial Physics, 70*, 803–819. http://dx.doi.org/10. 1016/j.jastp.2007.11.001

Tsiropouli, G. (2003). Signatures of solar activity variability in meteorological parameters. *Journal of Atmospheric and Solar-Terrestrial Physics, 65*, 469– 482. http://dx.doi.org/10.1016/S1364-6826(02)00295-X

Viereck, R. A., Puga, L., MacMullin, D., Judge D., Weber, M., & Tobiska W. K. (2001). The Mg II Index: A proxy for Solar EUV. *Letters, 28*(7), 1343-1346. http://dx.doi.org/10.1029/2000gl012551

4

Generalization of the Particle Spin as it Ensues from the Ether Theory

David Zareski[1]

[1] I.A.I, Yehud, Israel

Correspondence: David Zareski, I.A.I., Yehud, Israel. E-mail: zareski@inter.net.il

Abstract

In previous papers we generalized the ether waves associated to photons, to waves generally denoted ξ, associated to $\mathrm{Par}(m,e)$s, (particles of mass m and electric charge e), and demonstrated that a $\mathrm{Par}(m,e)$ is a superposition $\hat{\xi}$ of such waves that forms a small globule moving with the velocity $V_{m,e}$ of this $\mathrm{Par}(m,e)$. That, at a point near to a moving $\hat{\xi}$, the ether velocity $\partial_t\hat{\xi}$, i.e., the magnetic field **H**, is of the same form as that of a point of a rotating solid. This is the spin of the $\mathrm{Par}(m,e)$, in particular, of the electron. Then, we considered the case where e=0 and showed that the perturbation caused by the motion of a $\mathrm{Par}(m,0)$ is also propagated in the ether, and is a propagating gravitational field such that the Newton approximation (NA) is a tensor G_μ obtained by applying the Lorenz transformation for $V_{m,0}$ on the NA of the static gravitational potential of forces $G_{\mu,S}$. It appeared that G_μ is also of the form of a <u>Lienard-Wiechert potential tensor</u> A_μ created by an electric charge.

In the present paper, we generalized the above results regarding the spin by showing that the ether elasticity theory implies also that like the electron, the massive neutral particle possesses a spin but much smaller than that of the electron, and that the photon can possess also a spin, when for example it is circularly polarized. In fact, we show that the spin associated to a particle is a vortex in ether which in closed trajectories will take only quantized values.

Résumé. Dans nos précédents articles nous avons généralisé les ondes d'éther associées aux photons en les ondes généralement dénotées ξ associées aux $\mathrm{Par}(m, e)$s, (particules de masse m et de charge électrique e), et démontré qu'une $\mathrm{Par}(m, e)$ est une superposition $\hat{\xi}$ de ces ondes, formant un petit globule se mouvant a la vitesse $V_{m,e}$ de cette $\mathrm{Par}(m, e)$. Qu'en un point près d'un $\hat{\xi}$, la vitesse $\partial_t\hat{\xi}$ de l'éther, i.e., le champ magnétique **H** est de la même forme que celle d'un point d'un solide en rotation. Ceci est le spin du $\mathrm{Par}(m, e)$ en particulier, de l'électron. Puis nous considérâmes le cas où e=0 et montrâmes que la perturbation causée par le mouvement d'un $\mathrm{Par}(m, 0)$ est propagée dans l'éther et est un champ gravitationnel se propageant de telle sorte qu'à l'approximation de Newton (NA), ce champ est un tenseur G_μ obtenu en appliquant la transformation de Lorenz pour V_{m_0} à la NA du potentiel gravitationnel statique de forces $G_{\mu,S}$. Il apparaît que G_μ est aussi de la forme d'un <u>potentiel tenseur</u> A_μ <u>de Lienard-Wiechert</u> créé par une charge électrique.

Dans le présent article, nous généralisons les ci-dessus résultats en montrant que la théorie de l'élasticité de l'éther implique aussi que comme pour l'électron, la particule massive et neutre possède aussi un spin, mais beaucoup plus petit que celui de l'électron, et que les photons peuvent posséder aussi un spin, quand par exemple le photon est circulairement polarisé. En fait nous montrons que le spin associé à une particule est un vortex dans l'éther qui dans des trajectoires fermées ne pourra prendre que des valeurs quantifiées.

Keywords: The Lagrange-Einstein function, the generalized Lienard-Wiechert potential tensor, the spin of the massive neutral or electrical particle and of the circularly polarized photon, spin and vorticity

1. Introduction

Maxwell and Einstein assumed the existence of an ether, Cf. e.g., Zareski (2001), Zareski (2014). In this context, we showed there, that the Maxwell equations of electromagnetism ensue from the case of elasticity theory where the elastic medium is the ether, of which the field ξ of the displacements is governed by the Navier-Stokes-Durand Cf., e.g., Equation (4) of Zareski (2001), or Equation (1) of Zareski (2014). Then we extended this elastic interpretation to the case where the particles were not only photons, but can also be Par(m,e)s, (particles of mass m and of electrical charge e), that can be submitted to electromagnetic and or a gravitational fields. This extension which is well in accord with Einstein's opinion: *... the combination of the idea of a continuous field with the* *conception of a material point discontinuous in space appears inconsistent*...was achieved, in Zareski (2013). There we showed that the Lagrange-Einstein function L_G of such a Par(m,e) not only yields the particle four-motion equation, but also leads to the fact that φ, defined by $\hbar\, d\varphi/dt = L_G$, is the phase of a wave ξ [also denoted $\xi(m,e)$ when there is ambiguity], associated to these particles. Such a ξ is a field of the displacement of the points of the ether, i.e., propagated there, and is a solution of an equation that generalizes the Navier-Stokes-Durand equation where now a particle trajectory is a ray of such a wave that generalizes the light ray. Furthermore we showed there, that a specific sum of waves $\xi(m,e)$ forms a globule $\hat{\xi}(m,e)$ that moves like the Par(m, e) and contains all its parameters and that, reciprocally, a wave is a sum of such globules $\hat{\xi}(m,e)$, i.e., of particles. Then we showed that in its motion a $Par(m,e)$, i.e., a $\hat{\xi}(m,e)$ creates a Lienard-Wiechert potential tensor A_μ from which one deduces the electromagnetic field and in particular the magnetic field **H** which, in the elasticity theory is the velocity $\partial_t \xi$ of the ether points, Cf. Equation (6) of Zareski (2001) or Equation (3) of Zareski (2014). Then we demonstrated there that at a fixed observatory point $\mathbf{r}_{ob}$ near at a given instant to the moving electron, i.e., to the moving $\hat{\xi}(m,e)$, the velocity of the ether denoted there by $\partial_t \xi_{ob}$ is of the same form as that of a point of a rotating solid. This phenomenon is the electron spin which, as we showed, in a quantum state of an atom, can take only quantized values.

In the present article, we extend these results to the massive neutral particle and to the photon. It appears that the massive neutral particle possesses also a spin and that the circular polarization of the photon can be understood as an angular momentum that in fact is a spin. More generally, we show that this spin is in fact a vortex in the ether.

2. Notations

We denote by x^μ, ($\mu = 1,2,3,4$), the contravariant four- coordinates of a particle submitted to incident fields, the Greek indices taking the values 1,2,3,4, and the Latin, the values 1,2,3, these last refer to spatial quantities, while index 4 refers to temporal quantities. c denoting the light velocity in "vacuum", one always can impose $x^4 \equiv ct$. The Einstein summation will be used also with the Latin indices. As usual, we denote by $g_{\mu\nu}$ the co-covariant Einstein's fundamental tensor, by ds the Einstein infinitesimal element, by A_μ the Lienard-Wiechert electromagnetic potential tensor, by $\dot{f}$ the quantity defined by $\dot{f} \equiv df/dt$, in particular $\dot{x}^\mu$ denotes the four components of the velocity the particle, the expression for $\dot{s}$ is then $\dot{s} \equiv \sqrt{g_{\mu\nu}\dot{x}^\mu \dot{x}^\nu}$. The Newton approximation (**NA**), is the case where $V \ll c$.

3. Similitude of the "NA" of the Field Due to a Moving Neutral Massive Particle and the Lienard-Wiechert Field Due to a Moving Electric Charge

3.1 Introduction

The expression for the Lagrange-Einstein function $L_{G,EM}$ of a particle of mass m, of electric charge q and of velocity **V**, submitted only to a Lienard-Wiechert electromagnetic potential tensor A_μ is

$$L_{G,EM} = -mc\sqrt{c^2 - V^2} + qA_\mu \dot{x}^\mu / c. \tag{1}$$

where the expression for A_μ created by an electric charge q_0 of four-velocity $V_{q_0,\mu}$ located at the vectorial distance **R** from q, is, Cf. Equation (28) of Zareski (2014a),

$$A_\mu \equiv -\frac{q_0}{4\pi\varepsilon_0} \frac{V_{q_0,\mu}}{\left(Rc - \mathbf{R}\cdot\mathbf{V}_{q_0}\right)}, \tag{2}$$

in the static case, (1) becomes

$$L_{G,EM} = -mc\sqrt{c^2 - V^2} - \frac{qq_0}{4\pi\varepsilon_0}\frac{1}{r}. \tag{3}$$

On the other hand, the expression for the Lagrange-Einstein function $L_{G,G}$ of a particle of mass m, submitted to only a gravitational field $g_{\mu\nu}$, is

$$L_{G,G} = -mc\sqrt{g_{\mu\nu}\dot{x}^\mu \dot{x}^\nu}. \tag{4}$$

3.2 "NA" of the Lagrange-Einstein Function of a Massive Neutral Particle in a Gravitational Field

The explicit form of $g_{\mu\nu}\dot{x}^\mu \dot{x}^\nu$ is

$$g_{\mu\nu}\dot{x}^{\mu}\dot{x}^{\nu} = g_{\mu\mu}\dot{x}^{\mu}\dot{x}^{\mu} + 2g_{4_j}\dot{x}^4\dot{x}^j + 2\Delta_{i\neq j}, \tag{5}$$

where $\Delta_{i\neq j}$ is defined by

$$\Delta_{i\neq j} \equiv g_{12}\dot{x}^1\dot{x}^2 + g_{13}\dot{x}^1\dot{x}^3 + g_{23}\dot{x}^2\dot{x}^3, \tag{6}$$

and $g_{\mu\mu}$ as following

$$g_{\mu\mu} = g_{0,\mu\mu} + \delta g_{\mu\mu}, \tag{7}$$

where $g_{0,\mu\mu}$ denotes the free value of $g_{\mu\mu}$. With these notations (5) becomes

$$g_{\mu\nu}\dot{x}^{\mu}\dot{x}^{\nu} = c^2 - V^2 + \delta g_{\mu\mu}\dot{x}^{\mu}\dot{x}^{\mu} + 2g_{4_j}\dot{x}^4\dot{x}^j + 2\Delta_{i\neq j}, \tag{8}$$

and (4) can be written as following

$$L_{G,G} = -mc\sqrt{c^2 - V^2} - mc\frac{\left(\delta g_{\mu\mu}\dot{x}^{\mu}\dot{x}^{\mu} + 2g_{4_j}\dot{x}^4\dot{x}^j + 2\Delta_{i\neq j}\right)}{2\sqrt{\left(c^2 - V^2\right)}} + \cdots. \tag{9}$$

It appears that the "NA" of $L_{G,G}$ denoted then $L_{G,G,NA}$ is

$$L_{G,G,NA} = -mc\sqrt{c^2 - V^2} + mG_{\mu}\dot{x}^{\mu}/c, \tag{10}$$

where G_{μ} is the tensor defined by

$$G_4 \equiv -c^2 \delta g_{44}/2 \quad, \quad G_j \equiv -c^2 g_{4_j}, \tag{11}$$

and that Equations (1) and (10) are of the same form. Therefore, one can suppose that the tensor qA_{μ} plays the same role as the tensor mG_{μ}. This supposition is reinforced by considering that in the static case then $\delta g_{44} = -\alpha/r$, where $\alpha \equiv 2m_0 k/c^2$, and $g_{4_j} = 0$, i.e., the expression for $G_{4,S}$, (S for stationary), is

$$G_{4,S} = m_0 k/r. \tag{12}$$

Therefore, in the static case $L_{G,G,NA}$ denoted then $L_{G,G,NAS}$ becomes

$$L_{G,G,NAS} = -mc\sqrt{c^2 - V^2} + m\,m_0 k/r. \tag{13}$$

One sees that Equation (13) is of the same form as Equation (3), and that $-qq_0/4\pi\varepsilon_0$ and $mm_0 k$ are equivalent. So, the electrostatic field created by an immobile electric charge, i.e., Coulomb's field, differs from the "NA" of the gravitational field created by an immobile point mass, i.e., from the "NA" of Schwarzschild's field, by only a constant coefficient of proportionality. It appears that:

Electrostatic is of the same form as stationary gravitation where $-qq_0/4\pi\varepsilon_0$ and $mm_0 k$ are analogous.

Now, **as shown in Equations (23)-(27) of Zareski (2014a)**, the expression for G_{μ} created by a neutral massive particle of mass m_0 of covariant four-velocity $V_{m_0,\mu}$ located at the vectorial distance **R** from m, is

$$G_{\mu} \equiv m_0 k \frac{V_{m_0,\mu}}{\left(Rc - \mathbf{R}\cdot\mathbf{V}_{m_0}\right)}. \tag{14}$$

One sees that G_{μ} is of the same form as the Lienard-Wiechert potential A_{μ}, Cf. Equation (2), therefore G_{μ} will be called: gravitational Lienard-Wiechert potential. It is the gravitational potential field seen at an observation point R_{ob}, due to the particle of mass m_0 that moves with the velocity $\mathbf{V}_{m_0}$, and R is the distance between the position of m_0 at the time t' where the signal was emitted and reaches the point R_{ob} at the time t such that $(t - t')c = R$.

One sees the similarity of the electromagnetic Lienard-Wiechert potential tensor A_{μ} (2) with the gravitational Lienard-Wiechert potential tensor G_{μ} (14), that is the similarity of the electromagnetic Lagrange-Einstein function $L_{G,EM}$ (1) with the "NA" of the gravitational Lagrange-Einstein function $L_{G,G,NA}$ (10).

Since an electric charge possesses also a mass it follows that, in the NA, the expression for the total Lagrange-Einstein function $L_{G,GNA,EM}$ of a particle of mass m and electrical charge q submitted to the potential tensors G_μ and A_μ due respectively to m_0 and to q_0, is

$$L_{G,GNA,EM} = -mc\sqrt{c^2 - V^2} + \left(qA_\mu + mG_\mu\right)\dot{x}^\mu / c . \tag{15}$$

4. Review of the Ether Elastic Property of the Electron Spin

We recall how in Zareski (2014) we demonstrated that the spin of the moving electric charge is due to the ether elastic property. At the observatory point R_{ob}, let consider the electromagnetic field created by a moving electric charge q_0 of velocity V_{q_0}. This field derives from the Lienard-Wiechert potential tensor A_μ for which the expression is given in (2). If A denotes the vector of defined by the spatial components of the tensor A_μ, then the expression for the magnetic field H created by q_0 is

$$H \equiv (\mathbf{curl}\ A)/\rho_0 . \tag{16}$$

where ρ_0 denotes the density of the ether, one remind that ρ_0 is classically called, Cf. Zareski (2001), "coefficient of magnetic induction". On account of Equations (2) and (16) defined here above, and of Equations (63.8) and (63.9) of Landau and Lifshitz (1962), expressed in MKSA units, the explicit expression for H at the observation point R_{ob} of distance vector R from the charge at the retarded time, is

$$H = \frac{q_0 R}{4\pi c \rho_0 \varepsilon_0 R\left(R - R \cdot V_{q_0}/c\right)^3} \wedge \left\{ \frac{\left(R - V_{q_0} R/c\right)}{\beta^2} + \frac{R \wedge \left[\left(R - V_{q_0} R/c\right) \wedge \dot{V}_{q_0}\right]}{c^2} \right\}, \tag{17}$$

where $\beta^2 \equiv 1/\left(1 - V_{q_0}^2/c^2\right)$, and where $\dot{V}_{q_0} \equiv dV_{q_0}/dt$. As shown in Zareski (2014), when R_{ob} is very close to q_0 and then denoted r_{ob}, i.e., where R, denoted then r, is very small, then at r_{ob}, the expression for H denoted then H_{ob}, considering also that $c^2 \rho_0 \varepsilon_0 = 1$, Cf. Zareski (2014), is the following:

$$H_{ob} = \rho_{q_0} V_{q_0} \wedge r , \tag{18}$$

where ρ_{q_0}, for which the expression is

$$\rho_{q_0} \equiv q_0 / \left(4\pi r^3\right), \tag{19}$$

denotes the volumetric density of electrical charge. Yet, as shown in Zareski (2001, 2014), the magnetic field H is the field of the relative velocities $\partial_t \xi$ of the points of the ether, therefore, since at r_{ob} the expression for $\partial_t \xi$ denoted more specifically by $\partial_t \xi_{ob}$, is

$$\partial_t \xi_{ob} = \rho_{q_0} V_{q_0} \wedge r , \tag{20}$$

it follows that the velocity $\partial_t \xi_{ob}$ of a point of the ether at r_{ob}, i.e., of a vortex, Cf. Durand (1963), defined by (20), is of the same form as the velocity V_Ω of a point on a rotating solid of rotation vector Ω and of radius r, since V_Ω is of the form

$$V_\Omega = \Omega \wedge r . \tag{21}$$

One sees, by considering (18) and (21), that

$$\rho_{q_0} V_{q_0} = \Omega , \tag{22}$$

that is to say that $\rho_{q_0} V_{q_0}$ is a rotation vector, on can remark that this vector has the same dimension [1/T] as $[\Omega]$, indeed in Zareski (2012), we have shown that the ether elasticity theory implies that $\dim[\rho_{q_0}] = \dim[1/L]$. Now let us take $r = r_{q_0}$, where r_{q_0} is the radius of q_0, in this case Ω denotes the spin Ω_{q_0} of the electrically charged particle. And this is what we wanted to demonstrate, (CQFD). The quantum spin, e.g., of an electron in hydrogenous atom, is treated in Zareski (2014b).

5. The Ether Elasticity Implying that Like the Electron, the Massive Neutral Particle Possesses a Spin but much Smaller

From (1), (10), (2), and (14), it appears that $q_0 q /\left(4\pi\varepsilon_0\right)$ and $m_0 mk$ play the same role and as one can verify,

$$\dim\left[q_0 q/(4\pi\varepsilon_0)\right] = \dim\left[m_0 mk\right]. \tag{23}$$

It follows that for the neutral massive particle submitted to gravitation the coefficient which is equivalent to $q_0/(4\pi)$ of (19) is

$$m_0\sqrt{\varepsilon_0 k/(4\pi)}\,. \tag{24}$$

Therefore, if $A_{G,\mu}$ denotes the gravitational Lienard-Wiechert potential created by m_0 to which is submitted the massive neutral particle of mass m, then, considering (24) and (14), one has

$$A_{G,\mu} \equiv m_0\sqrt{\frac{k}{4\pi\varepsilon_0}}\frac{V_{m_0,\mu}}{\left(Rc - \mathbf{R}\cdot\mathbf{V}_{m_0}\right)}\,, \tag{25}$$

and by the same reasoning as made in **Sec. IV**, one has, denoting $\mathbf{H}_{G,ob}$ the equivalent of $\mathbf{H}_{ob}$,

$$\mathbf{H}_{G,ob} = \rho_{m_0}\sqrt{4\pi\varepsilon_0 k}\,\mathbf{V}_{m_0}\wedge\mathbf{r}\,, \tag{26}$$

where ρ_{m_0} is the mass volumetric density of the neutral massive particle defined by

$$\rho_{m_0} \equiv \frac{m_0}{4\pi r^3}\,. \tag{27}$$

As shown above, $\mathbf{H}_{G,ob}$ is the velocity $\partial_t\xi_{G,ob}$ of the points of the ether at $\mathbf{r}_{ob}$. Therefore, one has

$$\partial_t\xi_{G,ob} = \rho_{m_0}\sqrt{4\pi\varepsilon_0 k}\,\mathbf{V}_{m_0}\wedge\mathbf{r}\,, \tag{28}$$

and if we denote $\mathbf{\Omega}_{m_0} = \rho_{m_0}\sqrt{4\pi\varepsilon_0 k}\,\mathbf{V}_{m_0}$, then,

$$\partial_t\xi_{G,ob} = \mathbf{\Omega}_{m_0}\wedge\mathbf{r}\,. \tag{29}$$

That is, $\rho_{m_0}\sqrt{4\pi\varepsilon_0 k}\mathbf{V}_{m_0}$ denotes a rotation vector, i.e., a spin that plays the same role as the spin $\rho_{q_0}\mathbf{V}_{q_0}$. We show now that for the same velocity the proton spin is much smaller than that of the electron. Indeed, let us compare the coefficients ρ_{q_0} of (19) and $\rho_{m_0}\sqrt{4\pi\varepsilon_0 k}$ of (28) for the same r, that is, let us compare $m_0\sqrt{4\pi\varepsilon_0 k}$ and q_0. If m_0 is the mass of the proton and q_0 is the electric charge of the electron, then, in MKSA units, one has: $m_0 = 1.6\times10^{-27}$, $q_0 = 1.6\times10^{-19}$, $\varepsilon_0 = 8.9\times10^{-12}$, and $k = 7\times10^{-11}$, it follows that $m_0\sqrt{4\pi\varepsilon_0 k} = 5\times10^{-38}$ which is much smaller than $q_0 = 1.6\times10^{-19}$. This shows that the for the same velocity the spin of the electron is very very much greater than the spin of the proton, in fact the proton spin, i.e., the vorticity of the ether that it creates is negligible in front of that of the electron.

6. The Photon Circular Polarization as Identical to a Spin

Let $\mathbf{e}_x$, $\mathbf{e}_y$, $\mathbf{e}_z$ denote the unitary vectors along Resp. the x, y, z axes, and $\mathbf{E}$ the electrical field of an electromagnetic plane wave propagated along the x axe for which the expression is

$$\mathbf{E} = E\left(\mathbf{e}_y\cos\theta + \mathbf{e}_z\sin\theta\right), \tag{30}$$

where $E \equiv |\mathbf{E}|$, and where the expression for the phase θ is

$$\theta \equiv \omega\left(-t + x/c\right). \tag{31}$$

Since the vector potential $\mathbf{A}$ is related to $\mathbf{E}$ by the relation $\mathbf{E} = -\partial_t\mathbf{A}$, it follows that

$$\mathbf{A} = (E/\omega)\left(\mathbf{e}_y\sin\theta - \mathbf{e}_z\cos\theta\right). \tag{32}$$

Now, as one can verify

$$curl\mathbf{A} = (E/c)\left(-\mathbf{e}_y\sin\theta + \mathbf{e}_z\cos\theta\right), \tag{33}$$

therefore,

$$\partial_t \xi = \mathbf{H} = (1/\rho_0) \, curl\mathbf{A} = \left[E/(\rho_0 c) \right] \left(-\mathbf{e}_y \sin\theta + \mathbf{e}_z \cos\theta \right). \tag{34}$$

But since, considering (30), $\mathbf{e}_x \wedge \mathbf{E} = E\left(-\mathbf{e}_y \sin\theta + \mathbf{e}_z \cos\theta \right)$, it follows that

$$\partial_t \xi = \mathbf{H} = \left[1/(\rho_0 c) \right] \mathbf{e}_x \wedge \mathbf{E} \tag{35}$$

that can be written

$$\partial_t \xi = \left[E/(\rho_0 rc^2) \right] \mathbf{c} \wedge \mathbf{r}. \tag{36}$$

Let us show now that the dimension of $E/(\rho_0 rc)$ is 1/T that is, the vector $\left[E/(\rho_0 rc^2) \right] \mathbf{c}$ is a rotation vector $\mathbf{\Omega}$.

Indeed, as shown in Sec. VI of Zareski (2012), $\dim E = \dfrac{M}{LT^2}$ which implies $\dim \dfrac{E}{\rho rc} = \dfrac{1}{T}$, that is to say that

$$\left[E/(\rho_0 rc^2) \right] \mathbf{c} = \mathbf{\Omega}, \tag{37}$$

and

$$\partial_t \xi = \mathbf{\Omega} \wedge \mathbf{r}. \tag{38}$$

This shows that the photons created by the moving electron that itself possesses a spin, can possess also a spin this is the case where the photon is circularly polarized. Finally, an EM field is a vibration of the ether, this vibration can be linear, elliptic or circular, in this last case the photon possesses a spin, which in fact is a vortex in the ether.

6. Conclusions

We have shown that like the electron, the massive neutral particle possesses a spin, that the circular polarization of the photon is also a spin, and that the particle spin is in fact a vortex in the ether that in closed trajectories can take only quantized values.

Acknowledgments

I thank Professor Lawrence Horwitz from Tel-Aviv and Bar-Ilan Universities Israel, for his encouragements and his precious counsels.

References

Durand, W. F. (Ed.). (1963). *Aerodynamic Theory*. New York: Dover.

Landau, L. D., & Lifshitz, E. M. (1962). *The classical theory of field*. Pergamon Press.

Zareski, D. (2001). The elastic interpretation of electrodynamics. *Foundations of Physics Letters, 14*(5), 447-469.

Zareski, D. (2012). On the elasto-undulatory interpretation of fields and particles. *Physics Essays, 25*(2).

Zareski, D. (2013). Fields and wave-particle reciprocity as changes in an elastic medium: The ether. *Physics Essays, 26*(2), 288-295.

Zareski, D. (2014). The Electron Spin as Resulting From the Ether Elasticity. *Applied Physics Research, 6*(5), 41.

Zareski, D. (2014a). The ether theory as implying that electromagnetism is the Newtonian approximation of general relativity. *Physics Essays, 27*(4), 517-522.

Zareski, D. (2014b). The Quantum Mechanics as Also a Case of the Ether Elasticity Theory. *Applied Physics Research, 6*(4), 48.

5

Characterization of Water and Nitrogen Stress of Maize by Laser Induced Fluorescence

Adama Penetjiligue SORO[1], Emma Georgina ZORO-DIAMA[1], Kédro Sidiki DIOMANDE[2], Guy Euloge BANY[1], Yvon BIBILA MAYAYA BISSEYOU[1] & Adjo Viviane ADOHI-KROU[1]

[1] Laboratoire de Cristallographie et Physique Moléculaire (LaCPM), Université Félix Houphouët Boigny, Côte d'Ivoire

[2] Centre National de Recherche Agronomique (CNRA), Côte d'Ivoire

Correspondence: Adama Penetjiligue SORO, Laboratoire de Cristallographie et Physique Moléculaire (LaCPM), Université Félix Houphouët Boigny, Côte d'Ivoire. E-mail: adams.soro@gmail.com

Abstract

Water and nitrogen are essential for the optimal development of corn plants. A deficiency of these elements leads to lower crop production. Also, the health status of a plant influences the photosynthesis process. The photosynthetic diagnosis of a plant from the chlorophyll fluorescence spectrum induced by laser is non-destructive to the sample, reliable and fast method. As part of this work, we showed that it is possible to detect the nitrogen and water deficiencies of corn from the chlorophyll fluorescence ratio at 690 nm and 740 nm, when the measurements are performed before the senescence phase.

Indeed, we found that the R fluorescence ratio increases over time, for any stress on the plant. However, R decreases with the nitrogen stress and increases with increasing water loss.

The measures should be performed 51 Days After Planting (DAP) to detect water deficiency and the suitable date for nitrogen deficiency detection is 61 DAP.

Before each of these dates, the plants will be considered water deficient if the fluorescence ratio $R \leq 1.34$ and will be nitrogen stressed if $R > 1.36$.

Keywords: chlorophyll fluorescence, Maize, nitrogen deficiency, water deficiency, hydric stress

1. Introduction

Corn is the most consumed cereal in the world. Indeed, it represents 41 % of world cereal production (Perrier-Brusle, 2010). It is mainly used for feeding cattle in industrialized countries, but in sub- Saharan Africa and Latin America, it is used to feed the population (Bassalet, 2000). In Côte d'Ivoire, the 2014 national production of this cereal exceeded 600 000 tonnes (Réseau Non-Gouvernemental Européen sur l'Agroalimentaire le Commerce l'Environnement et le Développement [RONGEAD], 2014). However, the country is not self-sufficient in this cereal. It is forced to import corn to provide industrial and agro-pastoral needs. Therefore, it is necessary, even imperative to increase the production of this cereal to meet the existing needs and ultimately ensure food security of the Ivorian population. In Côte d'Ivoire, the corn growing areas are mainly located in Savannah, low rainfall area (Yéo, 2011).

Water and nitrogen are essential mineral elements for the optimal development of corn plants (Saccardy, 1997; Plénet, 1995; PNTTA, 1999). A deficiency in these elements leads to lower crop production. In agronomy, detecting the nutritional deficiencies of plants is usually carried out by foliar diagnosis which is a destructive method.

The fluorescence emission is directly related to the photosynthesis process (Papageorgiou & Govindjee, 2004; Stirbet & Govindjee, 2011; Wim & Andrej, 2013) in which are made of many biological exchanges. The health status of a plant influences this process. Thus, the study of the fluorescence spectrum can detect any stress on the plant at leaf and canopy scale (Chappelle, Wood, McMurtrey & Newcomb, 1984; Méthy, Olioso & Trabaud, 1994; Bourrié, 2007). The interest of this photosynthetic diagnosis is that it is non-destructive, fast and reliable. As part of this work we characterize the water and nitrogen deficiencies of corn plants by chlorophyll method of laser-induced fluorescence.

2. Material

2.1. Experimental Material

Data were collected in vivo and in situ using an USB4000 - type FL fluorescence spectrometer. This device can record plant chlorophyll fluorescence spectra whose wavelengths range is from 360 nm to 1 000 nm in steps of 0.22 nm. The samples were excited by a LED emitting at 450 nm through a bifurcated optical fiber. The acquisition, storage and processing of the collected spectral data were conducted using a laptop. The Figure 1 shows the experimental setup.

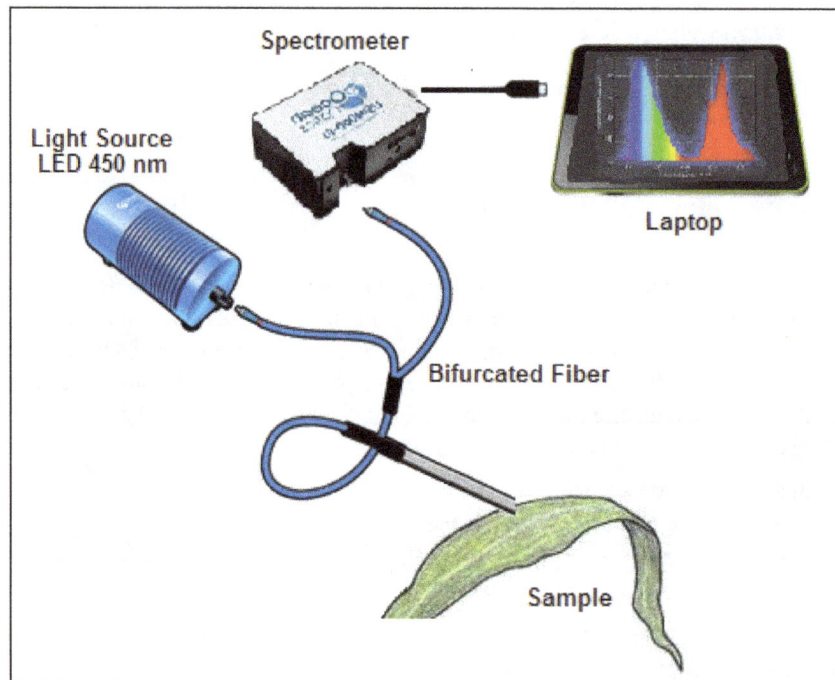

Figure 1. Experimental setup

2.2 Vegetal Material

The corn variety used in this study is called DMRESR-Y. It was provided by the Centre National de Recherche Agronomique (CNRA) in Côte d'Ivoire. Its maturation cycle is 90 to 95 days and its grains are yellow and horny (L. Akanvou , R. Akanvou, Anguété & Diarrassouba, 2006).

3. Methods

The plantations for both studies were divided into blocks. Each block contained the same deficiency levels (nitrogen or water) to ensure that the arrangement of the buckets in the field does not influence the results.

3.1 Water Stress Induction

In the case of the water stress study, the corn planting was made in a greenhouse as we did not want to be dependent on the weather and we had to control the water amount we had to provide to the corn plants.

A mineralogical analysis of the soil used to fill the buckets revealed that it had high content of all essential nutrients for the corn plants development.

We first sought the soil field capacity : it is the amount of water that the ground can retain. Knowing this value allowed us to determine the various doses to apply to the soil to induce the hydric stress. The field capacity of the used ground was 2 liters. We then generated four levels of water stress as listed below:

W12:   12.5 % of the soil field capacity per bucket (0.25 l of water)

W25:   25 % of the soil field capacity per bucket (0.5 l of water)

W50:   50 % of the soil field capacity per bucket (1 l of water)

W100: 100 % of the soil field capacity per bucket (2 l of water)

After a heavy watering the day before, we sowed corn grains the next day. The plantation in the greenhouse consisted of 72 buckets of 20 l capacity, left in 3 blocks (see figure 2).

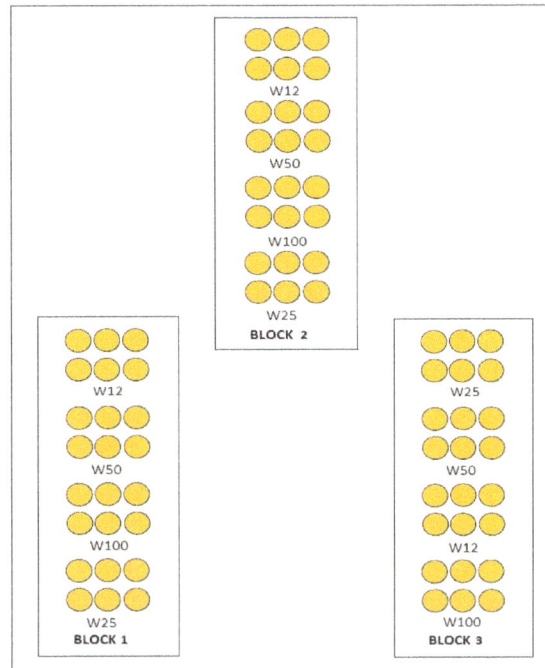

Figure 2. Water stress study planting plan

Every bucket contained two growth pouches at the rate of three grains per pouch. We removed corn plants from the bucket, to have a unique plant in a growth pouch 15 Days After Plantation (DAP). So, every bucket contained two corn plants.

We induced the water stress 30 DAP. Then, we started to collect spectral data 37 DAP. From this date and once a week, the fluorescence spectrum of every plant was recorded. This operation took place between 09:00 am and 1:00 pm. This phase ended 72 DAP when plants reached the senescence phase.

3.2 Nitrogen Stress Induction

The buckets were filled with poor nitrogenous soil, in order to control the intake of nitrogen fertilizer. Each bucket with a capacity of 30 liters had three growth pouches with three seeds per growth pouch.

We brought an amount of 2.27 g of nitrogen, phosphorus and potassium (NPK) to every hole to allow a good seeding of the corn grain. Then, the corn seedlings were thinned to one plant per hole 15 DAP. So, there were only three plants per pot.

The nitrogenous stress was led 30 DAP by providing various doses of urea. So, we generated five fertilization levels as listed below:

N0: 0 g of urea/plant (no nitrogen provided to the plant)

N1: 0.377 g of urea/plant (1/4 part of the nitrogen recommended dose)

N2: 0.755 g of urea/plant (2/4 part of the nitrogen recommended dose)

N3: 1.133 g of urea/plant (3/4 part of the nitrogen recommended dose)

N4: 1.510 g of urea/plant (nitrogen recommended dose).

This plantation consisted of 80 buckets of 30 l capacity, left in 4 blocks. Every block contained the 5 fertilization levels and each fertilization level included 4 buckets (see figure 3).

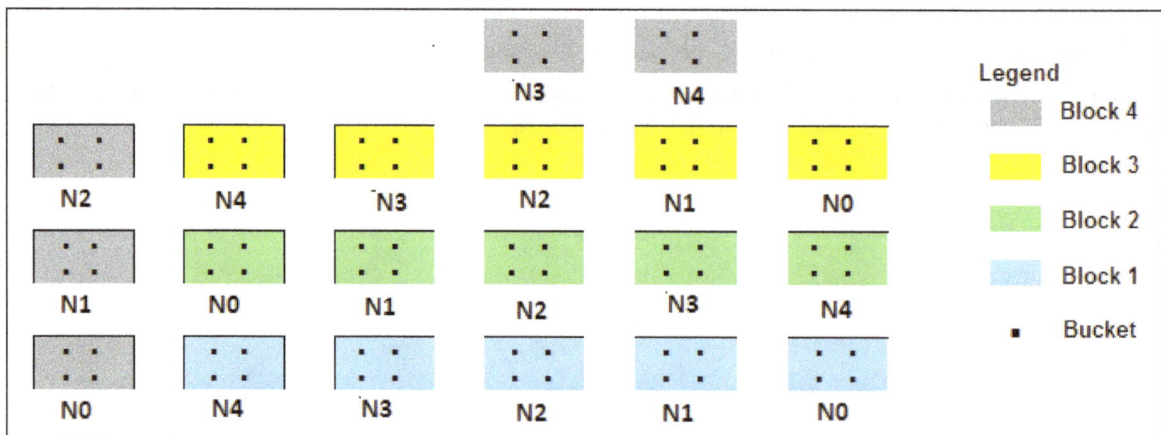

Figure 3. Plantation plan for the nitrogen stress study

Ten days after the application of nitrogen stress; that is 40 DAP, we began collecting spectral data. From that date, once every week, the fluorescence spectrum of each plant was recorded. This operation that ended 82 DAP took place between 09:00 am and 01:00 pm. A total of seven series of measurements were carried out during the different development stages of corn plants.

3.3 Data Processing

During the measures at leaf scale, every deficiency level (hydric or nitrogenous) applied to the plant in the same block, is characterized by the average fluorescence spectrum.

We noticed that for the same applied stress level, there is no significant difference between the measurements performed on the blocks. We then worked with the average values of the ratios R computed on the blocks, for the same given deficiency level.

For all the recorded spectra, the digital data are converted to text files and then imported into the MATLAB software to compute the ratios of intensities of the two characteristic chlorophyll fluorescence peaks ($R = F_{690}/F_{740}$). F_{690} corresponds to the intensity of the fluorescence peak at 690 nm and F_{740} is the intensity of the fluorescence peak at 740 nm. We used the fluorescence ratio in our study, among other fluorescence parameters as it is a pertinent indicator, widely used in plant stress detection (Tremblay, Wang & Cerovic, 2012).

The various graphics were edited with the software ORIGINPro 8. R values given on the charts are the average values for every deficiency level. Indeed, it is recommended to use the mean value of ratio *R* for many measurements for several plants to have reliable results instead of one single measurement (Fedotov, Bullo, Belov & Gorodnichev, 2016).

4. Results and Discussion

4.1 Water Stress Case

The graphs in Figure 4 show for each stress level, the intensities of fluorescence ratios depending on the development stage of corn plants. For all treatments, we notice an increase in the value of R depending on the plant development stage. Thus, the value of R increases with the plant age. From 51DAP to 65DAP each curve compared to other is a function of the nutritional stress: greater the stress is, smaller R is.

From the date 51DAP, we also find that the curves (W12, W25) and (W50, W100) are close to each other. The differences between the ratios values for considered couples treatments are weaker at 65DAP. We can then consider the two extreme curves : W12 represents stressed plants and W100 represents non- stressed ones. At 51DAP the gap between the extreme curves is maximum. It would be the indicated date for the water stress detection. The water deficiency detection measures are efficient in the time interval [51DAP, 65DAP].

As the histograms in Figure 5 show, until 51DAP, all water -deficient plants have a ratio R ≤ 1.34. But, for non-deficient plants the ratio is still greater than 1.34. In addition, as water is an essential element for plant survival, a lack of water causes early senescence. This is the case from 58DAP. Then, we notice that all plants have R > 1.34. So, the appropriate time to make the measures for the hydric stress detection would be 51DAP.

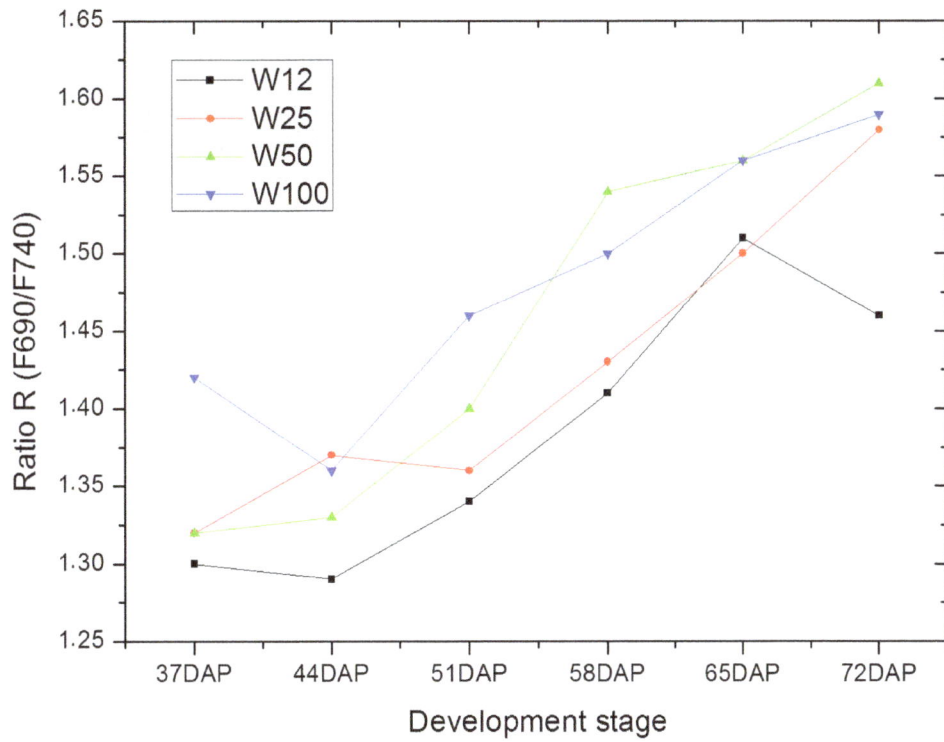

Figure 4. Temporal changes of the fluorescence ratio for the four hydric treatments

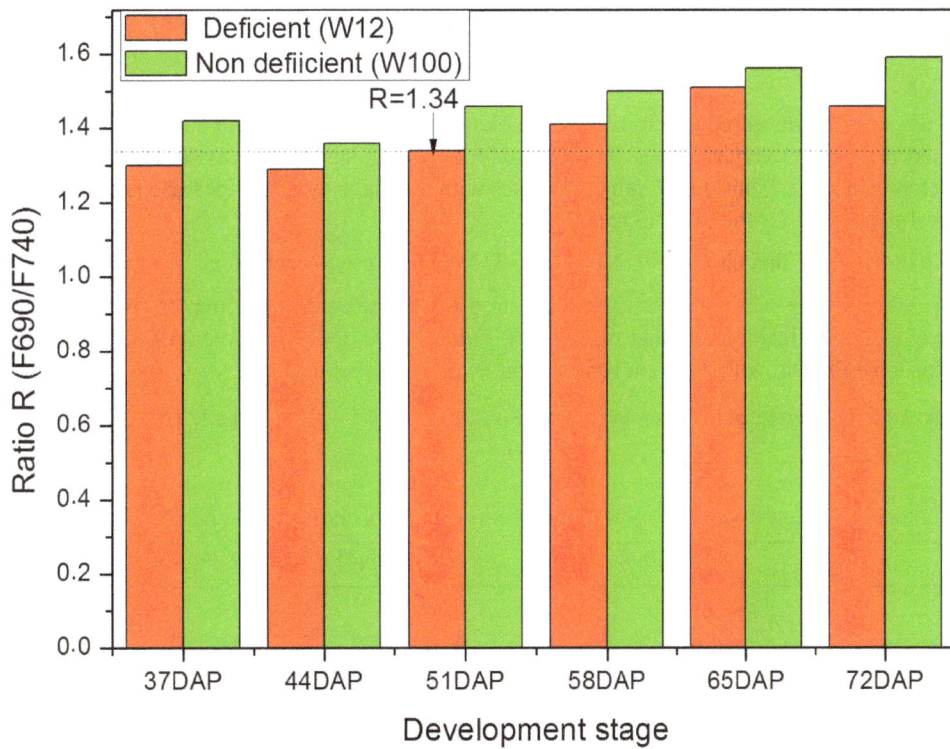

Figure 5. Histograms for the E12 and E100 water stress levels as a function of development stage

4.2 Nitrogen Stress Case

The charts in figure 6 show the fluorescence ratio over the treatment applied to corn plants. We find that at any plant development stage, the ratio R decreases when stress decreases.

According to Méthy, Olioso & Trabaud (1994), the amount of chlorophyll in the plant is proportional to the photosynthetic activity. However, the nitrogen deficiency causes the decrease of the activity. The ratio R therefore increases when the nitrogen stress decreases.

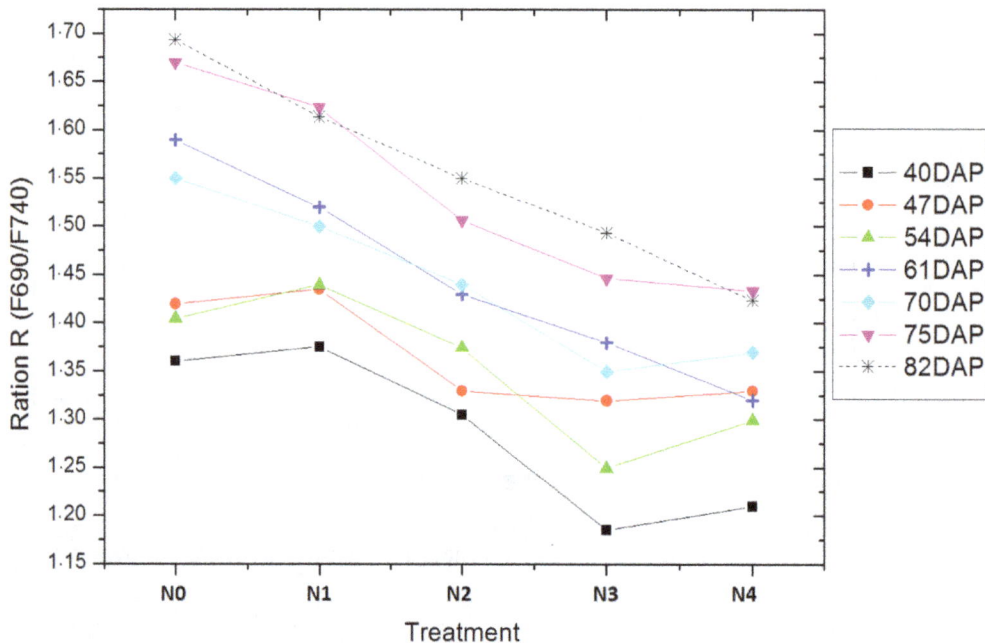

Figure 6. Changes in the relationship between the fluorescence ratio and the nitrogen treatment

The Figure 7 displays the fluorescence ratios over the plant development stage, for each nitrogenous stress level. Despite serrated evolution of certain values, we observe for all treatments, an increase of the R value depending on the plant development stage. Thus, the R value increases with the plant age. This increase is more pronounced for N0 and N1 treatments than for those of N3 and N4.

We also notice on Figure 7 that charts (N0, N1) and (N3, N4) are close to each other.

The median position of the N2 treatment chart compared to curves treatments couples (N0, N1) and (N3, N4) illustrates the average fertilization rate that we applied. Furthermore, the arrangement of each chart compared to other highlights a parallelism with different fertilization levels we generated.

The table 1 shows the differences between the pairs of curves (N0, N1); (N3, N4) and (N0, N4).

Table 1. Differences between R values for the pairs of curves (N0, N1); (N3, N4) and (N0, N4)

DAP	ΔR (N0-N1)	ΔR (N3-N4)	ΔR (N0-N4)
40 DAP	0.01	0.02	0.15
47 DAP	0.01	0.01	0.10
54 DAP	0.03	0.05	0.11
61 DAP	0.07	0.06	0.27
70 DAP	0.01	0.02	0.18
75 DAP	0.05	0.01	0.24
82 DAP	0.08	0.07	0.27

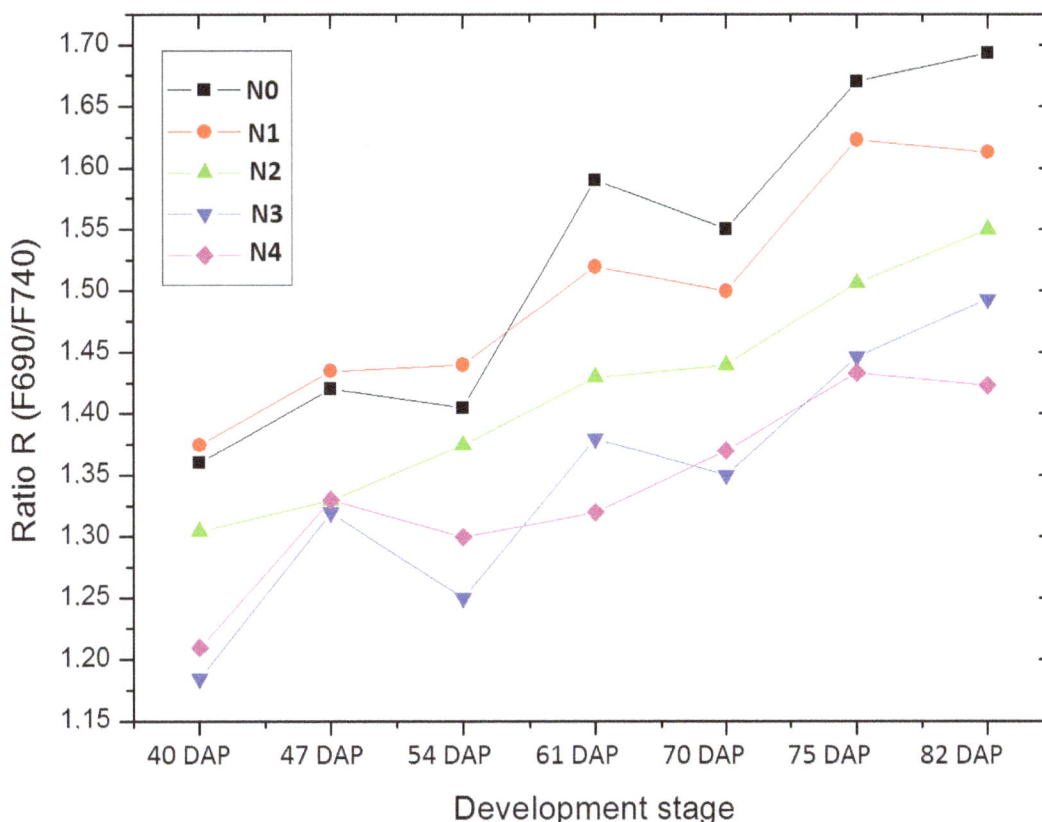

Figure 7. Temporal changes of the fluorescence ratio for the five nitrogen treatments

Table 1 shows that the greatest differences between the fertilization levels during the plant development, are obtained at 61 DAP and 82 DAP. We cannot consider the second date because it is already in the plant senescence phase. Thus, the best period to conduct early nitrogen stress detection measurements would be 61 DAP.

This table also shows that the differences between N0 and N1 on one hand and N3 and N4 on the other hand are very low compared to the differences between N0 and N4. N1 and N0 Fertilizations produce the same effect of stress on the plant. It would therefore be useless to bring 25 % of the nitrogen needs to the plant. However, it would be economical for the farmer to provide 75 % of the nitrogen needs of the plant because the N4 and N3 treatments produce the same stress effect.

We then considered both extreme fertilization levels :

- nitrogen deficient level   for N0 treatment plants;

- nitrogen fertilized level   for plants that have undergone the N4 treatment.

For both generated fertilization levels, figure 8 shows the fluorescence ratio over the development stage.

The maize variety used in this study has a short-cycle production (90-95 days). At 70 DAP, the culture is in the senescence process. Figure 8 shows that the fertilized plants have a ratio R = 1.36 on that date. All measurements performed before the senescence phase are such that:

-    $R \leq 1.36$ for fertilized corn plants;

-    $R > 1.36$ for deficient corn plants.

Moreover, the measures we took during the senescence phase provide R values greater than 1.36 whatever studied plants. In addition, the R value for fertilized plants is still lower than deficient plants.

Figure 8 also confirms that the date 61 DAP is convenient to perform nitrogen stress detection measures. These results are similar to those obtained by Soro, Adohi-Krou, Diomandé & Ebby (2004) in a study on oil palm trees.

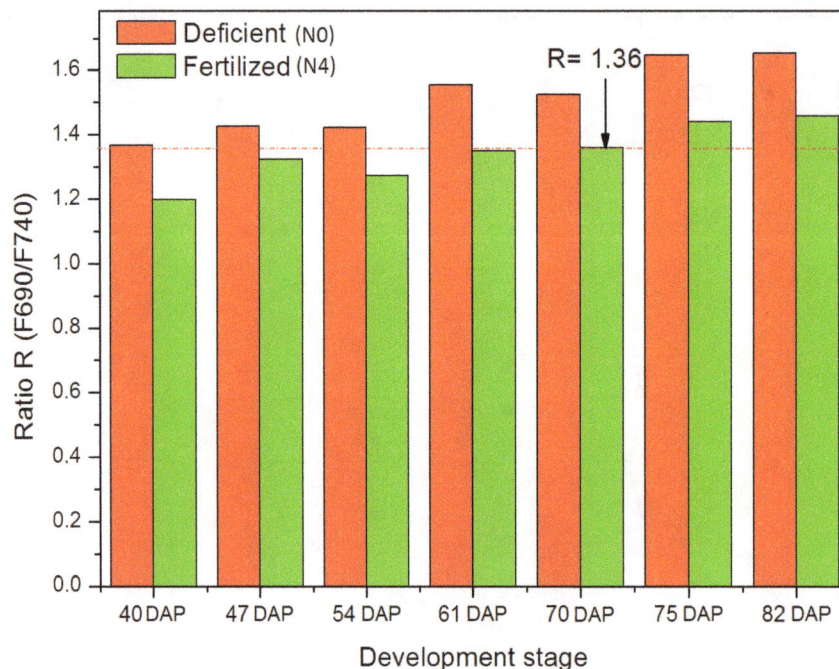

Figure 8. Fluorescence intensity ratios for N0 and N4 treatments over the development stage

5. Conclusion

This study allowed to show that it is possible to detect the nitrogen deficiency and the water deficiency of corn plants from the ratio of chlorophyll fluorescence intensities of at 690 nm and 740 nm, when the measurements are performed before the senescence phase. The senescence phase occurs earlier in plants for water stress:

- to detect water deficiency, the favorable date is DAP 51. Corn plants will be considered water deficient if $R \leq 1.34$ and water unstressed if $R > 1.34$ before the date specified for detection

- While to detect nitrogen deficiency, the convenient day would be 61 DAP. Corn plants will be considered nitrogen fertilized plants if $R \leq 1.36$ and nitrogen deficient plants if $R > 1.36$ before this date.

We find that the fluorescence ratio R increases over time, for any stress on the plant. However, R decreases with the nitrogen stress and increases with increasing water deficiency.

However, further experiments must be led to determine the amount of nitrogen and water to provide to corn plants to correct any detected deficiency.

Acknowledgements

Research reported in this paper was supported by the Programme d'Appui Stratégique à la REcherche Scientifique (PASRES): We also thank the Centre National de Recherche Agronomique (CNRA) and the Centre National de Floristique (CNF) for their scientific collaboration.

References

Akanvou, L., Akanvou, R., Anguété, K., & Diarrassouba, L. (2006). *Bien cultiver le maïs en Côte d'Ivoire*. fiche technique Centre National Recherche Agronomique, 4p. Retrieved from http://www.erails.net/images/cote-divoire/cnra/cnra/file/cultiver_mais.pdf

Bassalet, N. (2000). *Le maïs et ses avenirs*, Cahiers du LIPS, N°13, 47p. Retrieved from http://www.laprospective.fr/dyn:francais/:memoire/cahier_num13.pdf

Bourrié, B. (2007). *La fluorescence chlorophyllienne comme outil de diagnostic,* 8èmes Journées de la fertilisation raisonnée et de l'analyse de terre, GEMAS-COMIFER, Blois, 20-21 novembre 2007, 11p. Retrieved from http://www.comifer.asso.fr/images/pdf/8emes_rencontres/15%20-%20expos%20-%20%20la%20fluorescence%20chlorophyllienne%20comme%20outil%20.pdf

Chappelle, E. W., Wood F. M. Mcmurtrey J. E. & Newcomb W. W. (1984). Laser-induced fluorescence of green plants. 1:A technique for the remote detection of plant stress and species differenciation. *Applied Optics*, 23, 134-138. http://dx.doi.org/10.1364/AO.23.000134

Fedotov, Y., Bullo, O., Belov, M., & Gorodnichev, V. (2016). Experimental Research of Reliability of Plant Stress State Detection by Laser-Induced Fluorescence Method. *Int J Optics, 2016*, 1-6. http://dx.doi.org/10.1155/2016/4543094

Méthy, M., Olioso, A., & Trabaud, L. (1994). Chlorophyll fluorescence as a tool for management of plant ressources. *Remote Sensing of Environment, 47*, 2-9. http://dx.doi.org/10.1016/0034-4257(94)90121-X

Papageorgiou, G. C., & Govindjee. (2004). Chlorophyll a fluorescence: A signature of photosynthesis. *Advances in photosynthesis and respiration* (vol 19. Springer, Dordrecht, 795 p).

Perrier-Brusle, L. (2010). *Le maïs dans le monde: production, consommation et tension sur les marchés*. IRD France, 31p. Retrieved from http://laeti.perrierbrusle.free.fr/scpo_seances5_expose.pdf

Plénet, D. (1995). *Fonctionnement des cultures de maïs sous contrainte azotée - Détermination et application d'un indice de nutrition*. Thèse de doctorat, Institut National Polytechnique de Lorraine, Académie de Nancy-Metz, 247p.

PNTTA (Programme National de Transfert de Technologie en Agriculture). (1999). *Fertilisation azotée des céréales : cas des blés en Bour et en irrigué*. Bulletin mensuel d'information et de liaison du PNTTA, N° 62, 4p. Retrieved from http://agrimaroc.net/bul62.htm

RONGEAD – ONG CHIGATA. (2014). *Diagnostic de la filière maïs en Côte d'Ivoire*, 59p. Retrieved from http://www.rongead.org/IMG/pdf/Diagnostic_de_la_Filiere_Mais_en_Cote_d_Ivoire_RONGEAD_2014.pdf

Saccardy, K. (1997). *Effets de la contrainte hydrique sur le mécanisme photosynthétique des plantes en C4 ; interactions avec les fortes lumières*. Thèse, Université de Paris 11, Paris, 166 p.

Soro, P. A., Adohi-Krou, A., Diomandé, K., & Ebby, N. (2004). Caractérisation de la déficience en potassium des plants de palmier à huile par fluorescence induite par laser. *REVIST, 5*, 23-34.

Stirbet, A., & Govindjee. (2011). On the relation between the Kautsky effect (chlorophyll a fluorescence induction) and Photosystem II: Basics and applications of the OJIP fluorescence transient, *Journal of Photochemistry and Photobiology B, 104*(1-2), 236-257. http://dx.doi.org/10.1016/j.jphotobiol.2010.12.010

Tremblay, N., Wang Z., & Cerovic, Z. G. (2012). Sensing crop nitrogen status with fluorescence indicators- A review, *Agronomy Sust Developm, 32*(2), 451-464. http://dx.doi.org/10.1007/s13593-011-0041-1

Wim, V., & Andrej, P. (2013). Chlorophyll *a* fluorescence induction (Kautsky curve) in a Venus flytrap (*Dionaea muscipula*) leaf after mechanical trigger hair irritation, *Journal of Plant Physiology, 170*, 242–250. http://dx.doi.org/10.1016/j.jplph.2012.09.009

Yéo, Y. A. (2011). *Analyse de la compétitivité de la filière maïs en Côte d'Ivoire*, Programme de Renforcement et de Recherche sur la Sécurité Alimentaire en Afrique de l'Ouest (PRESAO), Composante SRAI, Michigan State University. *Rapport Final N°2-2011-11, 80p*. Retrieved from http://fsg.afre.msu.edu/srai/RCI_Rapport_competitivite_mais_RCI.pdf

6

On the Trigonometric Descriptions of Colors

Jiří Stávek[1]

[1] Bazovského 1228, 163 00 Prague, Czech republic

Correspondence: Jiří Stávek, Bazovského 1228, 163 00 Prague, Czech republic. E-mail: stavek.jiri@seznam.cz

Abstract

An attempt is presented for the description of the spectral colors using the standard trigonometric tools in order to extract more information about photons. We have arranged the spectral colors on an arc of the circle with the radius R = 1 and the central angle $\theta = \pi/3$ when we have defined cos $(\theta) = \lambda_{380}/\lambda_{760} = 0.5$. Several trigonometric operations were applied in order to find the gravity centers for the scotopic, photopic, and mesopic visions. The concept of the center of gravity of colors introduced Isaac Newton. We have postulated properties of the long-lived photons with the new interpretation of the Hubble (Zwicky-Nernst) constant $H_0 = 2.748\ldots * 10^{-18}$ kg kg^{-1} s^{-1}, the specific mass evaporation rate (SMER) of gravitons from the source mass. The stability of international prototypes of kilogram has been regularly checked. We predict that those standard kilograms due to the evaporation of gravitons lost 8.67 µg kg^{-1} $century^{-1}$. The energy of long-lived photons was trigonometrically decomposed into three parts that could be experimentally tested: longitudinal energy, transverse energy and energy of evaporated gravitons. We tested the properties of the long-lived photons with the experimental data published for the best available standard candles: supernovae Type Ia. There was found a surprising match of those experimental data with the model of the long-lived photons. Finally, we have proposed a possible decomposition of the big G (Newtonian gravitational constant) and the small kappa κ (Einsteinian gravitational constant) in order to get a new insight into the mysterious gravitational force and/or the curvature concept.

Keywords: spectral colors on an arc, trigonometrical tools, gravity centers of visions, specific mass evaporation rate (SMER), stability of the Urkilogram, long-lived photons, longitudinal energy, transverse energy, evaporation of gravitons, decomposition of the big G, decomposition of the small κ.

1. Introduction

The mathematical description of colors and the color ordering stay in the focus of researchers for centuries. There were published many valuable monographs, e.g. Isaac Newton in 1730, Johann Wolfgang Goethe in 1810, W.W. Abney in 1913, Vasco Ronchi in 1970, A.I. Sabra in 1981, Robert A. Krone in 1999, Klaus Hentschel in 2003, Rolf G. Kuehni in 2003, John A. Medeiros in 2006, Rolf G. Kuehni and A. Schwarz in 2008 (the very inspirative monograph), B. Stabell and U. Stabel in 2009, A. Mark Smith in 2015.

The first system of color ordering was the qualitative linear order positioned on the horizontal axis by Aristoteles. Marcilio Ficino in 1480 and Girolamo Cardano (very well known for his calculations with imaginary numbers) organized colors on the vertical axis with the first (pseudo) quantitative brightness of colors in 1563. The next innovation proposed Robert Fludd around 1630 with the circular form of the modified Aristotelian scale of simple colors. The breakthrough brought Isaac Newton with his invention of the semi-quantitative color wheel describing the prismatic spectrum in 1702.

The next important stimulus was the quantification of the wavelength scale of the spectrum by William Hyde Wollaston in 1802, and Joseph Fraunhofer in 1812. Hermann Günther Grassmann made the Newtonian color wheel more quantitative by adding the Fraunhofer lines into the wheel in 1853. Grassmann was working with colors as with vectors in order to predict the results of their additive mixing. Hermann Scheffler modified the color wheel into a spiral by inserting frequencies of the spectral light in 1883.

The search for the key to unlock the secret code of these strange Frauhofer lines in the solar spectrum attracted several researchers in the 19th century. A very original step made Johann Jakob Balmer in 1885 and 1897 when he used some trigonometric tools to derive his important formula later used by many researchers.

The next big breaktrough came with Planck (1900) and Einstein (1905) with their formula $h\nu = mc^2$ describing the energy of photon particles.

The main reason of this contribution is to revisit some forgotten concepts found during the light decomposition using a prism in the 17th and 18th centuries and to use standard trigonometric tools in order to extract more information about photon properties. We will propose to decompose the Planck - Einstein formula using several trigonometric tools in order to extract more information about photons. These newly predicted photon properties can be tested by instrumental techniques available to our generation and could bring to us a new glimpse into the realm of photons.

2. The Prismatic Rainbow – "Iris Trigonia"

The seventeenth century was the century in which the rainbow started to reveal its secret to several researchers in their laboratories. These scholars decomposed the white light using the prism and observed the prismatic rainbow called as "Iris Trigonia". Iris, in the Greece mythology the goddess of rainbow, inspired those scholars who slowly started to understand this phenomenom.

E.g., Marci (1648) in his forgotten book "Thaumantias liber de arcu coelesti ..." proposed that light passing the prism is compressed in the longitudinal direction and expanded in the transverse direction. This optical work of J.M. Marci was evaluated by Margaret D. Garber in 2005. The biography of J.M. Marci can be found in the book of Svobodný et al. (1998).

Boyle (1664) in his book "Experiments and Considerations touching Colours" speculated that light particles in the prism change their speed in the longitudinal and transverse directions. Robet Hooke (1665) speculated that the vibration of the light wave might cause the sense of colors. Isaac Newton in his "Optics" made many predictions for the future research. In one of his proposals he assumed that the vibration of light particles in the retina might cause the sense with various colors.

Many researchers described light particles as messengers with two properties – light energy and heat energy. E.g. Westfeld (1767) speculated that the retina could be stimulated by heat bringing with the light particles and the following increase in temperature might start the sense of colors. Johann Gottfried Voigt (1796) considered that light is a combination of light matter and caloric matter (two substances that appear on the top of Lavoisier´s table of elements).

It is possible to postulate an equation that could describe these qualitative arguments. We can mathematically describe two speeds of light particles passing through the prism by the proposal of Matzka (1850) that was rediscovered by Jiří Stávek in 2013. In this case we will postulate the new complex number – the Marci-Matzka complex number $¢$ (read $¢$ as cent):

$$¢ = \frac{c}{n} + i\, c \sqrt{1 - \frac{1}{n^2}} \tag{1}$$

where the modulus $|¢|$ has the size of the light speed constant c = 299 792 458 ms^{-1} and n is the index of refraction of the prism. We can experimentally test the norm of this complex number multiplied by the mass of the photon m (in this case there is no loss of photon mass by the attrition or no growth of photon mass in the prism):

$$E = m\,¢^2 = m\,\frac{c^2}{n^2} + m\,c^2\left(1 - \frac{1}{n^2}\right) \tag{2}$$

The first member of this expression – the longitudinal energy is visible during the photoelectric effect and was very well tested and confirmed. The second part of this expression – the transverse vibration of photons – can be tested quantitatively as heat – the increased vibrations of the molecules used in these experiments. This second effect is known qualitatively but was not tested quantitatively. E.g., Arnošt Reiser et al. (1996) found that some molecules served as local heaters after their activation by light beam absorption and started to promote the following chemical reactions.

The wave theory of heat was intensively studied in the period 1800 – 1850 as we can read in the great contributions of Brush (1970, 1986). Experiments on radiant heat by William Herschel, John Leslie, Macedonio Melloni, André Marie Ampère, and others showed that heat has most if not all of the properties of light: reflection, refraction, diffraction, polarization, interference, etc. This led to a belief that heat and light are essentially manifestions of the same physical agent.

Equation 2 can be tested for the scotopic, photopic and mesopic visions. As the index of refraction we can apply the value n = 1.337 for the vitreous humor. For the case of the scotopic vision the very well known Purkinje effect might be caused by the longitudinal photon energy (Jan Evangelista Purkyně discovered the "Purkinje effect" in 1819). The photopic vision might be caused by the transverse photon energy as was anticipated long time ago by several researchers. The mesopic vision is influenced by both mechanisms.

3. Trigonometric Description of Colors – the Duplicity Theory

The visible spectrum for a typical health human eye lays in the wavelength range from 380 nm to 760 nm. The wavelength boundaries slightly depend on the individual human eye and on the optical conditions during these vision experiments. The human vision is very complicated multi-stage event. Our modern duplicity theory of vision is based on the experimental work of many researchers. E.g., Max Schultze in 1866 discovered rods and cones in the retina, Franz Boll in 1876 discovered rhodopsin as a visual photopigment. The history of the development of the duplicity theory surveyed B. Stabel and U. Stabel in 2009.

We can arrange colors on the circle with the radius R = 1 as an arc with the central angle $\theta = \pi/3$ if we will define $\cos(\theta) = \lambda_{380}/\lambda_{760} = 0.5$. This color arc ("the color sextant") together with used trigonometric functions is depicted in Figure 1.

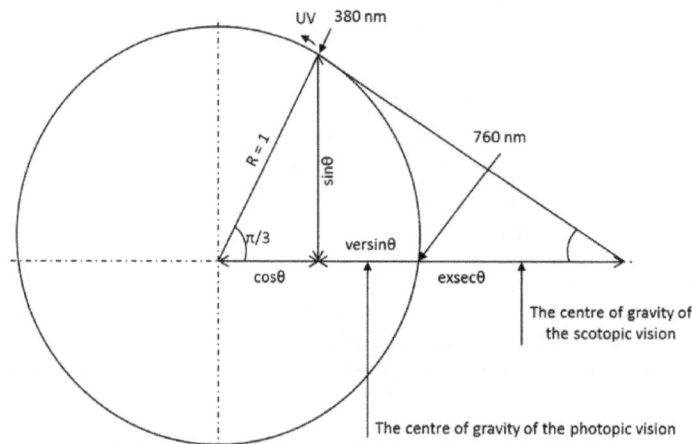

Figure 1. The "color sextant" of the visible light between 380 nm n mand 760 nm and some trigonometric functions for the determination of the centre of gravity of colors

From these relatively simple trigonometric functions we propose to formulate three Aristotelian horizontal linear color scales that describe the gravity centers of the scotopic, photopic and mesopic visions. Isaac Newton introduced in his color wheel the concept of the gravity center of colors. This topic is very well described by Briggs (2010) on his website. We will try to use this concept for the case of the trigonometric description of colors.

The scotopic vision will be modelled with the function exsec (θ) that describes the energy of individual colors where v stands for the frequency of that color:

$$ex\sec(\theta) = \sec(\theta) - 1 = \frac{1}{\cos(\theta)} - 1 = \frac{v}{v_{760}} - 1 \tag{3}$$

The gravity center of the exsecant (θ) = 0.5 can be identified with the wavelength 506.6667 nm. This is an interesting coicidence with the maximal rod sensitivity at 507 ± 7 nm for a typical human eye – the scotopic vision. The scotopic efficacy curve is officially assigned a value of unity at 507 nm by the Commision Internationale d´Eclariage (CIE). The quatitative data can be found in the Luminous Efficacy Tables.

The photopic vision can be modelled with the function versin (θ):

$$ver\sin(\theta) = 1 - \cos(\theta) = 1 - \frac{v_{760}}{v} \tag{4}$$

The gravity center of the versin (θ) = 0.25 can be identified with the wavelength 570 nm. It is known that the short-, medium-, and long-wavelength cone pigments add together to result in a psychophysical spectral sensitivity curve that peaks at 560 ± 10 nm. It is another interesting coincidence for the case of the photopic vision. In order to cover the whole color spectrum by the photopic vision the typical human retina contains about 2% of S-cones, 34% of M-cones and about 64% of L-cones. The relative number of S-, M-, and L-cones and their active surface area form together the quantitative response to individual colors.

The subjective impression of seeing has to be quantified for „normal"viewing conditions. In 1924, the Commision Internationale d´Eclairage (CIE) statistically determined with over one hundred observers the maximum photopic sensitivity under controlled conditions and defined the maximum at 555 nm. Since that time there have been numerous attempts to improve the standard photopic luminosity function to make it more representative of human vision. The great photographic chemist Edwin H. Land made an extensive research on the color vision influenced by the various illumination parameters in 1977.

The intermediate vision – the mesopic vision when both rods and cones are active in the vision process – is strongly influenced by the luminance in the range from 0.005 cd/m^2 to 5 cd/m^2. For the case of the mesopic vision we will use $\cos(\theta) = \lambda/\lambda_{760}$ in order to get probability relationships for the photopic vision ($\cos^2\theta$) and ($\sin^2\theta$) for the scotopic vision.

We can model the effect of photons on the sense of the mesopic vision as:

$$E = h\nu\left(\cos^2\theta + \sin^2\theta\right) = h\nu\left\{\left(\frac{\lambda}{\lambda_{760}}\right)^2 + \left[1 - \left(\frac{\lambda}{\lambda_{760}}\right)^2\right]\right\}$$

(5)

The first part of this expression describes the probability of the photopic vision and the second part describes the probability of the scotopic vision. The center of gravity for $\cos^2\theta = \sin^2\theta = 0.5$ can be identified with the wavelength 537.401 nm. The experimentally determined maximum for the mesopic vision depends strongly on the luminance and varies in the range 537 ± 30 nm. The future experimental study will be needed in order to investigate these coincidencies in more details. The UV light has been effectively blocked and do not enter the retina but has to be used for this statistical description as well.

Figure 2 shows the construction of three very well-known Pythagorean trigonometric identities. The trigonometric functions $\tan^2\theta$ and $\cot an^2\theta$ on the vertical axis were inspired by the Cardano´ formulas for probabilities. Geralomo Cardano was among the first researchers in the study of the calculus of probabilities. Cardano formulated two ratios r/s and s/r where r represents the favourable outcomes and s stands for unfavorable outcomes – see Gorroochum in 2012.

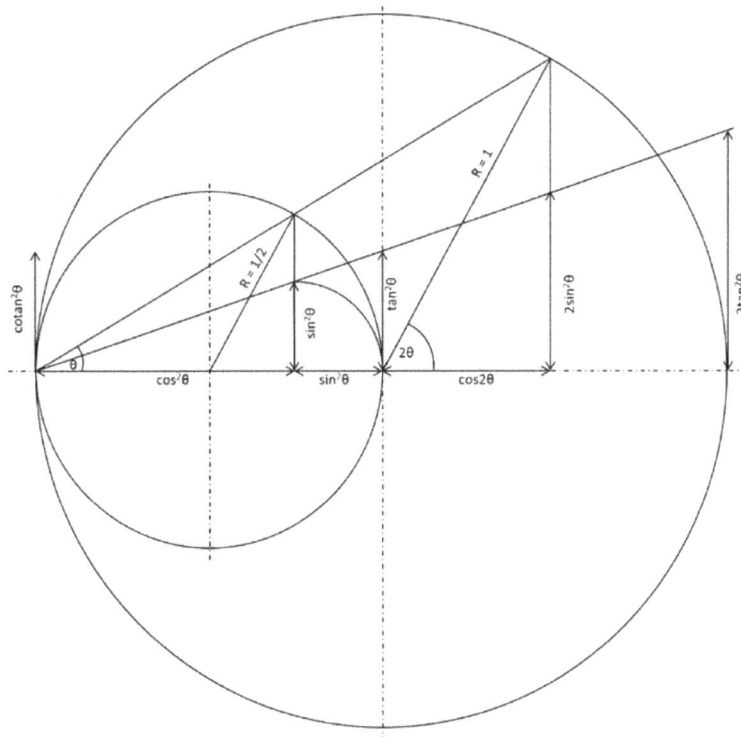

Figure 2. A trigonometric interpretation of $\tan^2\theta$ and $\cot an^2\theta$

The trigonometric functions on the vertical axis $\tan^2\theta$ and $\cot an^2\theta$ can be termed as Ficino-Cardano vertical orders of colors and are the measure of the scotopic – photopic vision probabilities for the colors in the mesoscopic vision range. Figure 3 depicts those Ficino-Cardano functions $\tan^2\theta$ and $\cot an^2\theta$ for the visible light.

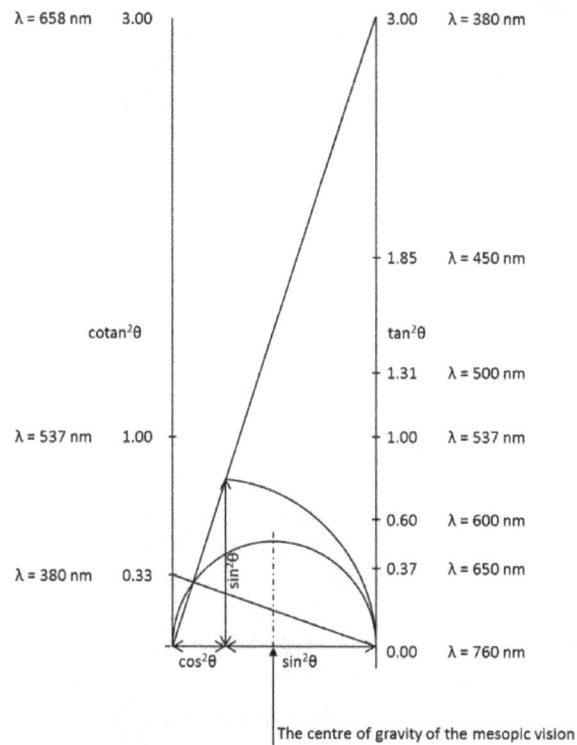

Figure 3. Ficino-Cardano functions $\tan^2\theta$ and $\cotan^2\theta$ describing the ratio of the probabilities of the scotopic–photopic visions

4. The Cosmological Transmutation of Colors

Christian Doppler (1842) predicted a color change of stars because of their relative motion to the observer. The observed redshifts and blueshifts in the analyzed spectra confirmed this Doppler prediction. However, during the next experimental investigation there were found bigger and bigger values of the observed redshifts. The dominant interpretation uses the characteristic Hubble constant with the dimension $kms^{-1}Mpc^{-1}$ describing the elasticity of the spacetime. There is one alternative concept proposed by Fritz Zwicky in 1929 and Walther Nernst in 1935. This model interprets the Hubble constant as a decay constant with the dimension s^{-1}. Nernst called the Hubbble constant as a quantum decay constant (Quantenzerfallskonstante). The development of this model slowly continued in stages through several generations. Marmet (2014) surveyed 59 interpretations of the cosmological redshift and his review might be a good starting base to this field of research.

In our trigonometric description of the cosmological transmutation of colors we will define several postulates:

(1) Hubble (Zwicky-Nernst) constant $H_0 = 2.748\ldots * 10^{-18}$ kg kg^{-1} s^{-1}, the specific mass evaporation rate of gravitons from the mass (SMER – in the Slovak language – "direction"), the numerical value will be derived later.

(2) Dependence of the photon wavelength, frequency, mass and complex longitudinal and transverse speed, and the average longitudinal speed on their life time.

(3) Matzka – Zwicky complex number describing the complex speed of the long-lived light.

(4) The cosmological hue angle $\cos(-\theta) = 1/(z+1)$ – see Figure 4.

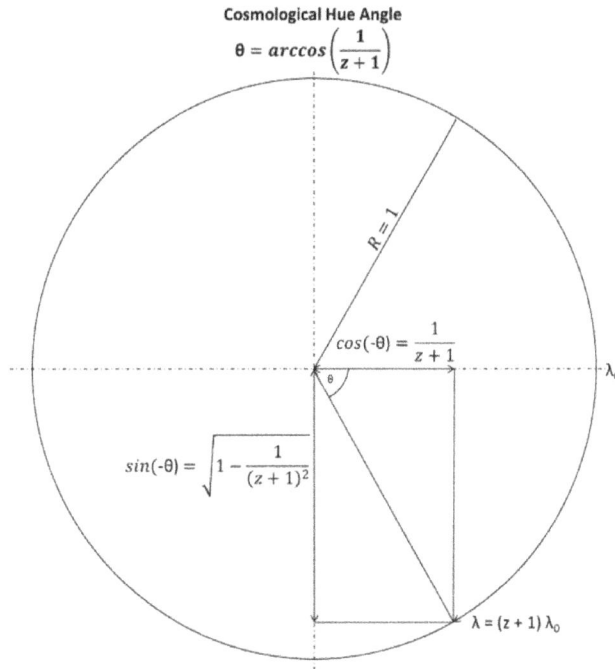

Figure 4. The cosmological hue angle $\theta = \arccos[1/(z+1)]$ describing the cosmological transmutation of colors

The photon parameters are postulated as:

$$\frac{m_1}{m_0} = \frac{\lambda_0}{\lambda_1} = \frac{v_1}{v_0} = \frac{V}{c} = e^{-H_0 t} = \frac{1}{z+1} \tag{6}$$

Parameters with the index 0 are at the source, parameters with the index 1 are at an observer, photon mass m, photon wavelength λ, photon frequency v, V is the longitudinal photon speed at the observer, c is the light speed at the source, t is time for the photon flight and z is the redshift.

The longitudinal photon speed exponentially decreases and thefore the average longitudinal speed V_{av} can be calculated as:

$$V_{av} = \frac{c}{(z+1)^{1/e}} \tag{7}$$

Matzka – Zwicky complex number č (read "č" as cheers!) describes the longitudinal and transverse photon speed:

$$\check{c} = \frac{c}{z+1} - i\,c\sqrt{1 - \frac{1}{(z+1)^2}} \tag{8}$$

The first hypothesis of the varying-speed of-light was proposed by Peter Wold in 1935. Since that time this hypothesis was used several times in later works. The big problem of this hypothesis is the constancy of the Rydberg constant, the fine structure constant and other constants where the value of the light speed constant c = 299 792 458 ms^{-1} has to be inserted. In our model we can insert in those constants the modulus of Matzka – Zwicky complex number $|\check{c}| = 299\ 792\ 458$ ms^{-1}.

The trigonometric decomposition of the expression $E = m_0 c^2$ for the long-lived photon is depicted on Figure 5.

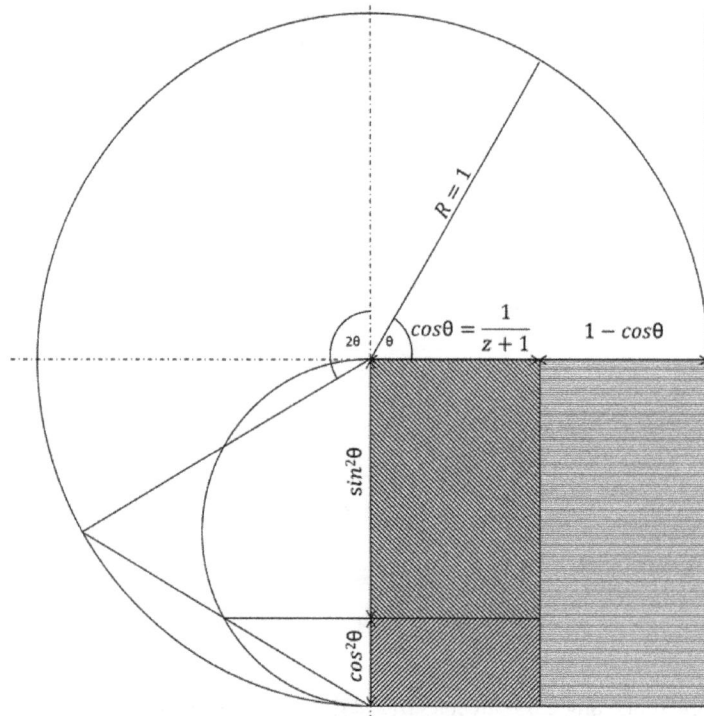

Figure 5. Trigonometric decomposition of the energy of the long-lived photon into the longitudinal energy, transverse energy, and graviton energy

This expression is composed from three members: the longitudinal energy and the transverse energy of the long-lived light and the third member stands for the graviton energy emitting from the long-lived light during its flight to an observer:

$$E = m_0 c^2 \left[\cos^3 \theta + \cos \theta \sin^2 \theta + \left(1 - \cos \theta \right) \right] =$$

$$= m_0 c^2 \left\{ \left[\frac{1}{\left(z + 1 \right)^3} \right] + \left[\frac{1}{\left(z + 1 \right)} \left(1 - \frac{1}{\left(z + 1 \right)^2} \right) \right] + \left[1 - \frac{1}{\left(z + 1 \right)} \right] \right\}$$

(9)

This trigonometric model of the long-lived light can be tested using the experimental data from the Hubble telescope:

(1) The longitudinal speed V = c/(z+1) – see data for the stretched temporal evolution of SNe Ia spectra: Leibundgut et al. in 1996, Goldhaber et al. in 1996 and 2001, Foley et al. in 2005, and Blondin et al. in 2008.

(2) The photoelectric effect of the long-lived light: in this experiment the longitudinal energy dependence on the redshift should be found with the exponent n = 3. The experimental data on the Tolman surface brightness test found the value n = 3.0 ± 0.4. For data see Lori M. Lubin and Alan Sandage (2001), Alan Sandage (2009) and Wikipedia (2016).

(3) Heat effect of the long-lived light – the transverse energy: the experimental data are not available.

(4) Specific mass evaporation rate (SMER – in the Slovak language „direction") – graviton evaporation from the long-lived photons – we should focus our attention on the value of the Hubble (Zwicky – Nernst) constant with the dimension kg kg^{-1} s^{-1} and to compare this value with the mass evaporation of the Urkilogram.

Figure 6 summarizes the dependence of these three parameters on the redshift of those photons.

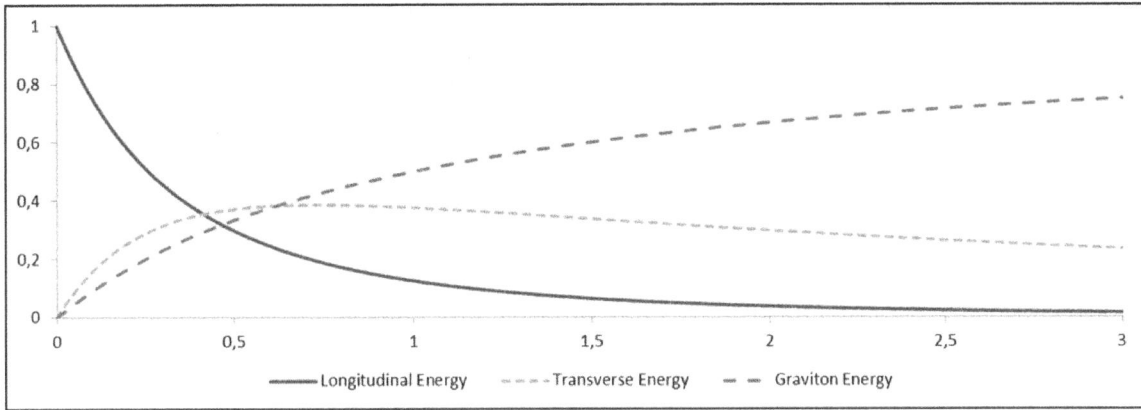

Figure 6. The predicted behavior of the long-lived photons on their redshift

We can test this long-lived light model using the experimental data published for the standard candles Supernova Type Ia. Firstly, we have to find the relation between the geometrical distance d of the measured supernova Type Ia and its luminosity distance d_L from the measurement of the longitudinal energy of long-lived photons:

$$E = h\frac{v_0}{z+1} = \frac{m_0 c^2}{(z+1)^3} \tag{10}$$

From this relation we find that the geometrical distance $d = d_L/(z+1)$. Then, we can express the luminosity distance from the postulates:

$$\ln(z+1) = H_0 t = H_0 \frac{d}{V_{AV}} = H_0 \frac{d_L}{(z+1)} \frac{(z+1)^{1/e}}{c} =$$
$$= \frac{H_0}{c} d_L (z+1)^{1/e-1} \tag{11}$$

$$d_L = \left[\frac{c}{H_0}(z+1)^{1-1/e} \ln(z+1)\right] \tag{12}$$

The Union2 Compilation (Amanullah et al., 2010) can be very well fitted with luminosity distance inserted into the standard formula for the supernova Type Ia:

$$m - M = 5 \log_{10}\left[\frac{d_L}{Mpc}\right] + 25 \tag{13}$$

$$m - M = 5 \log_{10}\left[\frac{c}{H_0}(z+1)^{1-1/e} \ln(z+1)\right] + 25 \tag{14}$$

where m is the apparent magnitude of the observed supernova, M is the absolute bolometric magnitude for the supernova Type Ia, c is the light speed, H_0 is the Hubble (Zwicky-Nernst) constant with the dimension kg kg^{-1} s^{-1} and z is the observed redshift, 25 enters the expression because the luminosity distance was expressed in Mpc.

The volumetric SN Type Ia rate has been recently studied in details using the Hubble telescope– see the data in Rodney et al. (2014) and Graur et al. (2014) with the maximum of the observed number of supernovae Type Ia in the spherical volume shell at around the redshift $z = 1.60 \pm 0.10$. With our model we have found the maximum spherical volume shell at the redshift $z = 1.59$:

$$V_{eff} = V_2 - V_1 = \frac{4}{3}\pi \frac{c^3}{H_0^3}\left[\frac{(\ln(z_2+1))^3}{(z_2+1)^{3/e}} - \frac{(\ln(z_1+1))^3}{(z_1+1)^{3/e}}\right] \tag{15}$$

where V_{eff} is the spherical volume shell between the redshift z_2 and z_1, c is the light speed, H_0 is the Hubble (Zwicky-Nernst) constant with the dimension kg kg^{-1} s^{-1} and z_1 and z_2 are two redshifts for the spherical volume shell calculation.

5. What is the Internal Structure of the Big G and the Small Kappa κ?

In order to penetrate more deeply into the structure of the Newtonian big G and the Einsteinian constant small kappa κ we should try to decompose the big G into several pieces that might bring to us more information about the mysterious gravitational force and/or the curvature concept.

We should fulfill four conditions:

(1)   Big G should contain the Hubble (Zwicky-Nernst) constant with the dimension kg kg^{-1} s^{-1} describing the specific mass evaporation rate (SMER) of gravitons from the source mass – the messengers of the gravitational force.

(2)   Big G should be composed from the square of the longitudinal speed of gravitons $[c/(z+1)]^2$ because the product $\Sigma m_g * [c/(z+1)]^2$ describes the longitudinal energy of gravitons. This member $[c/(z+1)]^2$ might explain the great success of the predictions of the Einsteinian curvature theory using his constant small kappa $\kappa = 8\pi G/c^2$ (see Wikipedia for two possible writings of the kappa with c^2 or c^4). This clever definition of the constant kappa eliminated the square of the longitudinal speed of gravitons $[c/(z+1)]^2$ from the big G. Since Einstein's constant κ had been evaluated by a calculation based on a time-independent metric, this by no mean requires that G and c must be constant themselves, but only their ratio G/c^2 must be the absolute constant.

(3)   Big G should contain the Planck constant h to create a bridge between the microworld and the macroworld.

(4)   The remaining member of the big G should match the dimension and the size the big G. As the best candidate we have found the size of the classical electron diameter $d_e = 2\, r_e = 5.63588 * 10^{-15}$ m.

$$G = \frac{d_e^3}{h} \left(\frac{c}{z+1} \right)^2 \frac{\Sigma m_{graviton}}{M_s T}$$

(16)

From the known constants G, d_e, h, c, for $z = 0$, we will get the value of the Hubble (Zwicky-Nernst) constant $2.74856 * 10^{-18}$ kg kg^{-1} s^{-1} (or $H_0 = 84.8116$ km s^{-1} Mpc^{-1}).

This value H_0 is higher than the recommended value from the astrophysical measurements (H_0 between 65 – 70 km s^{-1} Mpc^{-1}). The complexity of the experimental research on the Hubble constant can be found on the great web site of Huchra (2008). We should keep in mind that we get the deceleration parameter a_c in the form:

$$a_c = \frac{c\, H_0}{(z+1)} = \frac{c}{(z+1)} H_0$$

(17)

We have two choices for the calculation of the value H_0. We can use for those calculations the modulus of the light speed $|c|$ or the longitudinal light speed $c/(z+1)$. The distance of the redshifted objects plays a very important role, too.

We might approach this Hubble (Zwicky-Nernst) constant from some other field of the experimental research. The stability of international prototypes of kilogram has been regularly checked. Girard (1994) analyzed the stability of mass of those standard kilograms during the last century and his data can be interpreted in such a way that in the average one kilogram lost during one century about 30 ± 30 μg kg^{-1} century^{-1}. In our concept the Hubble (Zwicky-Nernst) constant $2.74856 * 10^{-18}$ kg kg^{-1} s^{-1} describing the evaporation of gravitons we predict that the standard kilogram should evaporate gravitons with the rate 8.67 μg kg^{-1} century^{-1}. In the recent time several researchers in this field continue to develop more sensitive instruments for the better determination of the stability of the international prototypes of kilograms.

Based on the known constants G, d_e, h, c, for $z = 0$, we can define the graviton flux $\Phi_{e,M}$ with the dimension W kg^{-1} as:

$$\Theta_{e,M} = \left(\frac{c}{z+1} \right)^2 \frac{\Sigma m_{graviton}}{M_s T} = 0.247...\, W\, kg^{-1}$$

(18)

(Comment: We can compare the Solar Luminosity $L_S = 3.846 * 10^{26}$ W with the Solar Gravitality $G_S = 4.914 * 10^{29}$ W).

6. Conclusions

(1)   The spectral colors were arranged on the arc of the circle with the radius $R = 1$ and the central angle $\theta = \pi/3$ when we have defined $\cos(\theta) = \lambda_{380}/\lambda_{760} = 0.5$.

(2) Trigonometric functions were used to find the gravity centers for the scotopic, photopic, and mesopic visions. This approach might bring new information about the mechanisms of those scotopic, photopic and mesopic visions.

(3) The properties of the long-lived photons were postulated with the new interpretation of the Hubble (Zwicky-Nernst) constant $H_0 = 2.748... * 10^{-18}$ kg kg^{-1} s^{-1}, the specific mass evaporation rate of gravitons from the source mass (SMER).

(4) It was predicted that international standard kilograms due to the evaporation of gravitons lost 8.67 µg kg^{-1} century^{-1}.

(5) The energy of the long-lived photons was trigonometrically decomposed into tree contributions that could be experimentally tested: the longitudinal energy, the transverse energy, and the energy of evaporated gravitons.

(6) The postulates of the long-lived photons were tested with the actual data sets found for the standard candles – supernova Type Ia. The match of predictions with the experimental data was very good.

(7) It was proposed a possible decomposition of the big G (Newtonian gravitational constant) and the small kappa κ (Einsteinian gravitational constant) in order to get a new insight into the mysterious gravitational force and/or the curvature concept.

Acknowledgments

This work was supported by the JP&FŠ Agency (Contract Number 25g/1963), by the VZ&MŠ Agency (Contract Number 16000/1989) and by the GMS Agency (Contract Number 69110/1992). We have found the valuable support from PP Agency that derived the Equation 7. The BŠ Agency translated the inspirative book "Thaumantias liber de arcu coelesti..." from the Latin language.

References

Abney, W. W. (1913). *Researches in Colour Vision and the Trichromatic Theory*. Longmans, Green and Co., New York.

Amanulah, R. (2010). *Spectra and HST Light Curves of Six Type Supernovae at 0.511 <z< 1.12 and the Union2 Compilation*. http://dx.doi.org/10.1088/0004-637X/716/1/712

Anonymous. (1979), J. W. Goethe: Farbenlehre, mit Einleitungen und Erläuterungen von Rudolf Steiner. Hrsg.: G. Ott und H. O. Proskauer, 3 Bände in Kassetten, 896 Seiten, 20 Farbtafeln, Verlag Freies Geistesleben Stuttgart 1979, kart. DM 38. *Physik in unserer Zeit, 10*, 194. http://dx.doi.org/10.1002/piuz.19790100610

Assis, A. K. T., & Neves, M. C. D. (1995). The redshift revisited. Astrophys. *Space Sci., 227*, 13-24. http://dx.doi.org/10.1007/BF00678063

Balmer, J. J. (1885). Notiz über die Spektrallinien des Wasserstoffs. *Annalen der Physik, 25*, 80-87.

Balmer, J. J. (1897). A new Formula for the Wave-lengths of Spectral Lines. *Astrophys. J., 5*, 199-209.

Blondin, S. (2008). Time Dilation in Type Ia Supernova Spectra at High Redshift. http://dx.doi.org/10.1086/589568

Boyle, R. (1664). *Experiment and Considerations touching Colours*. Reprinted Johnson Reprint. Corp., New York. Retrieved from http://www.gutenberg.org/ebooks/14504?msg=welcome_stranger

Briggs, D. (2010). *The Dimensions of Colour*. Retrieved from http://www.huevaluechroma.com/071.php; https://sites.google.com/site/djcbriggs/classes

Brush, S. G. (1970). The Wave Theory of Heat: A Forgotten Stage in the Transition from the Caloric Theory to Thermodynamics. *The British Journal for the History of Science, 5*, 145-167. http://dx.doi.org/10.1017/S0007087400010906

Brush, S. G. (1986). *The Kind of Motion We Call Heat: A History of the Kinetic Theory of Gases in the Nineteenth Century, Book 1: Physics and the Atomists*. North Holland, Amsterdam. http://dx.doi.org/10.1063/1.299504

Einstein, A. (1905). Über einen die Erzeugung und Verwandlung des Lichtes betreffenden heuristischen Gesichtspunkt. *Annalen der Physik, 17*, 132-148. Article first published online: 10 MAR 2006. http://dx.doi.org/10.1002/andp.19053220607

Einstein, A. (1915). Die Feldgleichungen der Gravitation. *Sitzungsberichte der Preussischen Akademie der Wissenschaften zu Berlin*, 844-847.

Foley, R. J., Filippenko, A. V., Leonard, D. C., Riess, A. G., Nugent, P., & Perlmutter, S. (2005). A definitive measurement of time dilation in the spectral evolution of the moderate-redshift type Ia supernova 1997ex. *The Astrophysical Journal Letters, 626*(1), L11. http://dx.doi.org/10.1086/431241

Garber, M. D. (2005). Chymical Wonders of Light: J. Marcus Marci´s Seventeenth-Century Bohemian Optics. *Early Science and Medicine, 10*, 478-509. http://dx.doi.org/10.1163/157338205774661843

Girard, G. (1994). The Third Periodic Verification of National Prototypes of the Kilogram (1988–1992). *Metrologia, 31*, 317-336. http://dx.doi.org/10.1088/0026-1394/31/4/007

Goldhaber, G., Boyle, B., Bunclark, P., Carter, D., Couch, W., Deustua, S., ... & Glazebrook, K. (1996). Cosmological time dilation using type Ia supernovae as clocks. *Nuclear Physics B-Proceedings Supplements, 51*(2), 123-127. http://dx.doi.org/10.1016/S0920-5632(96)00493-8

Goldhaber, G., Groom, D. E., Kim, A., Aldering, G., Astier, P., Conley, A., ... & Goobar, A. (2001). Timescale stretch parameterization of type Ia supernova B-band light curves. *The Astrophysical Journal, 558*(1), 359. http://dx.doi.org/10.1086/322460

Gorroochum, P. (2012). Some Laws and Problems of Classical Probability and How Cardano Anticipated Them. *Chance, 25*, 13-20. http://dx.doi.org/10.1080/09332480.2012.752279

Grassmann, H.G. (1853) Zur Theorie der Farbenmischung. *Annalen der Physik, 89*, 69-84. http://dx.doi.org/10.1002/andp.18531650505

Graur, O., Rodney, S. A., Maoz, D., Riess, A. G., Jha, S. W., Postman, M., ... & Strolger, L. G. (2014). Type-Ia supernova rates to redshift 2.4 from clash: The cluster lensing and supernova survey with Hubble. *The Astrophysical Journal, 783*(1), 28. http://dx.doi.org/10.1088/0004-637X/783/1/28

Hentschel, K. (2003). *Mapping the Spectrum. Techniques of Visual Representation in Research and Teachning.* Oxford, Oxford University Press. http://dx.doi.org/10.1093/acprof:oso/9780198509530.001.0001

Huchra, J. (2008). *The Hubble constant.* Retrieved from https://www.cfa.harvard.edu/~dfabricant/huchra/hubble/

J. W. Goethe: Zur Farbenlehre. Hrsg.: G. Ott und H. O. Proskauer, 3 Bände in Kassetten, Verlag Freies Geistesleben Stuttgart 1979.

Krone, R. A. (1999). *A History of Color.* Dordrecht, Kluwer Academic Publishers. http://dx.doi.org/10.1007/978-94-007-0870-9

Kuehni, R. G. (2003). *Color Space and Its Divisions: Color Order from Antiquity to the Present.* Hoboken,NJ, JohnWiley. http://dx.doi.org/10.1002/0471432261.ch9

Kuehni, R. G., & Schwarz, A. (2008). *Color Ordered.* A Survey of Color Order Systems from Antiquity to the Present Oxford, Oxford University Press. http://dx.doi.org/10.1093/acprof:oso/9780195189681.001.0001

Land, E. H. (1977). The Retinex Theory of Color Vision. *Scientific American, 237*, 108-128. http://dx.doi.org/10.1038/scientificamerican1277-108

Leibundgut, B., Schommer, R., Phillips, M., Riess, A., Schmidt, B., Spyromilio, J., ... & Kirshner, R. P. (1996). Time dilation in the light curve of the distant Type Ia supernova SN 1995K. *The Astrophysical Journal Letters, 466*(1), L21.http://dx.doi.org/10.1086/310164

Lubin, L. M., & Sandage, A. (2001). *The Tolman Surface Brightness Test for the Reality of the Expansion. Part IV. A Measurement of the Tolman Signal and the Luminosity Evolution of Early-Type Galaxies.* http://dx.doi.org/10.1086/322134

Luminous Efficacy Tables. (2016). Retrieved from http://hyperphysics.phy-astr.gsu.edu/hbase/vision/efficacy.html

Marci, J. M. (1648). Thaumantias liber de arcu coelesti..., Prague. THAVMANTIAS. LIBER DE ARCV COELESTI DEQVE Colorum Apparentium natura, ortu, et causis. In quo pellucidi Opticae fontes a suo scaturigine, ab his vero colorigeni rius deriuantur. DVCIBVS Geometria, et Physica Hermetoperipatetica / PRAGAE: TYPIS ACADEMICIS, 1648. Retrieved from http://aleph.nkp.cz/F/?func=direct&doc_number=000134631&local_base=STT

Marmet, L. (2014). *On the Interpretation of Red-Shifts: A Quantitative Comparison of Red-Shifts Mechanisms II.* Retrieved from http://www.marmet.org/cosmology/redshift/mechanisms.pdf

Matzka, W. (1850). *Versuch einer richtigen Lehre von der Realität der vorgeblich imaginären Grössen der Algebra, oder einer Grundlehre von der Ablenkung algebraischer Grössenbeziehungen.* Prague: J.G.Calve.

Medeiros, J. A. (2006). Cone Shape & Color Vision: Unification of Structure and Perceptions. Fifth Estate Publishers, Blountsville. http://dx.doi.org/10.1016/j.ajo.2006.12.026

Nernst, W. (1935). Einige weitere Anwendungen der Physik auf die Sternentwicklung. *Sitzungberichte der Preußischen Akademie der Wissenschaften, 28,* 473-479.

Newton, I. (1730, 4th Ed.). *Optics.* New York: Dover Publications.

Planck, M. (1900). Über irreversible Strahlungsvorgänge. *Annalen der Physik, 306,* 719-737. http://dx.doi.org/10.1002/andp.19003060410.

Purkyně, J. E. (1825). Neue Beiträge zur Kenntniss des Sehens in Subjektiver Hinsicht. Reimer, Berlin. Retrieved from http://aleph.nkp.cz/F/?func=direct&doc_number=004916257&local_base=SKC

Reiser, A., Shih, H.-Y., Yeh, T.-F., & Huang, J.-P. (1996). Novolak-Diazoquinone Resists: The Imaging Systems of the Computer Chip. Angew. *Chem., Int. Ed. Engl., 35,* 2428-2440. http://dx.doi.org/10.1002/anie.199624281.

Rodney, S. A., Riess, A. G., Strolger, L. G., Dahlen, T., Graur, O., Casertano, S., ... & Jha, S. W. (2014). Type Ia supernova rate measurements to redshift 2.5 from CANDELS: Searching for prompt explosions in the early universe. *The Astronomical Journal, 148*(1), 13. http://dx.doi.org/10.1088/0004-6256/148/1/13

Ronchi, V. (1970). *The Nature of Light. A Historical Survey.* Cambridge: Harvard University Press. http://dx.doi.org/10.1017/S0007087400011651

Sabra, A.I. (1981). *Theories of Light from Descartes to Newton.* Cambridge: Cambridge University Press. http://dx.doi.org/10.1063/1.2915547

Sandage, A. (2010). The Tolman surface brightness test for the reality of the expansion. V. Provenance of the test and a new representation of the data for three remote Hubble space telescope galaxy clusters. *The Astronomical Journal, 139*(2), 728. http://dx.doi.org/10.1088/0004-6256/139/2/728

Scheffler, H. (1883). *Die Theorie des Lichtes, physikalisch und physiologisch; mit spezieller Begrindung der Farbenblindheit.* Foerster, Leipzig.

Smith, A. M. (2015). *From Sight to Light. The Passage from Ancient to Modern Optics.* Chicago: The University of Chicago Press. http://dx.doi.org/10.7208/chicago/9780226174938.001.0001

Stabel, B., & Stabel, U. (2009). *Duplicity Theory of Vision. From Newton to the Present.* Cambridge: Cambridge University Press. http://dx.doi.org/10.1017/CBO9780511605413

Stávek, J. (2013). On the One Trigonometric Interpretation – The Decomposition Loophole. *Applied Physics Research, 5,* 131-144. http://dx.doi.org/10.5539/apr.v5n6p131

Svobodný, P. Ed. (1998). *Joannes Marcus Marci. A Seventeenth-Century Bohemian Polymath.* Prague: Charles University.

Westfeld, C. F. (1767). Die Erzeugung der Farben, eine Hypothese. Göttingen Universität, Göttingen. Retrieved from http://resolver.sub.uni-goettingen.de/purl?PPN640631835

Wold, P. (1935). On the redward shift of spectral lines of nebulae. *Physical Review, 47,* 21-29.

Zwicky, F. (1929). On the redshift of spectral lines through interstellar space. *Proc. Natl. Acad. Sci. USA, 15,* 773-779. http://dx.doi.org/10.1073/pnas.15.10.773

7

The Relationship Between the Possibility of a Hidden Variable in Time, Possible Photon Mass, Particle's Energy, Momentum and Special Relativity

Eyal Brodet[1]

[1] Michaeslon St, Jerusalem, 93707, Israel

Correspondence: Eyal Brodet, Michaeslon St, Jerusalem, 93707, Israel. E-mail: eyalbrodet@hotmail.com

Abstract

In this paper we will discuss the relationship between a possible hidden variable in time, f_r, and possible photon mass, particle energy, momentum and special relativity. One of the implications of the possibility of a hidden variable in time that may explains the origin of unstable particle decay time distributions, is the possible existence of f_r for stable particle such as photons. It will be discussed, that f_r may be linked to the photons spin and wave function, which may lead to the conclusion that the photon has a rest mass. More specifically, it will be argued that in order to explain the photon's large energy range, the photon may have a set of masses. Following the above, a correction to the energy and momentum's expressions given by special relativity will be presented. Possible experimental ways to test the above will be discussed.

Keywords: hidden variable, time, photon, spin, wave function, energy, special relativity

1. Introduction

In this paper we will discuss the relationship between a possible hidden variable in time, f_r, and possible photon mass, particle energy, momentum and special relativity. The hidden variable in time possibility, f_r, first suggested in Brodet (2010), attempts to explain unstable particle decay time distribution by a compatible distribution of f_r values. This is where f_r describes a range of frequencies for the unstable particle's internal virtual boson fluctuations. Although, stable particles, such as a photon, do not decay and therefore do not have a defined decay time, it is nevertheless suggested that also a photon should have such a hidden variable, f_r.

This suggestion originate from considering a process such as, $\gamma\gamma \to \mu^+\mu^-$. In this process the outgoing muons have a distribution of decay time. Therefore if the muons do have a hidden variable in time, in f_r, it is logical it originates from the process initial conditions, which in this case are the $\gamma\gamma$ and therefore the $\gamma\gamma$ should also have a hidden variable in time, f_r.

In Brodet (2016), it was suggested that f_r may be linked to the muon's spin and wave function. In this paper, it will be suggested that also in the case of the photon, f_r may be related to the photon's spin and wave function, and that in fact this suggests that the photon may have a non-zero rest mass. More specifically it will be suggested that the photon energy range may be explained by a set of photons with a set of different non-zero rest masses. Consequently, this means the expression for the photon's, energy and momentum, and in general any particle energy and momentum, given by special relativity should be modified. Therefore, a relevant correction is suggested to the energy and momentum expressions given by special relativity. At the end of the paper, general experimental ways are suggested to test the above.

The paper is organized as follows. Section 2 describes the theoretical background. Section 3 describes general experimental ways to test the existence of a non-zero photon mass or massive values and the consequence modification to special relativity. Section 4 contains the conclusions.

2. Theoretical Background

The initial conditions in the process $\gamma\gamma \to \mu^+\mu^-$ may generate two distributions in the final muons; one in space and one in time. As discussed in (1), the muon decay time distribution, may be generated by the possibility of a hidden variable in time, f_r. As discussed in Brodet (2015), the spatial distribution may be generated by a global hidden variable, H_{global}, which is suggested to be responsible for the muon's production angular distribution.

The experimental fact that the spatial and time distribution are not correlated, and the fact that a key ingredient in the $\gamma\gamma$ initial conditions is the $\gamma\gamma$ wave function, makes it logical to associate the final muon spatial and time distributions to the $\gamma\gamma$ initial wave functions which contains separate spatial and time parts.

Therefore, when we consider the spatial and time distributions of a single outgoing muon μ^-, we may argue that it is linked to the initial γ wave function. Since the wavy part of the γ wave function ($e^{i(E \cdot t + P \cdot x)}$) contains two parts, the energy-time phase, $e^{i(E \cdot t)}$ and the momentum-position phase, $e^{i(P \cdot x)}$, it would be logical to associate the muon decay time distribution to the energy-time phase and the muons spatial or angular distribution to the momentum-position phase.

In Brodet (2016) it was argued that one can relate the final muon decay time distribution to its intrinsic angular momentum value which can be identified as the muon's spin and to its wave function value at zero momentum ($P_\mu = 0$). Therefore if we consider such a muon production through a process such as, $e^+ e^- \rightarrow \mu^+ \mu^-$, we may associate, assuming angular momentum conservation, the final muon spin value to the initial electron spin value.

Therefore, comparing the $e^+ e^- \rightarrow \mu^+ \mu^-$ to the process, $\gamma\gamma \rightarrow \mu^+ \mu^-$, one can similarly relate the final muon decay time distribution to the initial photon spin and wave function value. This means that the photon spin is related to the energy-time phase of the photon's wave function which means that the photon must have a non-zero rest mass. That is:

$S_\gamma = E_\gamma \cdot t_\gamma$ where E_γ is the photon rest energy equals to $E_\gamma = M_\gamma \cdot c^2$ and $t_{\gamma(i)}$ is photon's i internal time, which may be given in terms of photon's i possible hidden variable in time $f_{r(i)}$:

$$t_{\gamma(i)} = \tau_\gamma \cdot \ln\left(\frac{M_\gamma}{f_{r(i)}}\right) \tag{1}$$

The above suggestion for a non-zero photon rest mass, poses an inconsistency with Einstein's special relativity as if a photon is an object with mass, according to special relativity its energy should be given by:

$$E = \gamma_{b\_original} \cdot M_\gamma \cdot c^2, \text{ where } \gamma_{b\_original} = \sqrt{1 - \frac{v_\gamma^2}{c^2}}.$$

However, since the velocity of a photon is always $v_\gamma = c$, the factor $\gamma_{b\_original} \rightarrow \infty$, and so the photon gets an infinite energy. This contradicts off Corse experimental observations where the photon's energy is measured with a range of values, from very low energy to very high but certainly not infinity.

Therefore, if we were to adopt the possibility of a hidden variable in time, f_r, the factor $\gamma_{\_b\_original}$ should be modified to agree with experimental observations.

It is suggested here to include, for every particle the value of its hidden variable in time f_r, in the expression for its boost factor, γ_b, such:

$$\gamma_{b(i)} = \sqrt{1 - \frac{v^2}{(c + \frac{a1}{f_{r(i)}})^2}} \tag{2}$$

Where $a1$ is a constant with the appropriate dimension to transform the term $\frac{a1}{f_{r(i)}}$ to be in velocity dimension.

It appears experimentally that a photon may have any energy value from very low to almost infinite energy values. If the above is correct, this would mean that low energy photons always have low f_r values and high energy photons have high f_r values and that the whole f_r spectrum is divided to parts according to the photon's energy spectrum. This would make the f_r idea not consistent with the experimental observation that even high energy photons(with large f_r values) may produce muons with long decay time values which should correspond according to the initial f_r idea to low f_r values (Brodet, 2010).

Therefore, it is suggested that the photon in fact comes with a discrete range of masses $j = 1, 2, 3.... \rightarrow n$ which are compatible to its energy range. Hence, a photon with large energy has a larger mass value and larger f_r and γ_b values and a photon with low energy has a low mass value and a low f_r and γ_b values. More specifically, it is suggested that each photon f_r shape variable, τ_γ ,(according to equation 1), is related to the mean photon mass, $M_{mean(j)}$ by:

$$\tau_{mean(j)} = f(M_{mean(j)}) \qquad\qquad (3)$$

The function $f(M_{mean})$ is unknown at this stage but maybe similar to the Standard Model function that calculated the mean lifetime of a particle as a function of its rest mass.

Therefore, the above means that behind each photon with a different given mean rest mass $M_{mean(j)}$ stands a different distribution of $f_{r(j)_i}$ values. Therefore, behind a photon with high energy and large $M_{mean(j)}$ mass value, stands a distribution of $f_{r(j,i)_i}$ values from $i \to n$ and $f_{r(j,i)_i}$ from zero to $M_{mean(j)}$ ($0 \to M_{mean(j)}$) according to its $\tau_{(mean)j}$ and equation 1. This means that even a photon with a large rest mass, $M_{mean(j)}$, that usually may be associated with high energies, could get low $f_{r(j,i)}$ and $\gamma_{b(j,i)}$ values and have low energies. However this situation is not expected to be very abundant as its $\tau_{(mean)j}$ would be very small and the number of such photon particles should be very small also.

The above may be related to the known photon's energy of: $E = h \cdot \nu$ by:

$$E_\gamma = a2 \cdot S_{\gamma(j,i)} \cdot \nu_j \qquad\qquad (4)$$

Where $S_{\gamma(j,i)} = M_j \cdot (c + \dfrac{a1}{f_{r(i)}})^2 \cdot \tau_j \ln(\dfrac{M_j}{f_{r(i)}})$, ν_j is the photon's frequency and $a2$ is a constant with yet an

unknown value.

3. Experimental Search for Photon Mass

In the case the above suggestion is correct; there should be experimental evidence for that. The above suggestion may affect the photon's spin and energy. Therefore, we should look at situations in which the photon's spin or energy may affect the experimental results.

For the spin case, it is not clear at this stage how to bias the photon's spin distribution and measure the possible effect it may have on measurements as the total cross section of a process such as: $\gamma\gamma \to \mu^+\mu^-$. So at this stage it may be easier to concentrate on the energy case.

For the energy case, if indeed the photon comes with different masses according to its energy, then there must be an evidence for that.

One may therefore suggest studying the process $\mu^+\mu^- \to \gamma\gamma$ more carefully in an experimental setup as a muon collider (2016). Therefore one may divide the above process into several data sets according to the initial muons decay time values; the first data set may be for muons that were just produced and we let them immediately to interact and produce the two photons, a second sample may consist muons that live a longer time after they are produced and a third sample may consist muons that are even more long lived before they annihilate to give the two outgoing photons. In every one of the three samples we measure the outgoing photons average energy. If the suggestion made in section 2 is correct, then we may observe differences in the average energies of the final photons. This is so since if the photons have a rest mass and a γ_b boost structure, then their energy depends on their hidden variable $f_{r(j,i)}$ which may be different in the 3 samples as muons with different average decay times may have different $f_{r(i)}$'s according to Brodet (2010). The above measurement may be carried out in different muon initial energies which may investigate this possible effect more thoroughly. Therefore if such final photon energy dependence is observed, then it will support the suggestion that the particle boost structure depends on $f_{r(j,i)}$ and that the photons have indeed a rest mass.

Another possibility, is that it could be argued that if the energy of the photon depends on its hidden variable value, f_r , and if indeed its energy spectrum contains different photon masses, then the photons energy spectrum depends on a sum of exponential distributions. Therefore one can attempt to measure the photons energy spectrum and try to fit to it the function:

$$E_{\gamma} = \sum_{j=1...n} W_j \cdot \gamma_{b(j)} \cdot M_{\gamma(j)} \cdot (c + \frac{a1}{f_{r(j,i)}})^2 \tag{5}$$

Where W_j is the weight associated with specific photon mass j. This procedure may reveal how many masses does the photon state contains, and whether it indeed fits a combination of exponential distributions.

The suggested measurement of the spectrum may be done by generating a part of the photon energy spectrum, from photons in the low radio frequency up to the visible range by an appropriate antenna. This photon energy generation may be done gradually from low energy (frequency) to the visible range where at each frequency the intensity of the generated radiation is measured. This will allow plotting a curve of the photons energy verses their intensity, and then attempt to fit equation 4 to this curve and test if it matches a combination of exponential distributions as suggested in this paper.

4. Conclusions

The relationship between the possibility of a hidden variable in time, f_r, the photon's mass ,particle's energy, momentum and special relativity were discussed. It was suggested, that as unstable particle's, also a stable particle such as a photon may have a hidden variable in time, f_r. More over it was suggested that as fermions, also a photons may have a link between their f_r and their spin and wave function values, and that in the case of a photon it may mean that the photon has a non-zero rest mass. More specifically, it was suggested that the photon's large energy range is explained by a set of different non-zero rest masses for the photons. Consequently, the relevant, modification was made for the energy and momentum expression given by special relativity. Finally, experimental ways to test the above were also suggested.

Refrences

Brodet, E. (2010). The possibility of a hidden variable in time. *Physics Essays*, *23*(4). 613-617. http://dx.doi.org/10.4006/1.3504889

Brodet, E. (2015). A possible framework for explaining the apparent randomness in particle angular distribution. *Physics Essays*, *28*(2), 161-163. http://dx.doi.org/10.4006/0836-1398-28.2.161

Brodet, E. (2016). The relationship between possible hidden variables in time and space and the particle's angular momentum and wave function. *Physics Essays*, *29*(3), 457-462. http://dx.doi.org/10.4006/0836-1398-29.3.457

Muon collider. (2016). In *Wikipedia, the free encyclopedia*. Retrieved September 2, 2016, from https://en.wikipedia.org/wiki/Muon_collider

8

Dipole Moment and Electronic Structure Calculations of the Electronic States of the molecular ion SiN⁺

Karam Hamdan[1], Ghassan Younes[1] & Mahmoud Korek[1]

[1] Faculty of Science, Beirut Arab University, P. O. Box 11-5020 Riad El Solh, Beirut 1107 2809, Lebanon

Correspondence: Mahmoud Korek, Faculty of science, Beirut Arab University, P.O.Box 11-5020, Beirut Lebanon. E-mail: mahmoud.korek@bau.edu.lb, fkorek@yahoo.com

Abstract

A theoretical investigation of the lowest electronic states of the molecular ion SiN⁺ has been performed via Complete Active Space Self Consistent Field (CASSCF) method with Multi Reference Configuration Interaction MRCI+Q (single and double excitations with Davidson correction) calculations. The potential energy curves of the low-lying 37 electronic states in the representation $^{2s+1}\Lambda^{(+/-)}$, up to 140000 cm⁻¹ have been investigated. The permanent dipole moment, the harmonic frequency ω_e, the equilibrium internuclear distance R_e, the rotational constants B_e and the electronic energy with respect to the ground state T_e have been calculated for these electronic states. The comparison between the values of the present work and those available in the literature for several electronic states shows a very good agreement. The permanent dipole moment, of the investigated 37 electronic states, have been calculated in the present work for the first time along with the investigation of nine new electronic states that have not been observed yet.

Keywords: *ab initio* calculation, permanent dipole moments, electronic structure, spectroscopic constants, potential energy curves

1. Introduction

The remarkable interest in silicon nitride SiN reside in its important role in the stellar atmosphere and in many properties such as strength, hardness, chemical inertness, good resistance to corrosion, high thermal stability, and good dielectric properties (Katz, 1980). In literature many spectroscopic investigations have been focused on the ground and the first excited states where some spectroscopic constants have been obtained (Linton, 1975, Bredohl et al., 1976, Saito et al., 1983, Foster, 1984, Foster et al., 1985, Yamada & Hirota, 1985, Yamada et al., 1988, Foster, 1989, Elhanine et al., 1992, Ito et al., 1993, Naulin et al., 1993).

In contrast, fragmented data on the molecular ion SiN⁺ have been published in literature. By using an ab initio calculation (MRD-CI), Bruna et al. (1980) found that the ground state is $X^3\Sigma^-$ for this cation. By using the second-order Möller-Plesset perturbation theory, Goldberg et al. (1994) determined the energy separation T_e between the $^3\Sigma^-$ and $^3\Pi$ electronic states along with the internuclear distance R_e of these electronic states for the SiN⁺ molecular ion. From CASSCF, CMRCI, CCSDT and density functional theoretical calculations, Cai and François (1999) calculated the spectroscopic constants of the SiN⁺ molecular ion and they found that the $^3\Pi$ is 460 cm⁻¹ above the ground state $^3\Sigma^-$ which is in disagreement with the calculated value of Bruna et al. (1980). Recently, Liu et al. (2016) investigated the low-lying 10 electronic states with spin orbit interaction of the SiN⁺ molecular ion by using MRCI+Q calculation.

By using a high-level ab initio MRCI+Q calculation, we investigate in the present work, the potential energy curves (PEC's) for 37 singlet, triplet and quintet electronic states of the SiN⁺ molecular ion up to 140000 cm⁻¹. The permanent dipole moment in terms of the internuclear distance R have been calculated for these electronic states along with the spectroscopic constants T_e, R_e, ω_e and B_e. The comparison of these results with those reported in the literature showed a very good agreement. Fourteen new electronic states are investigated here for the first time.

2. Method

The low-lying singlet, triplet and quintet electronic states of the molecular ion SiN^+ have been investigated by using the Complete Active Space Self Consistent Field (CASSCF) procedure followed by a multireference configuration interaction (MRCI+Q with Davidson correction). The entire CASSCF configuration space was used as the reference in the MRCI calculations. The calculation has been done via the computational chemistry program MOLPRO (MOLPRO package) taking advantage of the graphical user interface GABEDIT (Allouche, 2011). The silicon atom is treated as a system of 14 electrons by using the Ahlirchs-PVDZ;c basis set for s, p, and d functions. The 6 electrons of the carbon species are considered using the aug-cc-pVTZ;c basis set for s, p and d functions. Among the 20 electrons explicitly considered for SiN^+ molecule 6 valence electrons were explicitly treated, corresponding to 15 active Σ orbitals. Being an heteronuclear diatomic molecule, SiN^+ is of $C_{\infty v}$ point group symmetry; however, MOLPRO software can only make use of Abelian point groups which means that $C_{\infty v}$ will be treated using the subgroup C_{2v} point group placing the molecule along the + z-axis and keeping Si at the origin.

3. Results

3.1 Potential Energy Curves

The potential energy curves (PECs), in terms of the internuclear distance R, for the 36 singlet, triplet and quintet electronic states, in the representation $^{2s+1}\Lambda^{\pm}$ of the molecular ion SiN^+ were generated using the MRCI+Q for 101 internuclear distances. These curves are given in Figures 1-6.

The binding forces between the atoms of a molecule are the main origin of the physical properties of this molecule. The magnitude of these force are functions of the internuclear distance and they are either attractive or repulsive. The depth of a potential energy curve indicates the strength of the bond in order that the deepest potential energy curve belongs to the more stable molecule. From the drawn potential energy curves, in figures 1-6 for the molecular ion SiN^+, one can notice the deep potential wells for the low triplet and singlet electronic states and shallower wells for the higher excited electronic states. While most of the potential energy curves of the quintet electronic states are shallow. From the 37 investigated electronic states 26 are bounded while the other electronic states are unbound. For these states when the electronic clouds surrounding the two atoms start to overlap, the energy of the system increases abruptly in order that one repulses each other according to Pauli principle. Moreover, some crossings and avoided crossings have been obtained between the potential energy curves in Figures 1-6. Two potential energy curves belonging to states of different symmetry they cross each other where their wavefunctions remain unperturbed and they are the adiabatic solutions of the Schrödinger equation, while the solution becomes diabatic if these wavefunctions have the same symmetry. But by linear combinations of the diabatic wavefunctions, where the variation method is used to solve for the coefficients, one can have the adiabatic solution for these states. In this case the wave functions will mix with each other where no longer cross and the crossing become avoided.

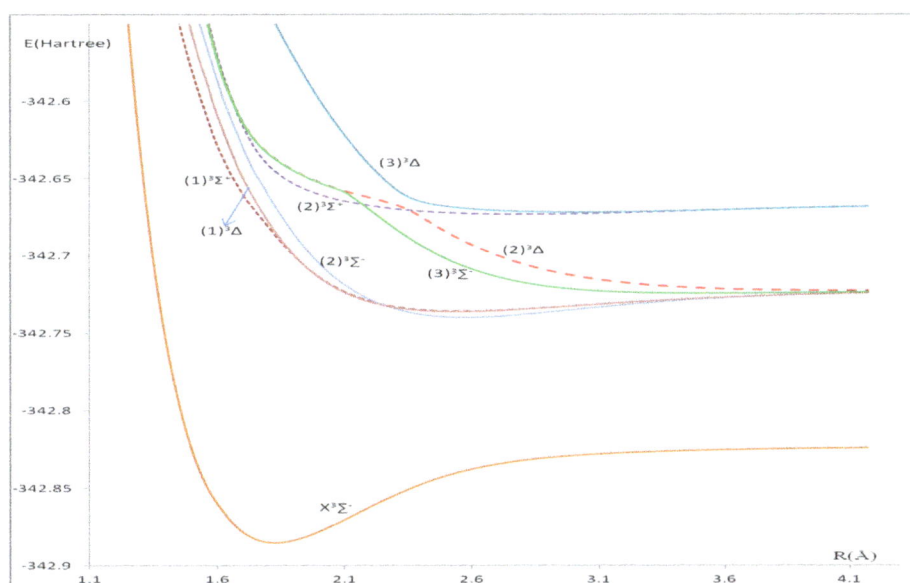

Figure 1. Potential energy curves of the electronic $^3\Sigma^{\pm}$ and $^3\Delta$ states of the molecule SiN^+

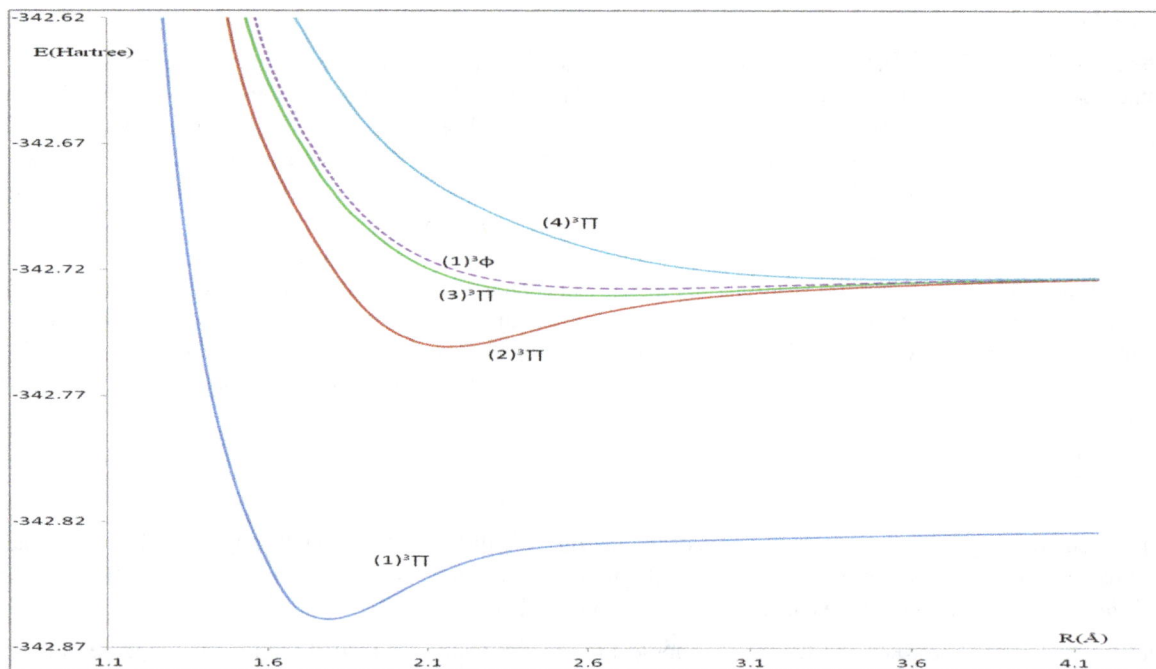

Figure 2. Potential energy curves of the electronic $^3\Pi$ and $^3\Phi$ states of the molecule SiN$^+$

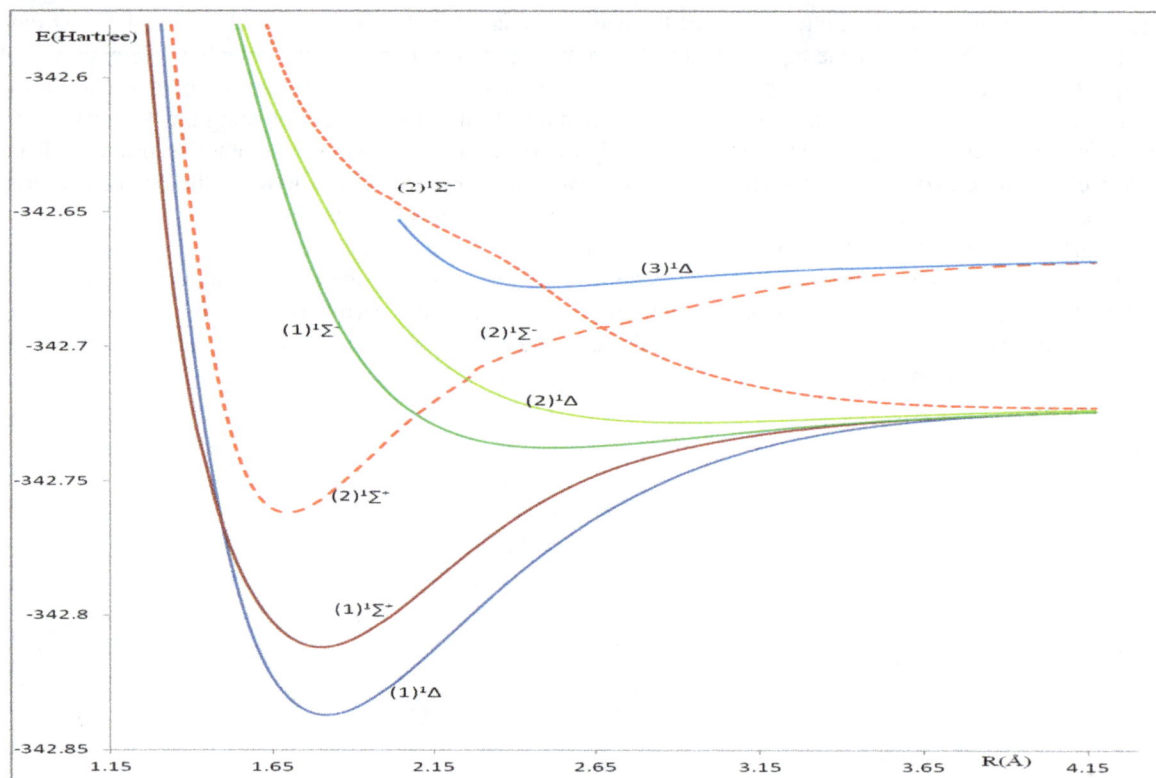

Figure 3. Potential energy curves of the electronic $^1\Sigma^\pm$ and $^1\Delta$ states of the molecule SiN$^+$

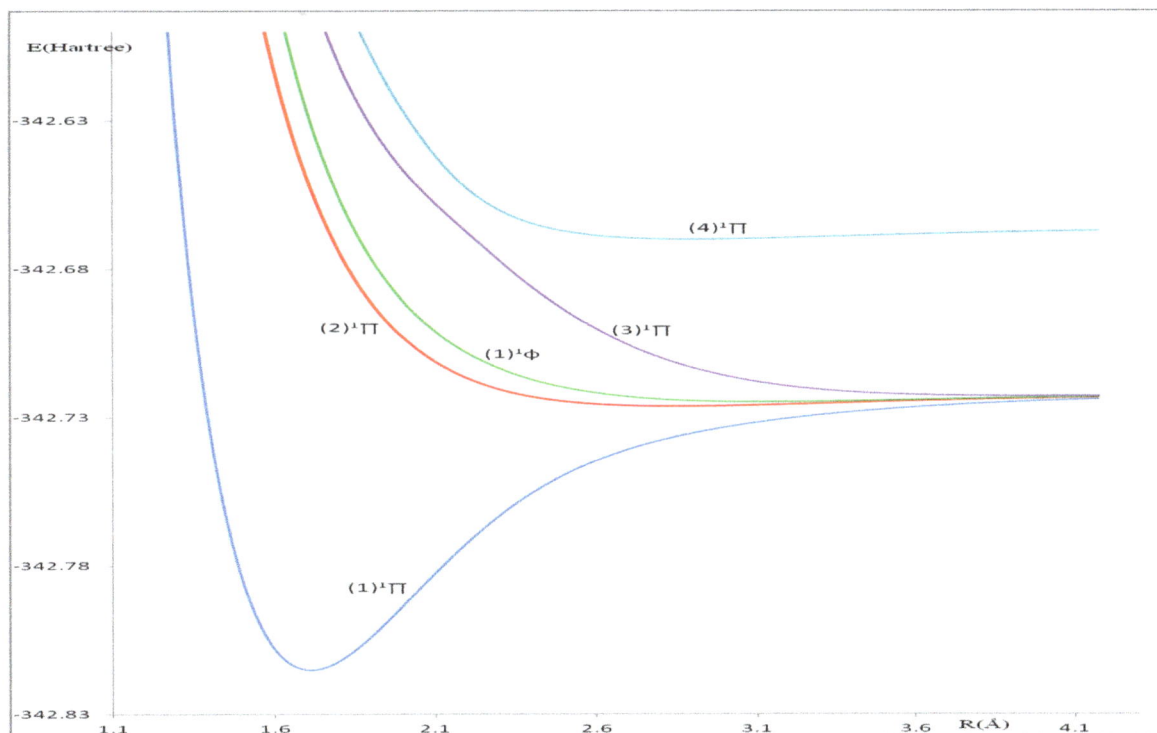

Figure 4. Potential energy curves of the electronic $^1\Pi$ and $^1\Phi$ states of the molecule SiN$^+$

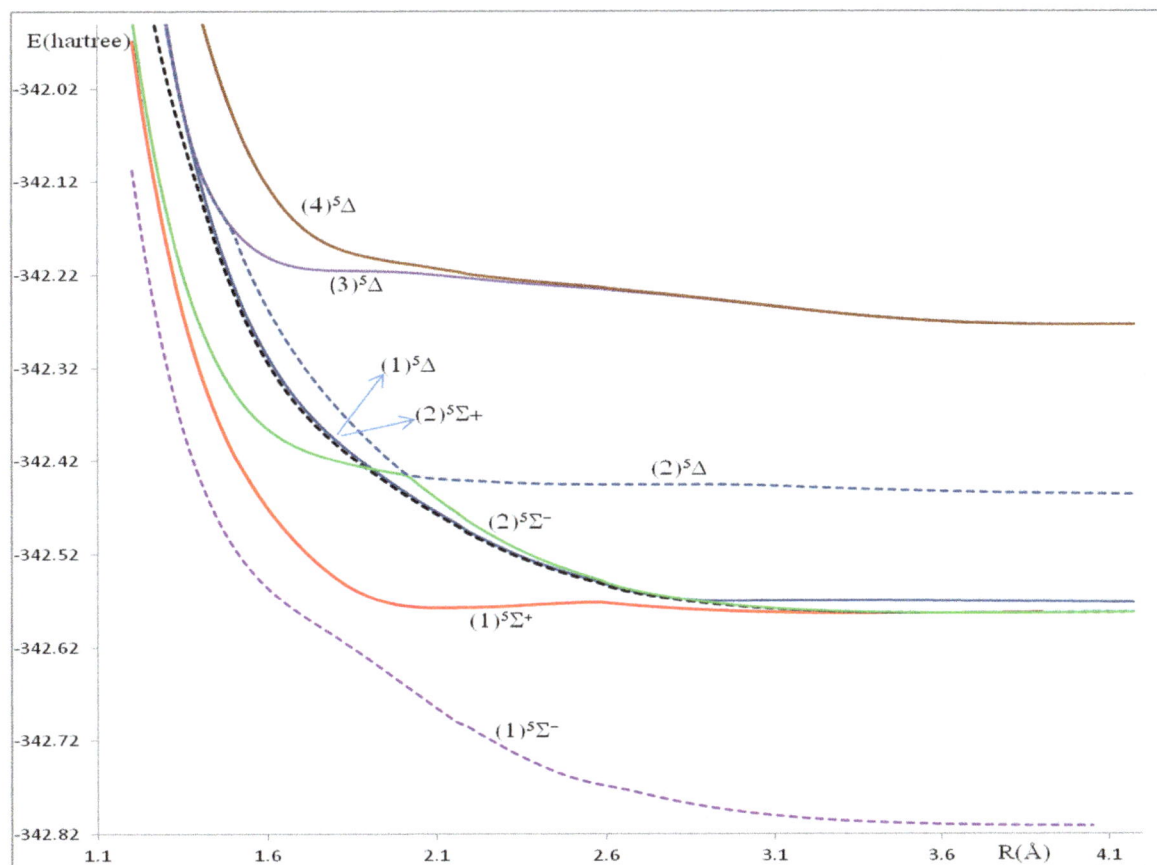

Figure 5. Potential energy curves of the electronic $^5\Sigma^\pm$ and $^5\Delta$ states of the molecule SiN$^+$

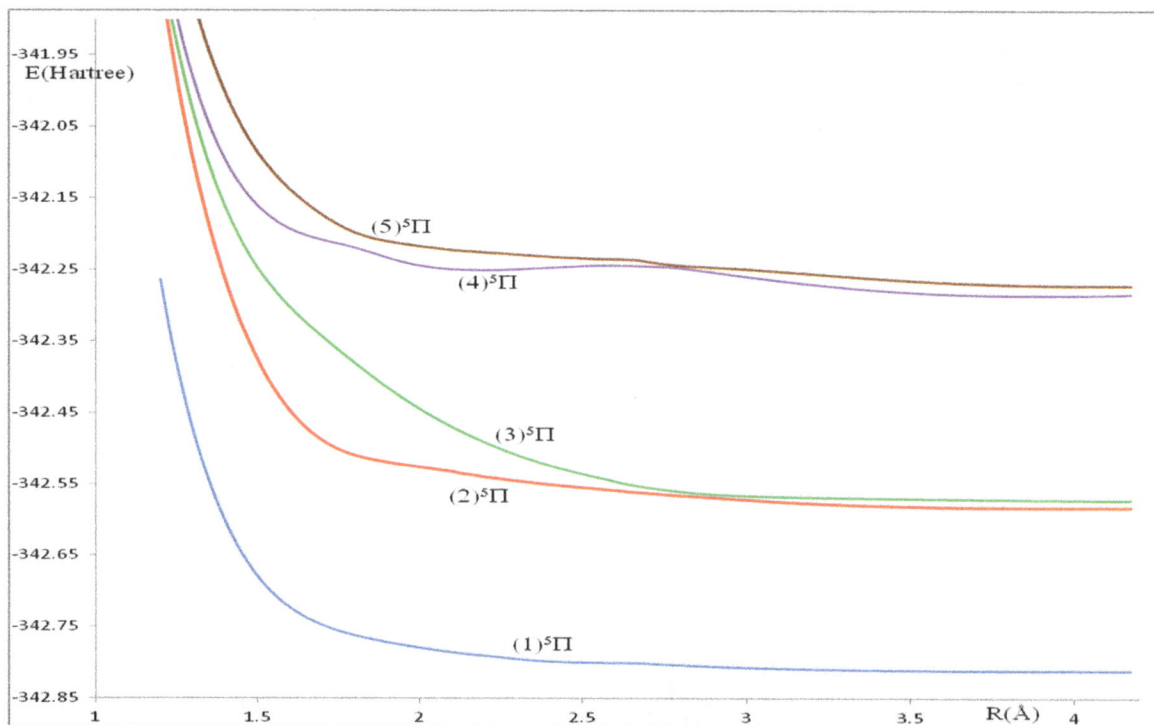

Figure 6. Potential energy curves of the electronic $^5\Pi$ states of the molecule SiN^+

3.2 Spectroscopic Constants

For the investigated bound electronic states of the molecular ion SiN^+ the transition energy with respect to the energy minimum for the ground state T_e, the equilibrium internuclear distance R_e, the harmonic frequency ω_e and the rotational constant B_e have been calculated by fitting the calculated energy values of the different investigated electronic states into a polynomial in terms of R around the internuclear distance at equilibrium R_e. These values along with those given in literature are given in Table 1. The comparison of our calculated values of T_e with those given by (Liu et al., 2016) shoes the good agreement for the investigated electronic states except for the three electronic states $(1)^3\Pi$, $(1)^3\Sigma^+$ and $(1)^1\Sigma^+$ where our potential energy curves are shifted up. Moreover this discrepancy can be noticed between the values given in literature (Liu et al., 2016) for the electronic states $(1)^3\Pi$ and $(1)^3\Sigma^+$. The comparison of our calculated values of R_e with those given in literature shoes a very good agreement, except the 2 electronic states $(1)^3\Sigma^+$ and $(3)^3\Pi$ which are displaced to the right by ~ 1Å. Similar good agreement is obtained by comparing the present calculated values of B_e with those given in literature except for the $(3)^3\Pi$, $(2)^1\Delta$ and $(1)^1\Phi$ where our values are systematically smaller than those of (Liu et al., 2016). Concerning the investigated values of ω_e, there is a good agreement with those given in literature for $T_e < 33000$ cm^{-1}, this agreement becomes less or deteriorate for some electronic states by comparison with those given by (Liu et al., 2016). The spectroscopic constants have not been calculated for the unbound electronic states and those where there is crossing or avoiding crossing near their minima of the potential energy curves R_e.

3.3 Permanent Dipole Moment.

The investigation of the dipole moment of a molecule is very useful in the study of the strength of the long range dipole-dipole forces and the understanding of the macroscopic properties of imperfect gases. From its calculation in terms of the internuclear distance, we can study the mobility of electrons through a polar gas. In the present work we place the silicon atom at the origin and the nitrogen atom along the positive direction of the internuclear axis. The investigated values of the dipole moment μ vary, in the region of calculation, between 1.44Å$\leq$R$\leq$4.14Å and 3.0 a.u.$\leq\mu\leq$5.0 a.u. for the singlet, triplet and quintet electronic states. At large internuclear distances, the dipole moment curves, of the singlet and triplet investigated electronic states, increase in absolute value in the negative region which is theoretically the correct behavior for a molecular ion. While for the quintet electronic state, one part tends in the positive region and the other part in the negative region. However, this behavior points out polarized states that dissociate into ionic fragments. An avoided crossing between the potential energy curves of two electronic states of the same symmetry is the origin of the abrupt gradient change of the DMCs.

Table 1. Speroscopic constants of the molecular ion SiN$^+$

States	T_e (cm^{-1})	ω_e (cm^{-1})	B_e (cm^{-1})	R_e (Å)
		749.6	0.5427	1.823
$(1)^3\Sigma^-$	0.0	777.6[b]	0.5628[b]	1.7972[b]
		717 [c]		1.81 [c]
	5734.2	852. 8	0.5667	1.784
$(1)^3\Pi$	723.4[b]	922.3[b]	0.6384 [b]	1.6825 [b]
	1693.8 [c]	856 [c]		
	10512.4	775.6	0.5487	1.813
$(1)^1\Delta$	9570.9 [b]	793.2 [b]	0.5628 [b]	1.7920 [b]
	11453.3 [c]			1.81 [c]
$(1)^1\Pi$	15352.6	855.86	0.6150	1.713
	16082.16[c]	915 [c]		1.69 [c]
	16090.0	733.6	0.5582	1.798
$(1)^1\Sigma^+$	7693.8 [b]			
	9840.19 [c]	1464 [c]		1.57 [c]
$(2)^1\Sigma^+$	27067.5	985.7	0.6300	1.692
$(2)^3\Pi$	29489.7	507.26	0.3832	2.169
	28330.7 [b]	700.25 [b]	0.5400 [b]	1.926 [b]
$(2)^3\Sigma^-$	31930.1	282. 9	0.2717	2.575
	34243.9 [b]	326.8 [b]	0.2928 [b]	2.4835 [b]
$(1)^1\Sigma^-$	32405.3	239.3	0.2865	2.511
	34033.5 [b]	320.9 [b]	0.3256 [b]	2.3560 [b]
$(1)^3\Delta$	32722.1	226.8	0.2803	2.537
	34541.1 [b]	308.6 [b]	0.3200 [b]	2.36 [b]
$(1)^3\Sigma^+$	32822.7	222.2	0.2774	2.551
	22770.2 [b]	1186.3[b]		1.5385 [b]
$(3)^3\Pi$	34011.5	165.9	0.2592	2.639
	30858.8 [b]	779.6 [b]	0.5189 [b]	1.8745 [b]
$(1)^3\Phi$	34597.4	130.9	0.2426	2.727
	36046.1 [b]	269.4 [b]	0.3976 [b]	2.1325 [b]
$(2)^1\Delta$	34439.7	136.3	0.2062	2.933
	37111.5 [b]	210.1 [b]	0.4947 [b]	2.6545 [b]
$(2)^1\Pi$	34856.6	62.6	0.2099	2.8494
$(1)^1\Phi$	35159.7	127.5	0.1828	3.125
	38266.0 b	144.1 [b]	0.2554 [b]	2.65 [b]
$(3)^3\Sigma^-$	35372.7	82.4	0.1447	3.5273
	38836.7 [b]	100.4 [b]	0.1732 [b]	3.2305 [b]
$(4)^3\Pi$	35446.1	69.9	0.1366	3.6300
	38282.9 [b]	860.9 [b]	0.510[b]	1.8715 [b]
$(3)^1\Delta$	35480.3	769.8	0.2793	2.542
$(3)^1\Pi$	35589.9	171.8		3.973
	39240.5 [b]	173.3 [b]		3.83 [b]
$(3)^1\Sigma^+$	45480.3	306. 2	0.2906	2.490
$(2)^3\Sigma^+$	46571.0	114.0	0.2245	2.806
$(3)^3\Delta$	46873.01	516.2	0.1709	2.994
$(4)^1\Pi$	47188.2	119. 4	0.2164	2.891
$(1)^5\Sigma^+$	67661.3	397.5	0.4018	2.114
$(4)^5\Pi$	139354.9	441.7	0.3759	2.190

For the considered electronic states of the molecular ion SiN$^+$ a reversed polarity of the corresponding atoms takes place at each avoided crossing region as indicated from the sharp change in the slopes of the DMCs of the states at which the dipole moment changes its gradient sign. There is no comparison of these investigated values of the dipole moments since they are calculated here for the first time. The variation of the dipole moment curves, in terms of the internuclear distance R, for singlet, triplet and quintet electronic states on SiN$^+$ cation are given in Figures 7-12.

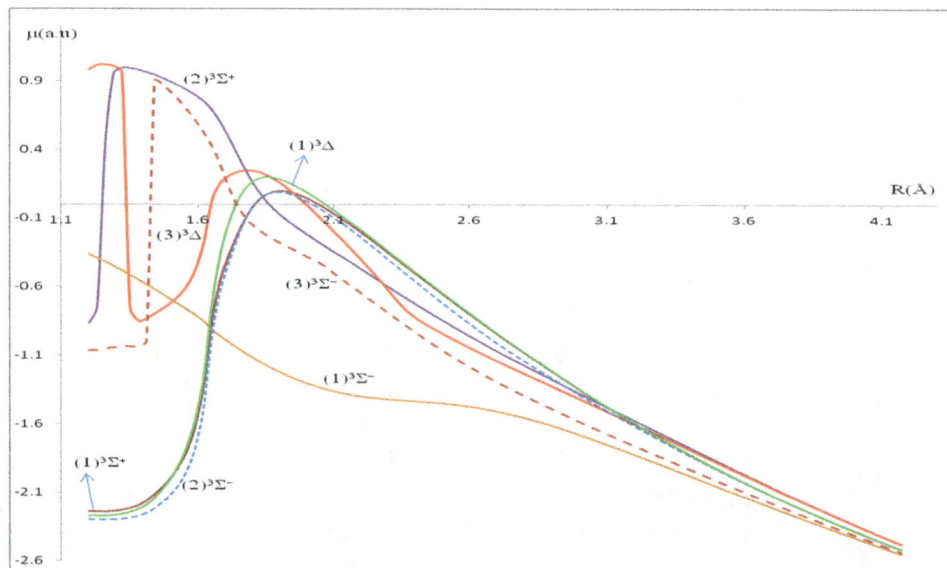

Figure 7. Dipole moment curves of $^3\Sigma^{\pm}$ and $^3\Delta$ electronic states of the molecular ion SiN$^+$

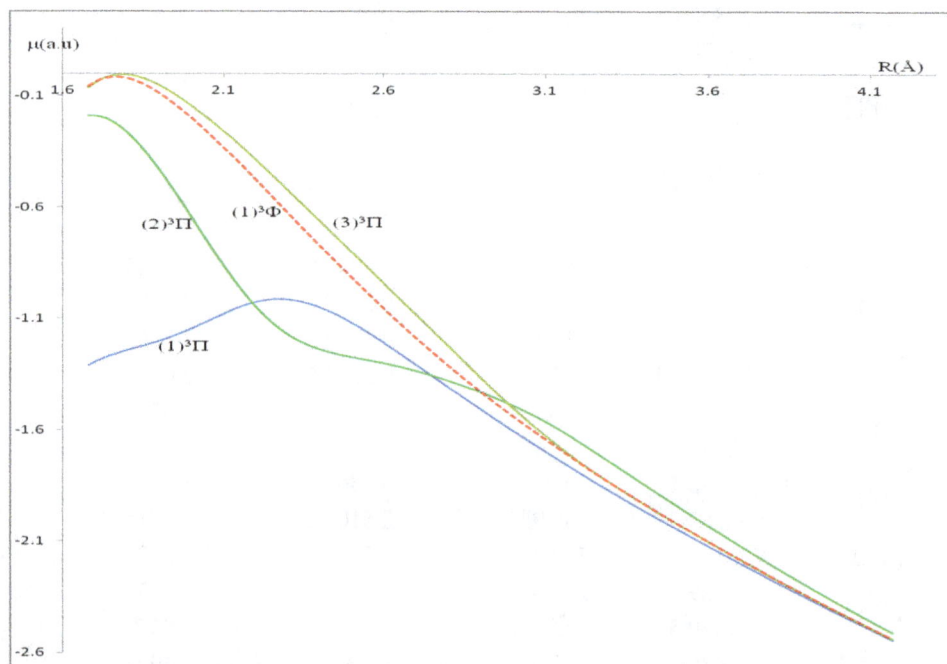

Figure 8. Dipole moment curves of $^3\Pi$ and $^3\Phi$ electronic states of the molecular ion SiN$^+$

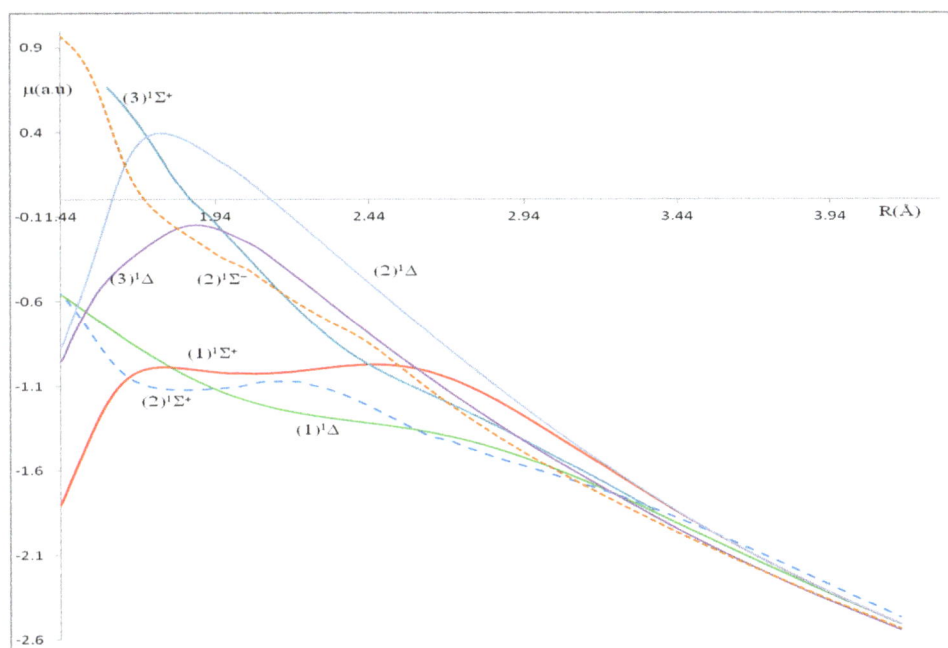

Figure 9. Dipole moment curves of $^1\Sigma^{\pm}$ and $^1\Delta$ electronic states of the molecular ion SiN$^+$

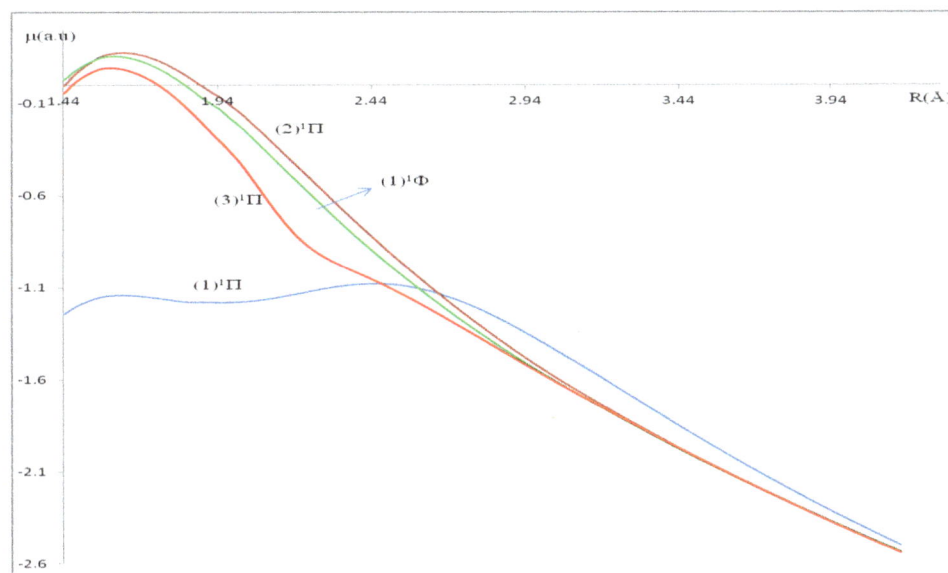

Figure 10. Dipole moment curves of $^1\Pi$ and $^1\Phi$ electronic states of the molecular ion SiN$^+$

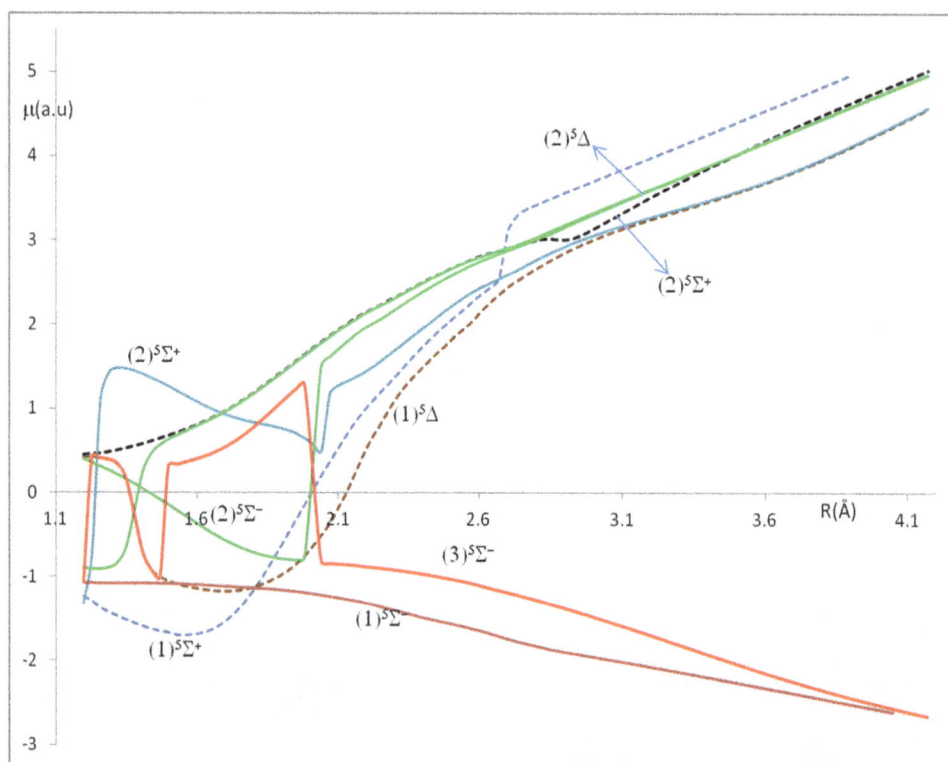

Figure 11. Dipole moment curves of $^5\Sigma^\pm$ and $^5\Delta$ electronic states of the molecular ion SiN$^+$

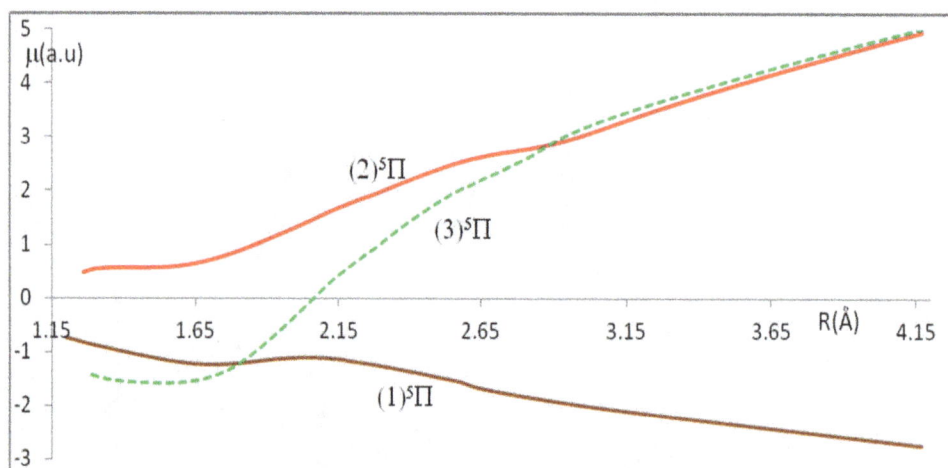

Figure 12. Dipole moment curves of $^5\Pi$ electronic states of the molecular ion SiN$^+$

4. Conclusion

The static dipole moment and the potential energy curves of 37 singlet, triplet and quintet electronic states of the molecule ion SiN$^+$ have been investigated by using a high precision level ab initio MRCI+Q calculation with Davidson correction. The spectroscopic constants T_e, R_e, ω_e and B_e have been calculated for the bounded states up to to 140000 cm^{-1}. The comparison of these calculated values with those given in literature shows an overall good agreement. At large internuclear distances, the dipole moment curves, of the singlet and triplet investigated electronic states, increase in absolute value in the negative region while for the quintet electronic state, one part tends in the positive region and the other part in the negative one. All these dipole moment curves along with nine electronic states have been studied in the present work for the first time. These new results may stimulate more investigation of experimental work on this molecular ion SiN$^+$.

Acknowledgements

The authors would like to thank the Lebanese National Council for Scientific Research for the financial support of the present work by the grant program for scientific research 2013-2015.

References

Allouche, A. R. (2011). Gabedit: A graphical user interface for computational chemistry software. *J. Comp. Chem. 32*, 174-182. http://dx.doi.org/10.1002/jcc.21600

Bredohl, H., Dubois, I., Houbrechts, Y., & Singh, M. (1976). The emission spectrum of SiN. *Can. J. Phys. 54*, 680-688. http://dx.doi.org/10.1139/p76-76

Bruna, P. J., Peyerimhoff, S. D., & Buenker, R. J. (1980). Theoretical prediction of the potential curves for the lowest - lying states of the isovalent diatomics CN+, Si2, SiC, CP+, and SiN+ using the ab initio MRD - CI method. *J. Chem. Phys., 72*, 5437. http://dx.doi.org/10.1063/1.439012

Cai, Z. L., & François, J. P. (1999). Ab initio study of the lowest 3Sigma-, 3Pi and 1Sigma+ states of the SiN+ cation. *Chem. Phys. Lett., 300*, 69-79. http://dx.doi.org/10.1016/S0009-2614(98)01329-3

Elhanine, M., Hanoune, B., Guelachvili, G., & Amiot, C. (1992). The infrared emission spectrum of SiN between 2.2 and 4.4 μm. *J. Phys. France, 24*, 931-938. http://dx.doi.org/10.1051/jp2:1992176

Foster, S. C. (1984). The B2Σ+→ A2Πi system of silicon nitride. *SiN. J. Mol. Spectrosc, 106*, 369-375. http://dx.doi.org/10.1016/0022-2852(84)90167-X

Foster, S. C., Lubic, K. G., & Amano, T. (1985). The 2-0 band of the A2Πi←X2Σ+ system of SiN near 3.3 μm. *J. Chem. Phys., 82*(2) 709-713. http://dx.doi.org/10.1063/1.448548

Foster, S. C. (1989). The vibronic structure of the SiN Radical. *J. Mol. Spectrosc, 137*, 430-431. http://dx.doi.org/10.1016/0022-2852(89)90185-9

Goldberg, N., Iraqi, M., Schwarz, H., Boldyrev, A., & Simons, J. (1994). A combined experimental and theoretical study of the neutral, cationic,and anionic Si3N cluster molecule. *J. Chem. Phys., 101*, 2871. http://dx.doi.org/10.1063/1.467601

Ito, H., Suzuki, K., Kondow, T., & Kuchitsu, K. (1993). An analysis of the B2Σ+−X2Σ+ emission of SiN. Dependence of the electronic transition moment on the SiN internuclear distance. *Chem. Phys. Lett., 208*(3-4), 328-334. http://dx.doi.org/10.1016/0009-2614(93)89084-U

Katz, R. N. (1980). High-temperature structural ceramics. *Science, 208*, 841-847. http://dx.doi.org/10.1126/Science.208.4446.841

Linton, C. (1975). Rotational analysis of some Mulliken D2Π-A2Π bands of the SiN molecule. *J. Mol. Spectrosc, 55*, 108-119. http://dx.doi.org/10.1016/00222852(75)90256-8

Liu, Y. F., Zhai, H., & Liu, Y. L. (2015). Extensive ab initio calculation on low-lying excited states of SiN+ cation including spin-orbit coupling. *Eur. Phys. J. D, 69*, 59. http://dx.doi.org/10.1140/epjd/e2015-50584-y

Naulin, C., Costes, M., Moudden, Z., Ghanem, N., & Dorthe, G. (1993). The dissociation energy of the SiN radical determined from a crossed molecular beam study of the Si+N2O→SiN+NO reaction. *Chem. Phys. Lett., 202*(5), 452-458. http://dx.doi.org/10.1016/0009-2614(93)90069D

Molpro Quantum Chemistry Software. (2008). *MOLPRO is a package of ab initio programs*. Retrieved from http://www.molpro.net/info/users

Saito, S., Endo, Y., & Hirota, E. (1983). The microwave spectrum of the SiN (2Σ+) radical. *J. Chem. Phys., 78*, 6447. http://dx.doi.org/10.1063/1.444682

Yamada, C., & Hirota, E. (1985). The A2Πi –X2Σ+ transition of the SiN radical by infrared diode laser spectroscopy. *J. Chem. Phys., 82*, 2547. http://dx.doi.org/10.1063/1.448304

Yamada, C., Hirota, E., Yamamoto, S., & Saito, S. (1988). The vibrational assignment for the A2Π–X2Σ+ band system of the SiN radical: The 0–0 bands of 29SiN and 30SiN. *J. Chem. Phys., 88*, 46. http://dx.doi.org/10.1063/1.454627

9

On the Nature of Space and the Emergence of the New Physics
Some Speculations

Philip J. Tattersall[1]

[1] Private researcher, 8 Lenborough St, Beauty Point, Tasmania, Australia

Correspondence: Philip J. Tattersall, 8 Lenborough St, Beauty Point, Tasmania, Australia. E-mail: soiltechresearch@bigpond.com

Abstract

In reflecting on the emergent ideas of Stochastic Electrodynamics (SED) the author suggests that the zero point activity, also termed 'vacuum noise,' is not only a fundamental aspect of space, but is space itself.

Recent experiments are discussed in which electromagnetic radiation has been shown to cause changes in gravitational effects. This essay speculates that the zero point energy may be responsible.

Keywords: space, virtual activity, stochastic electrodynamics, Zero Point, ZPE, ZPF, quantum, gravity, vacuum

1. Introduction

For many years there has been ongoing questioning as to the 'real' nature of space (Johnson & Walker, 2005). For its part classical physics assumes space is a vacuum; nothingness; a void. Quantum theory on the other hand posit that space, by necessity is full of 'virtual activity', with elementary particles being created and annihilated instantaneously, resulting in an overall void. By admitting such activity, even though virtual, the door was opened to ongoing questioning, experimental work and attendant theorizing. The subsequent schools of thought, fueled by the results of experimental work, have now taken us to the point where questions over the nature of space are, more than ever, ripe for consideration.

Increasingly researchers have embraced the idea that far from being merely a state of virtual activity space is actually full of *real* activity or quantum 'noise' (Davies, 2001). Researchers have termed the activity Zero Point Field (ZPF), meaning the complex and random radiation field, made up of zero point energy (ZPE) still present at absolute zero (de La Pena et al., 2012). Boyer (2015, pp. 13-14) provides a useful definition, *"...zero-point radiation and thermal radiation are treated on the same footing in classical electromagnetism, they have different effects because of the differences in the electromagnetic spectra at zero temperature and finite temperature. The spectrum of zero-point radiation holding at zero temperature is Lorentz invariant, scale invariant and invariant under adiabatic compression. The zero-point radiation spectrum is isotropic in any inertial frame. In free space, the zero-point spectrum cannot give rise to velocity-dependent forces on particles because of its Lorentz-invariant character...in contrast, the spectrum of thermal radiation at a non-zero temperature...has a preferred reference frame, that unique frame in which the spectrum is isotropic. At finite non-zero temperature in free space, the thermal spectrum gives velocity-dependent forces on any system having electromagnetic interactions which is moving through the thermal radiation."*

de la Pena et al. (2012, p.1) regard the ZPF as largely responsible for quantum phenomena and posit that inclusion of the ZPF in classical physics would lead to the emergence of quantum theory, thus securing the vital link between the macro and micro 'worlds'. They claim, *"In other words, that rather than being innate, the quantum phenomenon arises as a consequence of the interaction of the zero-point field with matter. The recognition of the reality of this field that fills the whole of space even at null absolute temperature, not only allows to better understand (rather than merely describe) the quantum phenomenon, but also opens the door to some new physics"*.

Although the idea is a relative newcomer to modern physics the theoretical underpinnings have been around since the 1960's when Marshall (cited in de la Pena et al., 2012) and Boyer (cited in de la Pena et al., 2012) developed the theoretical basis of Stochastic Electrodynamics (SED) (see Note 1). The key assumption germane to this essay is that SED takes the existence of the ZPF as real, seeing it as a classical background radiation field consistent with Maxwell's equations, with stochastic amplitude and it is assumed that its stochastic properties are the same for every inertial frame of reference (Hernandez, 2014).

In terms of 'wave-particle' phenomena, the SED theory maintains *inter alia* that an interaction between particles in motion and the ZPF causes the emergence of an electromagnetic wave that 'guides' particle trajectory. The apparent particle-wave duality and entanglement are thus explained. Marshall and Santos (1997) further elaborated on the role of the ZPF in the context of a wave description of light. They show how light waves and the ZPF interact under certain experimental conditions to yield correlated amplification outputs. In an analysis of the anticorrelation beam splitter experiments of Grangier, Roger and Aspect (1986), Marshall and Santos (1997, p.5) state, *"...It is easy to see, qualitatively, how the explanatory power of a purely wave theory is increased by the recognition of a real zero point field. A beam splitter does not simply split an incoming wave into parts; it **mixes** together **two** incoming waves, one of them from the 'vacuum', to give two outgoing waves..."* (see Note 2). (Bolded text is the original the author's emphasis).

de la Pena and Cetto (1996) have also elaborated the ZPF theory at some length over many years. Likewise Santos (2012, p.5), in keeping with his extensive and important contributions to the field, has recently presented cogent arguments regarding the role and nature of ZPF. In discussing the SED nature of the ZPF he states that fluctuation inherent in the ZPF, *"...gives rise to two characteristic traits of quantum physics, Firstly quantum theory should be **probabilistic**. Secondly it presents a kind of **wholeness**, quite strange to classical physics where the concept of **isolated** system is crucial. The fact that the vacuum fluctuations at different points maybe **correlated** is the origin of wholeness. Indeed **entanglement** may be a consequence of the correlation of quantum fluctuations at different points"*, (bolded text is the original author's emphasis).

The idea of wholeness opens the door to considerations of possible ZPF interactions with matter. Santos (2012) posits that atoms are in dynamic stationary state with the fluctuating ZPF, citing the electronic stability of the hydrogen atom (and atoms generally) as an example where the electron is essentially in a kind of resonance with an energy level within the ZPF. Boyer (2011), citing the theoretical calculations of other workers has elaborated on the idea that ZPF is responsible for electronic stability. Calculations on probability distribution have shown agreement (for hydrogen) with Schrödinger calculations.

These two key points relating to wholeness and dynamic energy source are thought to be fertile ground.

2. The ZPF and the Nature of Space

These speculations (along with the theoretical underpinnings of SED) would lead one to suggest that space is not full of ZPF, but rather that space *is* in fact ZPF. This is *not* a return to the aether theory of old, but is an acknowledgement of a mounting body of theoretical argument and experimental work suggesting that 'space' is in fact an active, all pervasive medium.

If we assume that space is full of activity and that the ZPF does interact with particles, fields and matter then it would seem plausible to extend its influence beyond the micro (quantum) to the macro world. In building on the ideas of Santos (2012), that mass is interacting with the ZPF it could be suggested that the effect of gravity may emerge from such interactions. Indeed there has been a considerable amount of research, discussion and theorizing regarding the ways in which ZPF energies (ZPE) interact with matter. Dirac (1951) concluded (rather provocatively), *"Thus with the new electrodynamics we are rather forced to have an aether"*. Dirac makes an interesting point and it has resonated with a number of researchers including Haisch, Rueda & Puthoff (1997). Haisch, Rueda and Puthoff immediately saw the ZPF as an electromagnetic component of Dirac's aether, stating that *"Ether is a loaded term today because it is so widely described as an outmoded, discredited idea. The 19th-century ether was an ill-defined 'substance' providing a universal frame of rest and a medium for propagation of light waves. The ZPF is not an ether of this sort. The ZPF **is** light waves... . Being Lorentz invariant it cannot and does not act as a universal frame of reference for rectilinear motion...However it does provide a universal frame of rest vis-a-vis acceleration..."* (Haisch, Rueda & Puthoff, 1997, p.8). In their recent work, Rueda and Haisch (2005) have shown how the electromagnetic quantum vacuum field (or ZPF) makes a contribution to inertial reaction force of an accelerated object. The authors present an elaborate set of theoretical arguments in support of their *Quantum Vacuum Inertia Hypothesis*, in which they have put forward detailed calculations and discussion relating to the connection of inertia and gravity to the vacuum fields.

This as it turns out is a most useful starting point in explaining the results of recent experiments involving the effects of gravity on selected masses.

3. A Speculation on the Cause of Gravity

While general relativity (GR) produces accurate predictions it remains silent as to the cause of gravity. Over many years there have been numerous attempts to elucidate the cause of gravity. For example, the recent efforts by Verlinde (2010) although seen as somewhat controversial, have sparked considerable interest and research activity. Verlinde posits that gravity emerges as an entropic force brought about by changes in the information associated with the position of bodies. While aspects of Verlinde's arguments strike chords with the key speculations in this paper, one senses that the cause still eludes us. The danger is that many of the candidate theories may effectively be surrogates for a deeper theory, in that they can make accurate predictions and are often in agreement with GR, but lack a clear explanation as to the cause of gravity.

Rueda and Haisch (2005) in drawing on the theoretical arguments of Sakharov and others pointed to the idea that gravity emerges from the vacuum, and in fact may arise from the effect of massive bodies on the permeability and permittivity of space itself (Puthoff, 2001). This view is close to the one I am suggesting, namely that space *is* ZPF; in a sense a real physical presence that due to its characteristics remains of itself undetectable, but whose effects emerge in the presence of matter.

Let us speculate that certain ZPE modes within the ZPF are continually moving into matter in order to provide an energy source for electron activity and possibly other activity within atoms. It would follow that the intensity of movement of the ZPE would be proportional to both mass and distance. The net result would be a distortion of space in the vicinity of massive bodies. On the face of it the mathematical approach used in GR would still apply.

The gravitational effect resulting from the ZPE-mass interaction would produce a slight thermal effect leading to an entropic emission, perhaps contributing to dark energy. The thermal effect is a significant thermodynamic consideration. As pointed out by Van Flandern (1996) the absorption of 'particle flux' by matter would result in heating effects. He noted that, *the six largest planets all appear to radiate more heat back into space than they take from the sun. It has been traditional to attribute this excess to radioactivity in the planetary cores, although no observation evidence supports this conjecture...".*

In a very real sense gravity would emerge from space and would appear to do so in the presence of matter. Therefore gravity is not a property of mass or space as such, but rather emerges as a consequence of the mutual interaction of ZPE modes within the ZPF (space) and matter.

4. An Interesting Experimental Development

The findings by Rancourt (2011), arising from experiments he conducted over a period of many years (in which he experimented with torsion pendulums and masses) suggest that light somehow interferes with the effect of gravity. Placing a laser light in the form of a curtain of light to one side of a mass on a torsion pendulum he was able to show that the mass moved toward the light. The elegant experiments, confirmed many times under a range of conditions led to the suggestion that gravity is in fact a pushing force (Rancourt, 2014). The movement of the test mass towards the light suggested that either the light was attracting the mass or that some kind of external 'influence' was blocked by the light thus creating a disequilibrium leading to movement of the mass, caused by an influence from the opposite (non-light) side. Professor Rancourt eliminated the effect of light as an attractor by housing it in a steel box. The effect was the same even though a number of different test mass materials were used.

If the movement of test masses is due to a pushing force then it is possible, based on my earlier suggestion, that the ZPF has been somehow affected by the light in the experiment (e.g. perhaps through correlative effects such as those discussed earlier in this paper (Marshall & Santos, 1997)) leading to a disequilibrium and the emergence of pushing forces. The pushing force hypothesis was further explored in experiments reported by Rancourt and Tattersall (2015). The authors speculated that zero point energy within the zero point field may be responsible for so called gravitational 'attraction'. As mentioned above, the movement of ZPE into matter may well be the cause of 'gravity' as bodies in proximity to each other are actually driven together by the pressure of space itself due to the movement of ZPE into matter in order to maintain atomic processes. Such a hypothesis would be contingent on stringent thermodynamic considerations.

5. Conclusions

This brief essay attempts to put forward a series of ideas in the form of speculations based on past and present developments in SED. Recent interesting experiments may prove to break new ground in our understanding of the nature of space as an active energetic influence that in the presence of matter leads to the emergence of gravity and inertia. The influence (zero-point energy) has been found to be Lorentz invariant, scale invariant and invariant under adiabatic compression and active at zero Kelvin (absolute zero). The emergence of gravity and inertial resistance are caused by the interaction of ZPE with leptons and possibly quarks at the atomic level (Haisch et al., 1997), thus the micro world of the quantum is linked to the macro world in a new concept of wholeness.

A quote from Santos (2012, p.1) is food for thought as we continue on jour journey toward the *new physics*, *"Understanding quantum mechanics presents big difficulties for many people, a paradoxical fact in view of the relevance and practical success of the theory. For physicists interested in applications understanding is not very relevant if it means getting an intuitive picture of the micro-world. In fact, for them the essential purpose of physics is to allow predicting the results of experiments. Other physicists, including most workers in foundations, think the real problem is that, in the attempt to understand quantum mechanics, we should not use concepts of classical physics, or we should adhere to a kind of 'weak' objectivity. In contrast with those opinions,...it is proposed that quantum mechanics is less different from classical physics than usually assumed, and might be understood in a similar manner."* To my mind the answers to many of the fascinating, and some say troubling questions facing physics may be closer than we think and may involve seeing what we have already seen in new ways.

It is without doubt that the growing field of SED is providing exciting and much needed opportunities for canvassing new ideas, and indeed old ideas anew!

By encouraging the publication of essays such as this the scientific community displays a willingness to entertain new ideas and intuitive speculations that may well help point toward a new way for physics. The joy is not so much in *knowing*, but is in the journey to the knowing.

References

Boyer, T. (2011). *Any classical description of nature requires classical electromagnetic zero-point radiation.* Retrieved from https://arxiv.org/abs/1107.3455v1

Boyer, T. (2015). *Understanding zero-point energy in the context of classical electromagnetism.* Retrieved from http://arxiv.org/pdf/1512.08033v1.pdf

Davies, P. C. W. (2001). Quantum vacuum noise in physics and cosmology. *Chaos, 11*(3). http://dx.doi.org/10.163/1.1378796

de la Pena, L., & Cetto, A. M. (1996). *The quantum dice. An introduction to stochastic electrodynamics.* Dordrecht, The Netherlands: Kluwer Academic Publishers.

de la Pena, L., Cetto, A. M., & Hernandez, A. V. (2012). *Whence the quantum?* Retrieved from http://fqxi.org/community/forum/topic/1384

Dirac, P. A. M. (1951). Is there an aether? *Nature, 168.*

Grangier, P., Roger, G., & Aspect, A. (1986). Experimental evidence for a photon anticorrelation effect on a beam splitter: a new light on single-photon interference. *Europhysics Letters, 1*(4).

Haisch, B., Rueda, A., & Puthoff, H. E. (1997). Physics of the zero point field: implications for inertia, gravitation and mass. *Speculations in Science and Technology, 20.*

Hernandez, C. A. D. (2014). *Wave duality from stochastic electrodynamics.* Retrieved from https://arxiv.org/pdf/1403.0016v1.pdf

Johnson, G. W., & Walker, M. E. (2005). Sir Michael Atiyah's Einstein lecture: "The Nature of Space". *Notices of the AMS, 53*(6).

Marshall, T., & Santos, E. (1997). *The myth of the photon.* Retrieved from http://arxiv.org/pdf/quant-ph/9711046v1.pdf

Puthoff, H. E. (2001). *Polarizable-vacuum (PV) representation of General Relativity.* Retrieved from http://lanl.arxiv.org/ftp/gr-qc/papers/9909/9909037.pdf

Rancourt, L. (2011). Effect of light on gravitational attraction. *Physics Essays, 24*(4). http://dx.doi.org/10.4006/1.3653936

Rancourt, L. (2014). Why I think gravity is a pushing force. *Gravityforces.* Retrieved from http://www.gravityforces.com/?author=1

Rancourt, L., & Tattersall, P. J. (2015). Further experiments demonstrating the effect of light on gravitation. *Applied Physics Research, 7*(4), 4-13. http://dx.doi.org/10.5539/apr.v7n4p4

Rueda, A., & Haisch, B. (2005). Gravity and the quantum vacuum inertia hypothesis, *Ann. Phys. (Leipzig), 14*(8). http://dx.doi.org/10.1002andp.200510147

Santos, E. (2012). *Vacuum fluctuations, the clue for a realistic interpretation of quantum mechanics.* Retrieved from https://arxiv.org/pdf/1208.4431.pdf

Van Flandern, T. (1996). Possible new properties of gravity. *Meta Research Bulletin, 5*(38). (Note 3)

Verlinde, E. (2010). *On the origin of gravity and the laws of Newton.* Retrieved from https://arxiv.org/abs/1001. 0785

Notes

Note 1. Boyer (2015) provides a detailed and useful discussion of the zero point energy that is accessible and presents cogent mathematical arguments

Note 2. Marshall and Santos (1997, p.5) summarize the findings of the Grangier, Roger and Aspect beam splitter experiments in terms of the ZPF as, *"If we consider the 'vacuum' as empty then it seems almost unavoidable to assume that the intensity of any incoming classical signal is equal to the sum of the intensities in the outgoing channels...with such a description it is not possible to explain the anticorrelation data; these were interpreted by Grangier, Roger and Aspect as evidence that the whole 'photon' goes onto one or the other of the outgoing channels."*

Note 3. Reprinted from Astrophys. & Space Sci. *244.* (1996): http://dx.doi.org/10.1007/BF00642296

10

The Fabrication of the Infrared CdSe Doped With Cu Photodetector

Hassan H. Mohammed[1] & Salwan K. J. AL-Ani[2]

[1] Department of Computer Engineering Technology, Iraq University College, Alestiqlal Street, Basrah, Iraq

[2] Department of Physics, College of Science, Al-Mustansirya University, Baghdad, Iraq

Correspondence: Hassan H. Mohammed, Department of Computer Engineering Technology, Iraq University College, Alestiqlal Street, Basrah, IRAQ. E-mail: dr.hassanh50@gmail.com

Abstract

In this work, the implementation method of the CdSe doped with Cu (CdSe: Cu) photodetector is presented. This detector is prepared by vacuum evaporation of CdSe films on glass substrate followed by vacuum annealing under an argon atmosphere for doping with copper. This detector is found, for the first time, to cover a wide range of the infrared besides the visible region of the electromagnetic spectrum. This finding of the wavelength tuning is due to the localized energy states of copper atoms inside the band gap of the CdSe. This tuning is compared with recent work in the corresponding colloidal CdSe-ZnS core shell quantum dots and with the quantum well (QWIR) and quantum dots (QDIR) infrared detectors. The major significance of this developed detector is in its synthesis simplicity and its fabrication processes costs in comparison with that of the (QWIR) and (QDIR) detectors. The structural analysis results demonstrated that the vacuum annealing in competition with the doping concentration improves significantly the film structure. Better crystalline structure was reported at 5 wt% of Cu concentration and at annealing temperature of 350 ºC. Besides the measured detectivity at room temperature is $D*=2.31\times10^8$ cm Hz$^{1/2}$W^{-1}. This value approaches the detectivity of the state of art mercury cadmium telluride (MCT). This result paves the way for further investigations and improvements.

Keywords: CdSe:Cu, Photodetector, Infrared, Tuning range, detectivity

1. Introduction

In the last two decades there are great potential interests in the design and implementation of photodetectors. The motivation of such interests is the expansion of absorption spectrum and increase of detectivity at room temperature. Expansion of electromagnetic spectrum provides a wide range of applications including high speed data communication Zhang et al. (2015), medical diagnosis Maiti et al. (2013), solar cells in the visible range Kashyout et al. (2012), while the infrared (IR) detectors are used in thermal imaging, night vision, heat seekers. In the present work we have found that the bulk CdSe semiconductor material when doped with Cu impurity leads to the electromagnetic spectrum expansion. Besides, the implementation of this material as a photoconductive detector could reach a detectivity value comparable with that of the state of art mercury cadmium telluride (MCT) at room temperature. Recently CdSe quantum dots detector is developed in the visible region of the electromagnetic spectrum (400-700 nm) with detectivity $D*=10^8$ cm Hz$^{1/2}$W^{-1}. The quantum dots infrared photodetector (QDIP) have emerged as a potential alternative to MCT and QWIP (Oertel et al., 2005). Asgari and Razi (2010) have designed a novel infrared quantum photodetector using a cubic shaped 6nm GaN quantum dots capped with AlGaN. The $D*$ of this detector was calculated and found to be 3×10^8 cmHz$^{1/2}$W^{-1} at room temperature. Diedenhofen et al. (2015) have demonstrated that the integration of colloidal quantum dot photodetector with color tunable plasmonic nanofocussing lenses offer a significant color selectivity. Such lenses facilitate light concentration at the nanoscale and enhance light matter-interactions. The aim of the present work is to investigate an alternative approach for wavelength tuning extending from the visible to the IR spectrum region. For this purpose, we focus in particular on the CdSe doped with different concentrations of Cu atoms in which photoconductive detector plays a major role. This detector has found for the first time to cover a wide spectral range and operating at room temperature.

2. Experiment

Thin films of CdSe (purity= 99.999%) are prepared by vacuum evaporation on a glass substrate (7.6×2.6 cm²) using Balzers-OE-8 evaporation unit. High purity (99.999%) aluminum is deposited on the CdSe film to act as ohmic contacts. Copper atoms of a different weight percentage are also introduced into the CdSe lattice by dipping the films in a CuCl solution, complemented by annealing using a vacuum furnace and flowing argon gas. The experimental layout is shown in Figure 1. The sample is annealed at variable temperature between 100 and 350°C for a period of 2.5 hours. It has been found that a better film structure was obtained at the temperature of 350°C. The doped CdSe films prepared by this method have enjoyed characteristics similar to the CdSe single crystal doped with impurity atoms. The mask was fabricated from Mo material using a CNC cutting machine. The shape and dimensions of the mask are given in Figure 2. The photocurrent and spectral response were measured by the detector test bench DSR-500 supplied by Optronic laboratories. For lasers wavelength response test, the mounted test bench shown in Figure 3a was used. This test bench consists of the laser source, mechanical chopper for chopping the CW CO_2 laser, CdSe:Cu detector and storage oscilloscope. The detector bias circuit is shown in Figure 3b which presents a DC power supply (100V, 5 mA) and a load resistor. In the present experiment a bias voltage of 20V was used. For responsivity measurements, the power incident on the detector area is measured by a calibrated thermopile.

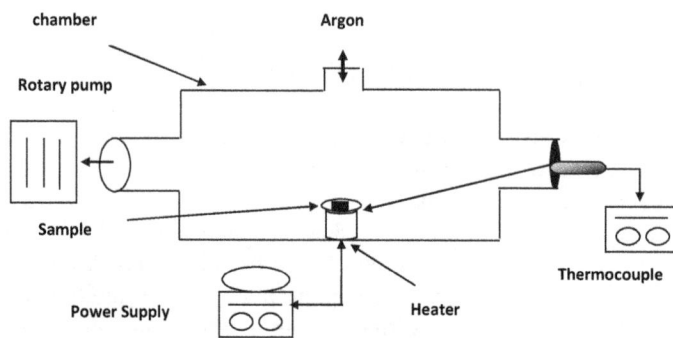

Figure 1. Film annealing setup

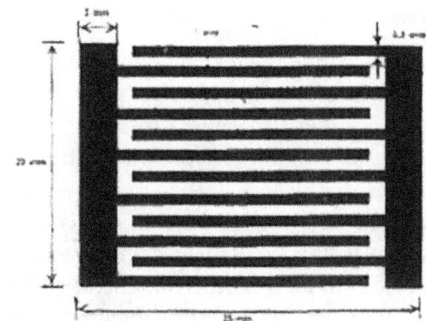

Figure 2. The detector mask

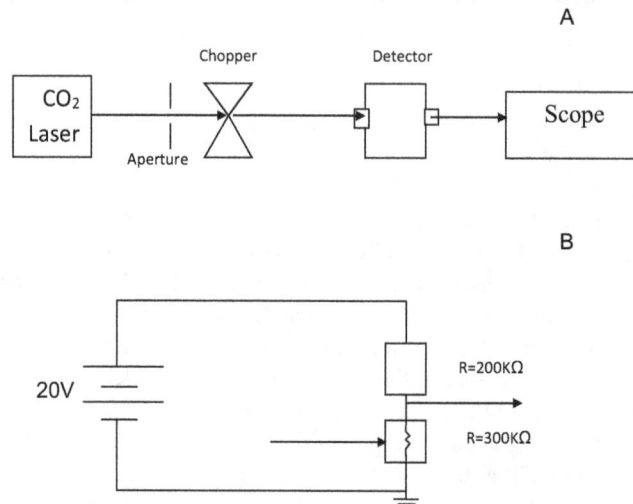

Figure 3. (A) Detector response test bench, (B) Detector bias circuit

3. Results and Discussion

3.1 Structural Properties

The doping of pure CdSe with Cu at different concentrations (film thickness= 1μm) improves significantly the structure. Better crystalline is reported at 5wt% Cu concentration. The X-ray diffraction of Cu doped samples depicts that only reflection from the (100) plane appeared and indicated that a single crystal grew on this plane. Figure 4 shows the diffraction pattern of CdSe doped with 5 wt% of Cu and at 2wt% for comparison. The results of the X-ray diffraction pattern are confirmed by the SEM morphology which indicates the single crystalline with hexagonal (wartzite) of the film structure at 5wt% of doping (Figure 5).This large crystalline structure change with doping indicates the incorporation of this dopant within substitusional or interistitial site in the CdSe lattice structure. It was shown that the grain size increases from 0.4μm to 2.87μm with an increase in Cu concentrations (Figure 6). A great improvement of the films structures was achieved at doping with 5wt% of Cu complemented by 2.5 hours of annealing at 350°C. The improvement in the crystal structure leads to a better optical sensitivity than that found in the case of the pure CdSe. The crystal structure is found to be hexagonal and cubic when Cu concentrations are below 3wt%. The increase in grain size can be related to the atoms thermal energy gained by annealing in presence of argon gas which facilitates the recrystalization. The SEM morphology results of CdSe:Cu (5wt% of Cu) and the X-ray diffraction of the same material can be considered to be identical.

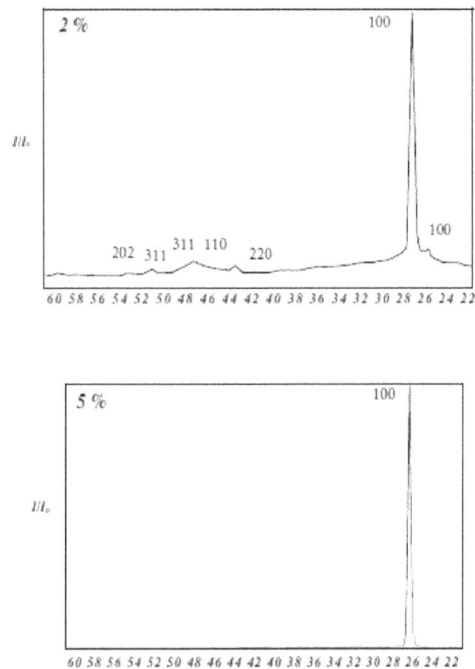

Figure 4. the diffraction pattern of CdSe doped with 5 wt% of Cu and 2 wt%

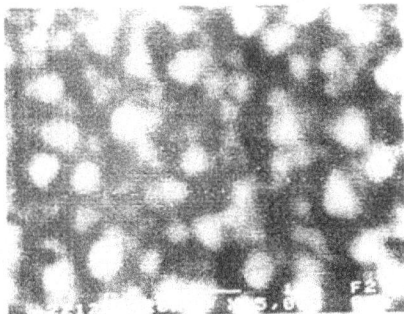

Figure 5. SEM morphology of CdSe doped with Cu at 5 wt% concentration

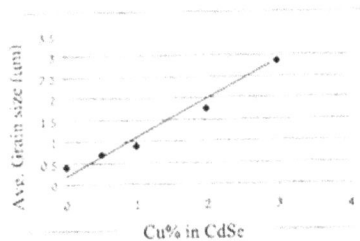

Figure 6. the average grain size as a function of Cu concentration

3.2 The Optical Properties

The absorption coefficient increases with the increase in the Cu doping concentration (Figure 7). Red shift in photon energy was found to be relative to the pure CdSe. This shift can be related to the presence of localized Cu levels inside the band gap of CdSe. The Cu levels induced field effect that shift the band edge towards energy lower than that of the pure CdSe. The change in the absorption coefficient $\Delta\alpha = (\alpha_{doped} - \alpha_{pure})$ was calculated for various doping concentrations. Figure 8 depicts the values of $\Delta\alpha$ as a function of photon energies (hf). $\Delta\alpha$ represents the electronic transitions for various copper levels inside the band gap to the conduction band. The positions of the copper levels were estimated from the full width at Maximum (FWHM) for each doping concentrations. As predicted by Wei et al. (2000), the CdSe can easily be n-doping in consistent with observation of deep levels below the conduction band minimum. These levels were localized relative to the conduction band at (0.67, 0.825 and 1.1eV) corresponding to the Cu concentrations (1, 2 and 5wt% respectively). Other levels were detected in the near and mid infrared region of the spectrum. The first level is situated at 0.116eV just below the conduction band. The position of this level is accurately ascertained when CdSe: Cu detector is selectively excited by a modulated CO_2 laser (10.6 μm), and the output signal was detected by the storage oscilloscope (Figure 9). Another accurate determination of Cu level is by selective excitation using the GaAs semiconductor laser pulse. This pulse is shown in Figure 10 with a rise time of 0.2 μs.

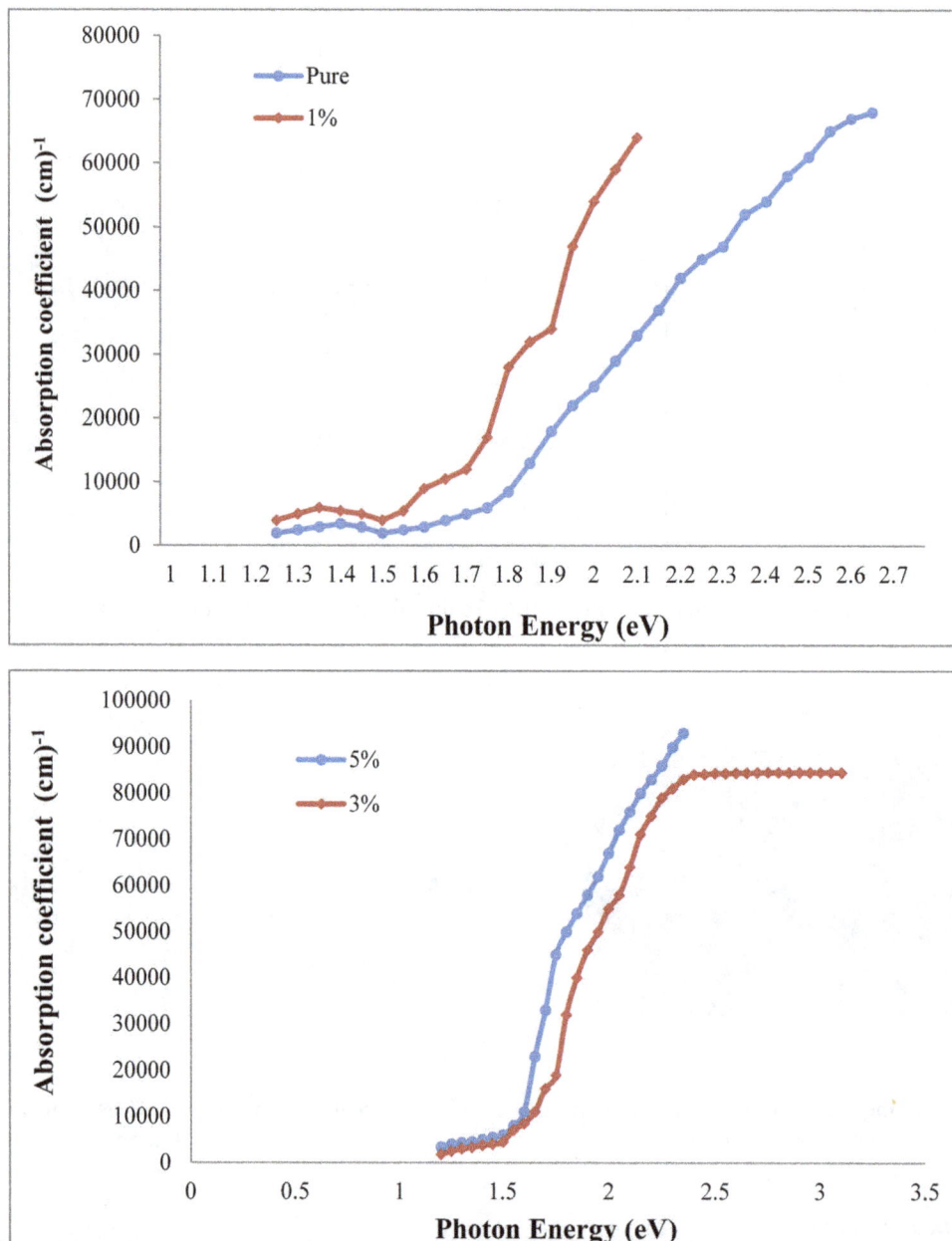

Figure 7. The absorption coefficient as a function of photon energy at different Cu wt%

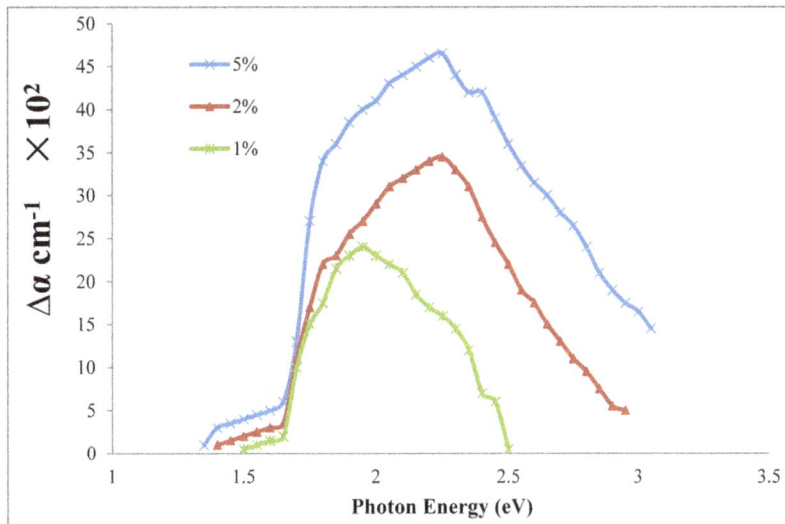

Figure 8. The change in absorption coefficient for various doping concentrations as a function of photon energies

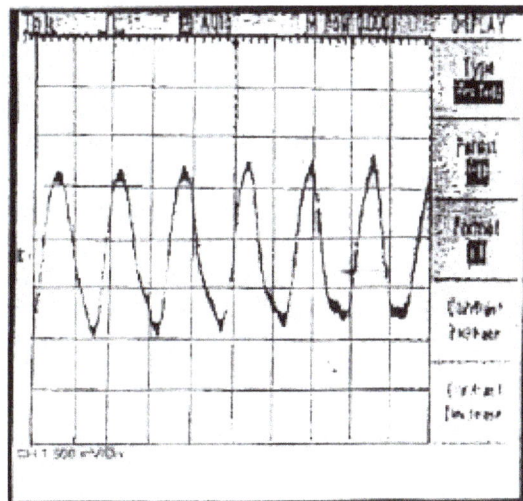

Figure 9. The output signal of the modulated CO_2 Laser

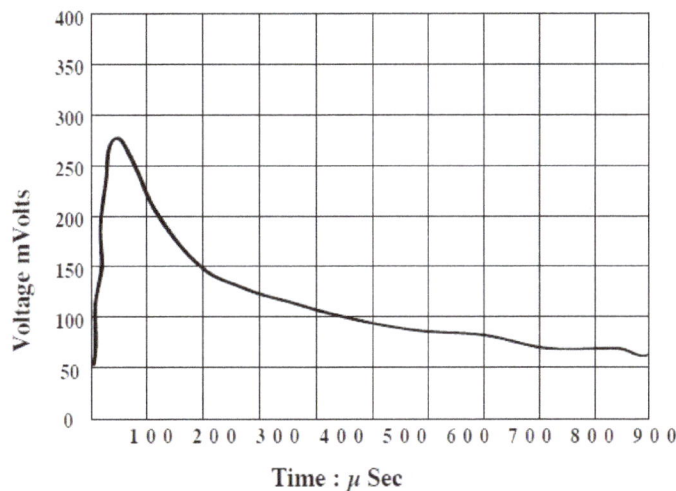

Figure 10. The output signal of the GaAs semiconductor laser

Figure 11a depicts the energy levels involved in the absorption of CdSe:Cu film. This schematic diagram shows a wide wavelength tuning range extending from the visible to the infrared spectrum. This tuning spectrum of the copper doped bulk CdSe can be compared with colloidal CdSe- ZnS core-shell quantum dots (Figure 11b). The common index of the bulk CdSe:Cu and the colloidal CdSe core-shell quantum dots is the single crystalline structure. As a matter of fact the bulk crystal covers the micorscale range, whereas the quantum dots is in the nanoscale. However, this tuning spectrum was detected by using a surface Plasmon's technique. Five distinguishable transitions were detected by the absorption and fluorescence spectra of the CdSe quantum dots (Mohammed, 2013). The populations of these levels were explained by the multiple exciton generation (MEG), in full agreement with the theoretical predictions. It can be remarked that the traditional bulk CdSe doped Cu and the CdSe –ZnS quantum dots are manifested in wavelength tuning. The advantages of the bulk CdSe over the size dependent confinement energy of the same semiconductor material are easier to prepare and at much less fabrication cost than the corresponding size confinement quantum dots. Besides, the CdSe quantum dots tuning range never extends over the visible region of the spectrum, whereas this reported tuning of the bulk CdSe:Cu does extend to the infrared region. Recently, researchs were done in the synthesis and characterization of CdSe doped Cu nanoparticles. Raj et al. (2014) were observed that the increase in the concentration of copper shifts the emission towards the higher wavelength. They explained such blue shift was caused by strong confinement effect. Meulenberg et al. (2004) have studied the structure and composition of Cu doped CdSe nanocrystals using soft X-ray absorption near edge spectroscopy (XANES). They indicated changes in the Se density of states with Cu doping, due to a local bonding environmental effect. Bear et al. (2014) were studied copper doped CdSe-ZnS core-shell quantum dots (QD) and have shown that doping small amounts of Cu into the ZnS shell partially quenches the QD core luminescence and a blue shifts in luminescence peak with respect to pure ZnS shell was detected. This is in contrast to a pure ZnS shell, where a significant increase in quantum yield and a red shift in luminescence were observed. The pioneer work of Türe et al. (1986) have investigated by photoconductivity and space-charge region capacitance technique multiple levels of copper centers in single crystal of CdSe. They demonstrated that the only center referring to copper was situated at 1eV with respect to the valence band. Photoluminescence blinking dynamics for Cu^+:CdSe was analyzed by Whitham et al. (2015) and compared to undoped CdSe nanocrystal. They identified the effect of Cu^+, which selectively traps photogenerated holes and revealed that the Cu^+:CdSe nanocrystal off state dynamics are statistically identical. A theoretical and experimental studies implemented by Wright and Meulenberg (2013) have demonstrated the effect of dopant into the CdSe quantum dots. They predicted that the dopant concentration causes a lowering of band energy as compared to the bulk band gap energy. This red shift was explained in terms of the contribution of both the hybridization energy and the confinement energy. They demonstrated that the dopant can affect not only the electronic structure but also the optical properties.

It can be concluded from the previous studies in CdSe:Cu as a bulk or a nanostructure, no evidences were reported for the existence of Cu localized levels with an accurate determination of their positions in the band gap of CdSe, particularly, in the infrared region of the spectrum.

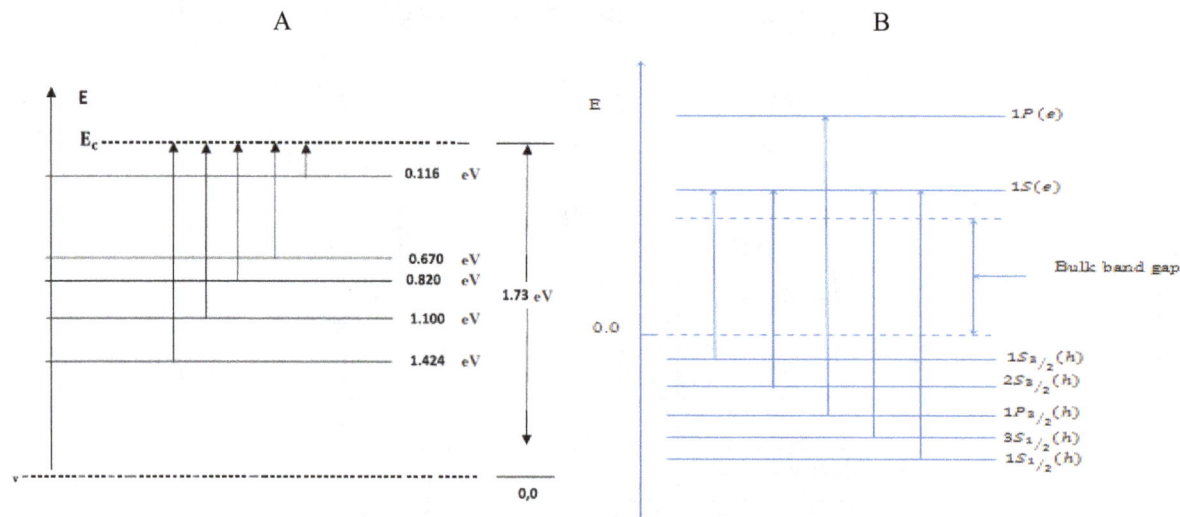

Figure 11. A) Energy levels involved in the absorption of CdSe:Cu film B) Energy Levels involved in the absorption and fluorescence of CdSe-ZnS core-shell quantum dots (Mohammed, 2013)

3.3 Detection Characteristics

Figure 12 shows that the photoconductivity of the CdSe grows exponentially with the increase in Cu wt% concentration. This growth feature is due to the improvement of the film structure. This improvement is confirmed by the increase in the crystal grain size as stated earlier. The noise voltage and the responsivity of this detector are measured and found to be ~1nV and 0.73V/W respectively. The detectivity was calculated by using the following relationship:

$$D^* = (A \Delta f)^{1/2} R / V_n \tag{1}$$

Where A is the detector active area (A=0.1 cm^2), (Δf=1Hz) the electrical bandwidth, V_n is the noise voltage and R is the responsivity. Using the Eq.1, a value of D*=2.31×10^8 cm Hz$^{1/2}$W^{-1} is obtained. Comparison of this D* value of the CdSe bulk detector at room temperature with the specific detectivity across a different technology is reasonable. The room temperature detectivity of the state of art MCT is 2×10^8 cm Hz$^{1/2}$W^{-1} (Hamamatsu 2015) which approaches the measured D* value for the CdSe:Cu detector. However, the predicted detectivity can be increased by an optimization methodology which includes surface plasmons effect by deposition of the CdSe films into gold coated substrates. Besides, other dopant than Cu like Ag and Au can be used for comparison. The present CdSe:Cu detector paves the way for further investigations and improvements in the field of a dual range antiaircraft missile seekers.

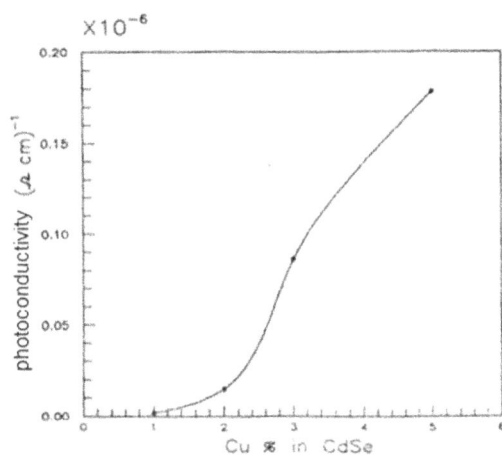

Figure 12. The photoconductivity of CdSe as a function of Cu wt% concentration

4. Conclusions

The following conclusions are drawn:

(i) The tuning range of the bulk CdSe:Cu is found greater than the recent technological CdSe quantum dots. This may open the door for more applications, particularly in the IR region of the electromagnetic spectrum.

(ii) The detectivity of this detector at room temperature is comparable with the corresponding state of art MCT detector and the recent designed GaN QDs. Therefore, the CdSe:Cu detector can be considered as an additional prototype to the traditional MCT as an IR detector operating at room temperature.

References

Asgari, A., & Razi, S. (2010). High performances III-nitride quantum dot infrared photodetector operating at room temperature. *Optics Express, 18*(14), 14604-14615.

Bear, J. C., Hollingsworth, N., McNaughter, P. D., Mayes, A. G., Ward, M. B., Nann, T., ... & Parkin, I. P. (2014). Copper ‑ Doped CdSe/ZnS Quantum Dots: Controllable Photoactivated Copper (I) Cation Storage and Release Vectors for Catalysis. *Angewandte Chemie International Edition, 53*(6), 1598-1601. http://dx.doi.org/10.1002/anie.2013308778

Diedenhofen, S. L. D., Kufer, D., Lasanta, T., & Konstantatos, G. (2015). Integrated colloidal quantum dot with color-tunable plasmonic nanofocussing lenses. *Light Science Applications, 4*, e234. http://dx.doi.org/1038/lsa.2015.7

Hamamatsu, Infrared detectors, Selection Guide-March. (2015). Retrieved from http://www.hamamatsu.com/resources/pdf/ssd/ingaos_kird0005epdf

Kashyout, A. B., Soliman, H. M. A., Fathy, M., Goma, E. A., &.Zidan, A. A. (2012). CdSe Quantum dots for solar cells devices. *International Journal of Photoenergy*, 1-7. http://dx.doi.org/10.1155/2012/952610

Mohammed, H. H. (2013). CdSe-ZnS core-shell quantum dots: surface plasmons effect and optical spectra. *Applied Physics Research, 5*(6), 15-22. http://dx.doi.org/10.5539/apr.v5n6p15

Maiti, A., & Bhattacharyya, S. (2013). Quantum dots and applications in medical science. *International Journal of Chemical Science and Chemical Engineering, 3*(2), 37-42. Retrieved from http://www.r.publication.com

Meulenberg, R. W., Buuren, T. V., Hanif, K. M., Strouse, G. F., & Terminello, L. (2004). Structure and composition of Cu-doped CdSe nanocrystals using soft X-ray absorption spectroscopy. *Nano Letters*, *4*(11), 2277-2285.

Oertel, D. C., Bawendi, M. G., Arango, A. C., & Bulovic, V. (2005). Photodetector based on treated CdSe quantum dots films. *Applied Physics Letters, 87*(21), 213505. http://dx.doi.org/10.1063/1.2136227

Raj, D. J. V., Linet, J. M., & Das, S. J. (2014). Synthesis and characterization of 1- thioglycerol capped CdSe and Cu (copper) doped CdSe nanoparticles at room temperature. *International Journal of ChemTech Research, 6*(3), 2042-2044. Retrieved from http://www.sphinxsai.com/framesphinxsaichemtech.htm

Türe, I. E., Claybourn, M., Brinkman, A. W., & Woods, J. (1986). Copper centers in CdSe. *J. Appl. Phys., 60*(5), 1670- 1675. http:// dx.doi.org/10.1063/1.337256

Wright, T. W., & Meulenberg, R. W. (2015). Effect of dopant on the band structure of quantum dots: A theoretical and experimental study, *Physical Review B., 88*, 045432-1-8. http:// dx.doi.org/10.1103/Phys Rev B.88.045432

Whitham, P. J., Knowles, K. E., Raid, P. G., & Gamelin, D. G. (2015). Photolumiescence blinking and reversible electron trapping in copper-doped CdSe nanocrystal. *Nano Lett., 15*(6), 4045-4051. http://dx.doi.org/10.1021/acs.nanolett.5b01046

Wei, S. H., Zhang, S. B., & Zunger, A. (2000). First-principles calculation of band offset, optical bowings, and defects in CdS, CdSe, CdTe, and their alloys. *Journal of Applied Physics, 87*(3), 1304-1311. Retrieved from http://scitation.aip.org/termsconditions

Zhang, J., Itzler, M. A., Zbinten, H., & Pan, J. W. (2015). Advances in InGaAs/InP single- photon detector systems for quantum communications. *Light Science and Applications, 4*, e286. http://dx.doi.org/10.1038/Isa.2015.59

11

Fluorescence and Reflectance Spectroscopy for Early Detection of Different Mycorrhized Plantain Plants

Wilfried G. Dibi[1,2], Beaulys Fotso[3], Casimir Y. Brou[3], Jérémie T. Zoueu[1], Adolphe Zeze[3] & Jocelyne Bosson[2]

[1] Laboratoire d'Instrumentation Image et Spectroscopie, Unité Mixte de Recherche et d'Innovation en Electronique et Electricité Appliquée, Ecole Supérieure d'Industrie, Institut National Polytechnique Houphouët Boigny BP 1093 ESI Yamoussoukro, Côte d'Ivoire

[2] Laboratoire de Physique Fondamentale et Appliquée, Unité de Formation et de Recherche des Sciences Fondamentales et Appliquées, Université Nangui Abrogoua, 02 BP 801 Abidjan 02, Côte d'Ivoire

[3] Laboratoire des Biotechnologies Végétale et Microbienne, Unité Mixte de Recherche et d'Innovation en Sciences Agronomiques et Génie Rural, Ecole Supérieure d'Agronomie, Institut National Polytechnique Houphouët-Boigny BP 1313 ESA Yamoussoukro, Côte d'Ivoire

Correspondence: Jérémie T. Zoueu, Laboratoire d'Instrumentation Image et Spectroscopie, Institut National Polytechnique Felix Houphouët-Boigny Yamoussoukro, Cote d'Ivoire.
E-mail: Jeremie.zoueu@inphb.edu.ci

Abstract

Sustainable agriculture with use of Arbuscular Mycorrhizal Fungi (AMF) is an emerging farm management that improves crops nutrient and water use efficiency. Decision making on the effect of AMF is still dependent on agronomic diagnosis which is long, tedious, expensive and destructive. This study demonstrates the applicability of proximal fluorescence and reflectance spectroscopy for evaluating and detecting at early stage distinct types of mycorrhized plantain from two cultivars (*Musa paradisiaca*).

Visible-near infrared (400-1000 nm) reflectance and fluorescence data were collected from control and three levels mycorrhized plants designed in randomized and complete block under greenhouse conditions. Two spectral measurements at a week interval were performed on plant leaves by using an USB spectrometer mounted with an Arduino-based LED driver clip.

A new normalized reflectance water NWI5 index shows with Datt5 alone highly significant differences at P<0.001 respectively for Orishele and fhia21 cultivars. dNIRmin920_980, NDVI3 and GI reflectance index are significant at P<0.01. Seven other reflectance and 3 fluorescence indices ANTH, FRF_R and NBI_R are significant at P<0.05. The two first principal components for each cultivar spectral features explaining 94.1 % of variance were used to build predictive classification models. LogitBoost algorithm indicates accuracy of 90.27% on stratified cross-validation and 87.5% on test split. Our results confirm that fluorescence and reflectance spectroscopy is a valuable tool for early assessment of mycorrhization success rate evaluation and pattern recognition. They also show promise for the development of non-destructive and cost-effective detectors in monitoring crops under biofertilizers with arbuscular mycorrhizae.

Keywords: classification, fluorescence and reflectance spectroscopy, LogitBoost algorithm, mycorrhized plantain plants.

1. Introduction

Agricultural remote sensing has gained great importance in sustainable and intensification farming management. It generates digital data from sensors that help in reducing the risk and minimize damage of inappropriate fertilization. Indeed, low or excessive uses of chemical fertilizers can have high adverse impact on the environment and grain production system. Unused nitrogen released to the environment can have detrimental effects (Cameron, Di, & Moir, 2013). Large application of phosphate fertilizers and by-products has been practiced on arable lands, to improve crop production, induced soil nutrients deficiency, and increasing the levels of available S and P (Kassir, 2014). In the increasing demand of environmental sustainability, more attention of agricultural scientific community is being turned on biofertilizers particularly Arbuscular Mycorrhizal Fungi

(AMF) (Weber, 2014; Sahraoui, 2013). Thus investigating the potential output of such biological microorganisms on crops with use of remote sensing tools is highly relevant.

Controlled endomycorrhization is a current problem that stayed long time ago in agriculture, horticulture and arboriculture (Gianinazzi, 1981). It was recently reported that diversity of plant response to mycorrhizal symbiosis essentially depends on isolates spores of species, soil fertility and environmental culture conditions (Garbaye, 2013). Optical sensing systems are potential and suitable tools for assessing mycorrhization success rate since they gather information on plant response, rather than costly, time-consuming, and destructive laboratory analyses. In this field, many efforts are done with various acquisitions sensing system and several spectral features developed (Mulla, 2013; Pinheiro & Gusmo Dos Anjos., 2014) .However, there is an apparent lack of study in remote sensing crop investigation with arbuscular mycorrhizae. New spectral indices that simultaneously allow assessment of multiple crops characteristic and reveal better understanding of plant molecular mechanism could be retrieved.

An incident photosynthetically active radiation (PAR) interaction with plant leaves induced various spectral response in absorbance, reflectance and transmittance. Reflectance expresses light use efficiency of plant (Barton & North, 2001). Reflectance properties of leaves are determined by the concentration of chlorophyll, other pigments (carotenoids, xanthophylls and anthocyanins) and absorbing biochemicals in the visible (400–700 nm) wavelength region, by mesophyll structure in the near infrared (700–1200 nm) region and by amount of water in the near infrared (950-970 nm) and in the middle infrared (1200–2400 nm) region (Peñuelas, 1998). When a fluorescent molecule absorbs the energy of a given wavelength, a part of it is dissipated by light emission at longer wavelengths within a very short time. This small but variable amount of energy loss is known as the fluorescence emission which is fingerprint of leaf photosynthetic activity. Coupling the measurements of reflectance and fluorescence with high spectral resolution has been suggested for improving crop productivity evaluation (Baret, Guyot, & Major, 1988), stress diagnosis (Lichtenthaler, Wenzel, & Buschmann, 2006) and nutrient diagnosis (Belanger, Viau, Samson, & Chamberland, 2005). Decades of research in vegetation reflectance and fluorescence-based methods has gone into finding multiple indices precisely related to several leaf pigments, stresses, and mineral content. Combination of fluorescence and reflectance could reveal some functional properties of AMF in crop production particularly in plantain and banana case where their effects have been studied in recent Africa's agricultural researches (Nwaga et al., 2011; Gaidashova et al., 2013; Jefwa et al., 2013).

Multi-class classification in crops remote sensing is performed by means of data mining techniques. In this approach, several methods of machine learning have been utilized for plant characteristics detection. Support Vector Machine are important advanced techniques intensively used (Mountrakis & Ogole, 2011; Moshou, Pantazi, Kateris, & Gravalos, 2014). Others efficient tools are concerned with fuzzy logic, neural network and extreme learning machine (Moreno, Corona, Lendass, Graña, & Galvão 2014; Murmu & Biswas, 2015; Cvetković, Stojanović, & Nicolić, 2015).The boosting meta-algorithm is relatively new, efficient, simple, and easy to manipulate additive modeling technique that can use potentially any weak learner available. LogitBoost is another variant of the boosting algorithm that performs additive logistic regression (Friedman, Hastie, & Tibshirani, 2000) and has shown better performance than many other machine learning algorithm especially in protein data structure classification (Krishnaraj & Reddy, 2009). At present, new improvements of this algorithm are proposed (Kanamori & Takenouchi, 2013; Sun, Reid & Zhou, 2014).

In this study, we aimed at evaluating early detection of mycorrhized plantain using active fluorescence and reflectance spectrometry. The specific objectives were to (1) identify the most significant fluorescence and reflectance vegetation indices explaining mycorrhizae effects on plantains and to (2) build predictive classification model for each mycorrhizal treatment using spectral features discovered.

2. Material and methods

2.1 Plant Material and Experimental Design

The experiment was conducted under greenhouse conditions at the Laboratoire des Biotechnologies Végétale et Microbienne in National Polytechnic Houphouet-Boigny Institute of Yamoussoukro (Côte d'Ivoire). Two plantain varieties namely Orishele (sensitive to water stress) and Fhia 21 (tolerant to water stress) were cultivated in pots containing sterilized substrate composed of loam and sand. The plants were inoculated with a mycorrhizal morphotype of *Glomus* (T1), a complex of spores from the rhizosphere of plantain in Bouaflé (T2) and another complex of spores from the rhizosphere of plantain in Azaguié (T3). All treatments including control plants without any mycorrhizae (T0) were split into a randomized complete block design consisting of three replicates per treatment in each of the three blocks.

2.2 Formulation and Preparation of Inoculum

Spores are extracted at the Laboratoire des Biotechnologies Végétale et Microbienne in National Polytechnic Houphouet-Boigny Institute of Yamoussoukro (Côte d'Ivoire) according to the method described by Gerdemann & Nicholson (1963). They were disinfected using bleach 2% and 0.2% streptomycin. The banana suckers were firstly cleaned, washed and acclimated in coco peat and then inoculated with these spores at 150 spores per pot.

2.3 Optical Instrumentation and Spectral Measurements

A hand-held clip LEDs was built in our laboratory and mounted with USB4000 spectrometer Ocean Optics for acquiring fluorescence and reflectance spectra on the upper leaf surface. The spectrometer can collect data in the 350–1000 nm spectral region, with a sampling interval of 0.22 nm. Three Light-emitting-diodes excited the fluorescence at 375 nm (UV), 520 nm (green) and 630 nm (red) while the emitted fluorescence light was detected in the red (RF: 680–690 nm) and far-red (FRF: 720–755 nm) spectral regions using a low-pass filter at 650 nm. A white LED emitting (400-700 nm) was utilized for reflectance measurement. All four LEDs were controlled using the computer and Arduino, a cost effective, open source and easy programmable microcontroller board (Arduino - Home). Reflectance spectra were carried out after comparing with a 99% white reflectance standard (Labsphere®, Edmund Optics Inc. — 101 East Gloucester Pike, Barrington, NJ 080071380 USA). After three scans, an average spectrum is recorded using SpectraSuite® Software. Spectra acquisitions occurred between 29 November and 06 December 2014 with experimental setup showed in Figure 1.

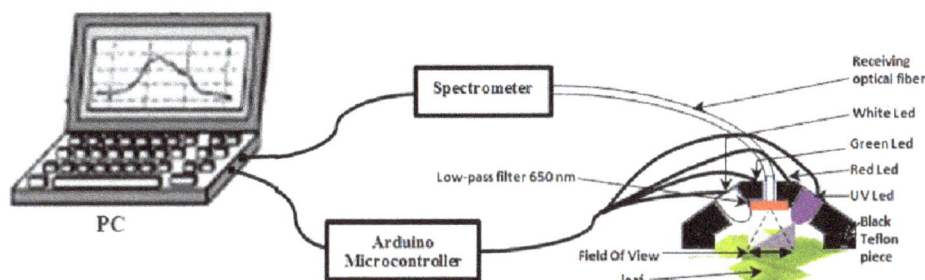

Figure 1. Schematic drawing of the experimental setup used for spectra acquisition

2.4 Spectra Preprocessing and Normalization

The spectra dataset recorded are firstly filtered using Gaussian normal probability density function in matlab. Then, each spectrum is standardized by its own average and standard deviation for baseline and quantity correction (Barnes, Dhanoa & Lister., 1993). Finally, in the hypothesis that the number of photon at any wavelength should be constant across shots, we proceeded to min/max normalization. Each spectrum is normalized to have a minimum value of 0 and a maximum value of 1. This normalization was accomplished according the formula:

$$x_{norm} = \frac{(x - \min_i x_i)}{(\max_i x_i - \min_i x_i)} \tag{1}$$

2.5 Fluorescence and Reflectance Indices Calculation

Vegetation indices are widespread in remote sensing scientific literature. A database composed of 18 fluorescence and 90 reflectance indices were implemented in matlab for spectra features calculation. We tried to extend to almost reviewed vegetation index at leaf and canopy level for enabling maximum crop characteristic detection. These indexes are also supplemented with our proposed one.

2.6 Statistical Analysis

2.6.1 One-Way Variance Analysis with Step-Down Dunnett Post Hoc Test

ANalysis Of VAriance (often referred to as ANOVA) is a technique for analyzing the way in which the mean of a variable is affected by different types and combinations of factors. One-way analysis of variance is the simplest form in which we are interested in comparing the dependent variable means of two or more groups defined by a categorical grouping factor (Bewick, Cheek, & Ball, 2004). Analysis of variance is initially performed on two groups (control and all mycorrhized plants together) then on four groups (control and each of three mycorrhizal treatments). In order to understand subgroup differences among the different experimental and control group, we chose to perform step-down Dunnett test which is more suitable and powerful (Bretz, Hothorn & Westfall, 2010). R software version 3.1.2 (R Core Team, 2014) was used for ANOVA analysis and for multiple comparisons.

2.6.2 Principal Components and Vegetation Index Correlation Analysis

Principal components analysis (PCA) is a data reduction technique that transforms a larger number of correlated variables into a much smaller set of uncorrelated variables called principal components. The weights used to form the linear composites are chosen to maximize the variance each principal component accounts for, while keeping the components uncorrelated. We generated principal components for each cultivar with their respective significant vegetation indices.

Correlation analysis was also used for establishing relationship between new indice developed and existing ones.

2.6.3 LogitBoost Classifier

Boosting implements forward stagewise additive modeling. This class of algorithms starts with an empty ensemble and incorporates new members sequentially. At each stage the model that maximizes the predictive performance of the ensemble as a whole is added, without altering those already in the ensemble. Additive logistic regression use similar adaptation by modifying the forward stagewise modeling method (Witten, Frank, & Hall, 2011).

For K-class classification ($K \geq 2$), consider an N example training set $\{x_i, y_i\}_{i=1}^N$ where x_i denotes a feature value and $y_i \in \{1, ..., K\}$ denotes a class label. Class probabilities conditioned on x denoted by $p = (p_1, ..., p_K)^T$ are learned from the training set. For a test example with known x and unknown y, we predict a class label by using the Bayes rule: $y = argmax_k p_k, \ k = 1, ..., K$.

Instead of learning the class probability directly, one learns its "proxy" $F = (F_1, ..., F_K)^T$ given by the so-called Logit link function:

$$p_k = \frac{\exp(F_k)}{\sum_{j=1}^K \exp(F_j)} \tag{2}$$

with the constraint $\sum_{k=1}^K F_k$ (Friedman et al., 2000). For simplicity and without confusion, we hereafter omit the dependence on x for F and for other related variables.

The F is obtained by minimizing a target function on training data:

$$Loss = \sum_{i=1}^N L(y_i x_i), \tag{3}$$

where F_i is shorthand for $F(x_i)$ and $L(y_i, F_i)$ is the Logit loss for a single training example:

$$L(y_i, F_i) = -\sum_{k=1}^K r_{ik} \log p_{ik}, \tag{4}$$

Where $r_{ik} = 1$ if $y_i = k$ and o otherwise. The probability p_{ik} is connected to F_{ik} via (2).

To make the optimization of (3) feasible, a model is needed to describe how F depends on x. For example, linear model $F = W^T x$ is used in traditional Logit regression, while Generalized Additive Model is adopted in LogitBoost:

$$F(x) = \sum_{m=1}^M f_m(x), \tag{5}$$

where each $f_m(x)$, a K dimensional sum-to-zero vector, is learned by greedy stage-wise optimization. That is, at each iteration $f_m(x)$ is added only based on $F = \sum_{j=1}^{m-1} f_j$. Formally,

$$f_m(x) = argmin_f \sum_{i=1}^N L(y_i, F_i + f(x_i)), \tag{6}$$
$$s.t. \sum f_k(x_i) = 0, i = 1, ..., N,$$

This procedure repeats M times with initial condition $F = 0$. Owing to its iterative nature, we only need to know how to solve (6) in order to implement the optimization (Sun et al., 2014).

Models were performed in WEKA 3.6.13 software Explorer interface (Hall et al., 2009). They were built separately for each cultivar using their own significant fluorescence and reflectance indices according to ANOVA results. The four output classes used in the analysis were T0, T1, T2 and T3 distributed on 72 observations in each model. In holdout evaluation, the data were randomly separated into training and test datasets such that 66% of the data was utilized for training the classifier model; while 33% of the data was used for testing the developed classifier model. In order to mitigate any bias by the particular sample chosen for holdout, we performed a tenfold stratified cross-validation. The outcomes metrics are automatically generated in WEKA. Two-class outcomes are presented in figure 2 for interpretation needs.

		Predicted Class	
		yes	no
Actual Class	yes	true positive (**TP**)	false negative (**FN**)
	no	false positive (**FP**)	true negative (**TN**)

$$TP\ rate(\%) = \frac{TP}{TP+FN} \times 100$$

$$FP\ rate(\%) = \frac{FP}{FP+TN} \times 100$$

$$Precision(\%) = \frac{TP}{TP+FP} \times 100$$

$$Recall(\%) = \frac{TP}{TP+FN} \times 100$$

Figure 2. Confusion matrix and associated classification measures in two-class case

3. Results and Discussion

3.1 Data Representation after Preprocessing and Normalization

Fluorescence spectra using UV (375 nm), Green (520 nm) and Red (630 nm) essentially show two peaks in red band (685- 690nm) and far –red band (735-740 nm) known as chlorophyll a fluorescence (Buschmann Langsdorf & Lichtenhaler, 2008; Misra & Singh, 2012). Intensity at these two bands vary when using an UV (375 nm) or Green (520 nm) light because of respective absorption at these wavelengths of flavonols (Cerovic *et al.*, 2002) and anthocyanins pigments (Agati et al., 2005). Analysis of reflectance curve presents a deep through in 850-1000 nm band and two others at 778-788 nm and 834-840 nm bands. These observations may be attributed to water absorption.

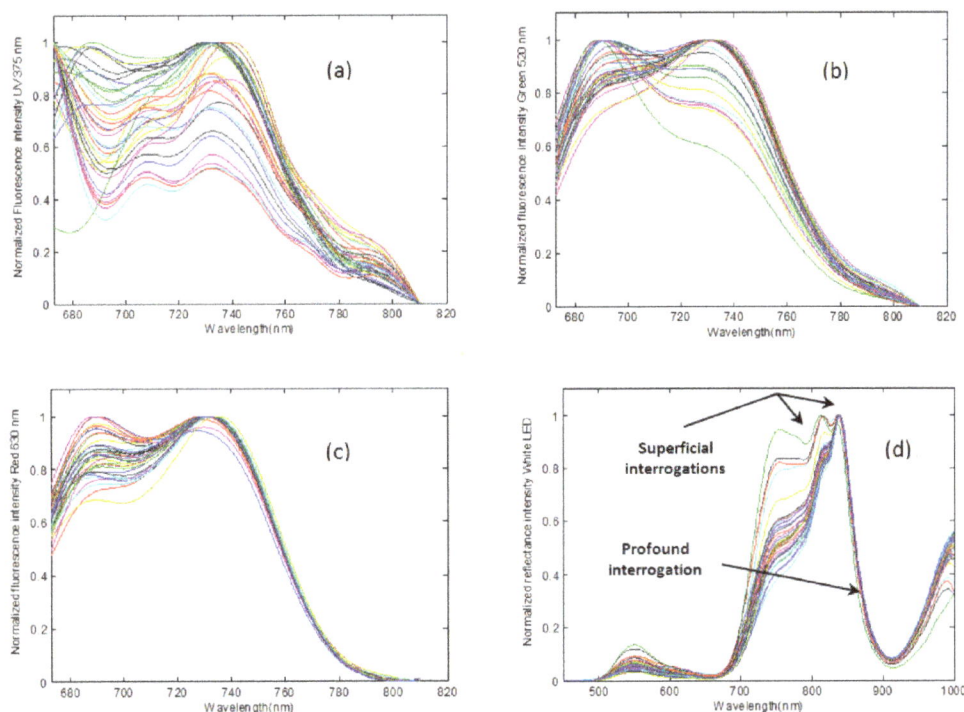

Figure 3. Normalized fluorescence spectra using excitation Led source at 375 nm (a), excitation Led source at 520 nm (b), excitation Led source at 630 nm (c) and normalized reflectance spectra using white light Led source 400-700 nm (d) in experimental setup

3.2 Spectral Responses to Plantain Mycorrhizal Treatment

Analysis of variance of 108 vegetation indexes reveals 3 significant fluorescence indices and 12 significant indices for reflectance spectra. ANOVA results in Table 1 and 2 show that differences between control and each of three mycorrhizal treatments are revealed by reflectance indexes. Two reflectance indexes are highly significant ($P<0.001$): Datt5 for Fhia 21 cultivar and the new index NWI5 for Orishele cultivar.

Table 1. Effect of AM fungal inoculation on calculated vegetation indexes of plantain cultivar Fhia 21 under greenhouse conditions

ANOVA	Vegetation Index											
	Fluorescence		Reflectance									
	ANTH	FRF_R	Carter5	Datt5	SR5	GI	NDVI3	NWI	NWI4	NWI5	dNIRMin920_980	min950_970
All mycorrhized treatment	*	*	NS	**	NS	**	**	NS	NS	*	*	*
Three level mycorrhized treatment	NS	NS	*	***	*	**	**	*	*	**	NS	*

Significance code: * Significant at 0.05, ** significant at 0.01, *** significant at 0.001, NS Not significant

Table 2. Effect of AM fungal inoculation on calculated vegetation indexes of plantain cultivar Orishele under greenhouse conditions

	Vegetation Indexes						
	Fluorescence	Reflectance					
ANOVA	NBI_R	DPI	RatiodRE_703	NWI	NWI5	dNIRmin920_980	min950_970
All mycorrhized treatment	*	*	NS	*	**	**	*
Three level mycorrhized treatment	NS	*	*	*	***	*	*

Significance code: * Significant at 0.05, ** significant at 0.01, *** significant at 0.001, NS Not significant.

Table 3 gives us an overview on their definition and relationship with bioparameters. From most to least importance order, these spectral features are related to water content (NWI, NWI5, NWI4, dNIRMin920_980, min950_970), chlorophyll content (Datt5, NDVI3, SR5, and RatiodRE_703), anthocyanin content (ANTH), foliar density (FRF_R), nitrogen content (NBI_R), photosynthetic activity (DPI) and leaf area index (GI).

As expected, most of these spectral features appear to be relevant in plantain mycorrhization studies. On the one hand, we can see that indices linked to water content appear with great occurrence in both varieties: NWI, NWI4, NWI5, dNIRmin920_980, min950_970. These water-based indexes demonstrate that mycorrhized plantains have better water content in comparison to control. This result agrees with Nwaga et al. (2011) study which has shown good correlation between water use efficiency and drought tolerance of mycorrhized plantains. As an example, Orishele water content indices given in Figure 4 show that water-based indexes means for all mycorrhized classes are higher than control ones.

Table 3. Overview of significant fluorescence and reflectance indexes in Fhia 21 and Orishele plantain cultivars

Vegetation index	Related to	Equation	Reference
Fluorescence indexes			
Anthocyanin (ANTH)	Anthocyanin content	$\log(FRF\_R/FRF\_G)$	Agati et al. (2005)
Far-Red Fluorescence with Red light (FRF_R)	Foliar density, leaf strength	FRF_R	Bramley et al. (2011)
Red Nitrogen Balance Index (NBI_R)	Nitrogen content	$\log(FRF\_UV/FRF\_R)$	Cartelat et al. (2005)
Reflectance indexes			
SR5	Chlorophyll a	R_{675}/R_{700}	Chappelle, Kim & McMurtrey
Datt5	Chlorophyll b	R_{672}/R_{550}	Datt (1998)
Normalized Difference Vegetation Index (NDVI3)	Chlorophyll total	$(R_{682} - R_{553})/(R_{682} + R_{553})$	Gandia, Fernandez, Garcia & Moreno (2004)
Ratio of first derivative maximum value in red-edge with reflectance one at 703 nm (RatiodRE_703)	Chlorophyll, Water content	$maxD_{680-780}/D_{703}$	Filella & Peñuelas (1994)
Greenness Index (GI)	Chlorophyll, Leaf Area Index	R_{554}/R_{677}	Smith, Adams, Stephen & Hick (1995)
Carter5	Stress	R_{695}/R_{670}	Carter (1994)
Double Peak Index (DPI)	Photosynthetic activity	$(D_{688} \times D_{710})/D_{697}^2$	Zarco-Tejada, Pushnik, Dobrowski & Ustin (2003)
Normalized Water Index (NWI)	Water content	$(R_{970} - R_{900})/(R_{970} + R_{900})$	Barbar et al. (2006)
Normalized Water Index 4 (NWI4)	Water content	$(R_{970} - R_{920})/(R_{970} + R_{920})$	Prasad et al. (2007)
Normalized Water Index 5 (NWI5)	Water content	$\dfrac{(minR_{778-788} - minR_{834-840})}{(minR_{778-788} + minR_{834-840})}$	In this study
Minimum value of the first derivative in 920-980 range (dNIRMin920_980)	Water content	$minD_{920-980}$	Peñuelas, Fillela, Biel, Serrano & Salvé(1993)
Minimum value in 950-970 range (min950_970)	Water content	$minR_{950-970}$	Peñuelas, Fillela, Biel, Serrano & Salvé(1993)

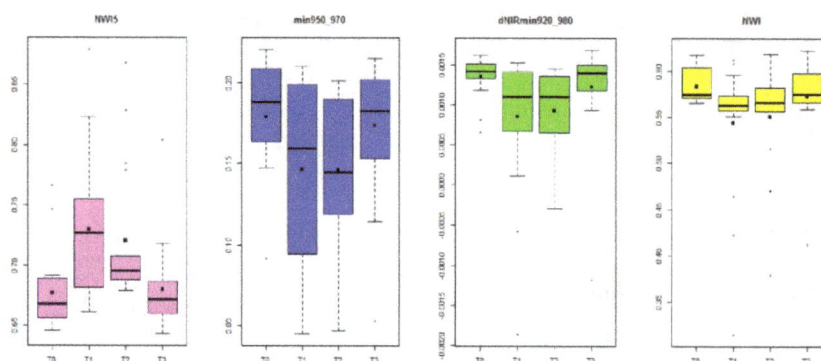

Figure 4. Boxplots of water-based vegetation indices (NWI5, min950_970, dNIRmin920_980 and NWI) for each treatment group for Orishele variety. Means are shown by fat dots, boxes indicate median and interquartile range, with vertical lines depicting the range. Dots outside vertical range are outliers' data

As suggested from the reflectance curve analysis, we computed a new normalized water index NWI5. This index shows significant differences between treatments for both varieties and is strongly correlated with leaf water content as it appears in correlation matrices (Table 4 and 5) particularly with water index content dNIRmin920_980 and min950_970 ($-0.84 < r < -0.88$). Therefore this parameter may stand as an indicator of water status. The negative correlation with the two literature indexes mentioned above suggests that water content in leaf is high when NW5 is high. Indeed dNIRmin920_980 and min950_970 are negatively correlated with relative water content. In addition, NWI5 presents more significance in Orishele variety than in Fhia 21 because Orishele is more sensitive to water stress. As a result, NWI5 could better track plant water stress sensitivity.

Table 4. Pearson correlation coefficients (all with p value<0.01) for water content reflectance indexes of Fhia 21 variety

Reflectance indexes	dNIRmin920_980	min950_970	NWI4	NWI5	NWI
dNIRmin920_980	-				
min950_970	0.842	-			
NWI4	0.572	0.508	-		
NWI5	-0.843	-0.879	-0.511	-	
NWI	0.445	0.358	-0.860	-0.452	-

Table 5. Pearson correlation coefficients (all with p value <0.01) for water content reflectance indexes of Orishele variety

Reflectance indexes	dNIRmin920_980	min950_970	NWI5	NWI
dNIRmin920_980	-			
min950_970	0.826	-		
NWI5	-0.859	-0.838	-	
NWI	0.780	0.559	-0.754	-

On the other hand, related works (Tropical Soil Biology and Fertility [TSBF], 2007; Declerck, Cevos, Devos, & Plenchette, 1994) have reported that AMF improves nutrient uptake and growth of bananas particularly an enhancement of P uptake. In this context, our study provides additive information especially about mycorrhization effects on plantain nutrition, leaf pigment content and photosynthetic activity. Indeed, in variety Fhia 21, ANTH index which is negatively correlated with anthocyanin content as demonstrated by J. Baluja (2012), indicates that all mycorrhizal treatment induces an increase of anthocyanins. FRF_R index is related to foliar density. Bramley et al. (2011) showed that NDVI provided by GreenSeeker has a positive relationship with index FRF_R obtained using the Multiplex long distance to flowering when the vegetation is less developed. So we deduce low rate of chlorophyll content in Fhia 21 variety for mycorrhized plants. This is also illustrated in figure 5 by chlorophyll index (Datt5, SR5, NDVI3 and GI) and is due to higher anthocyanin content.

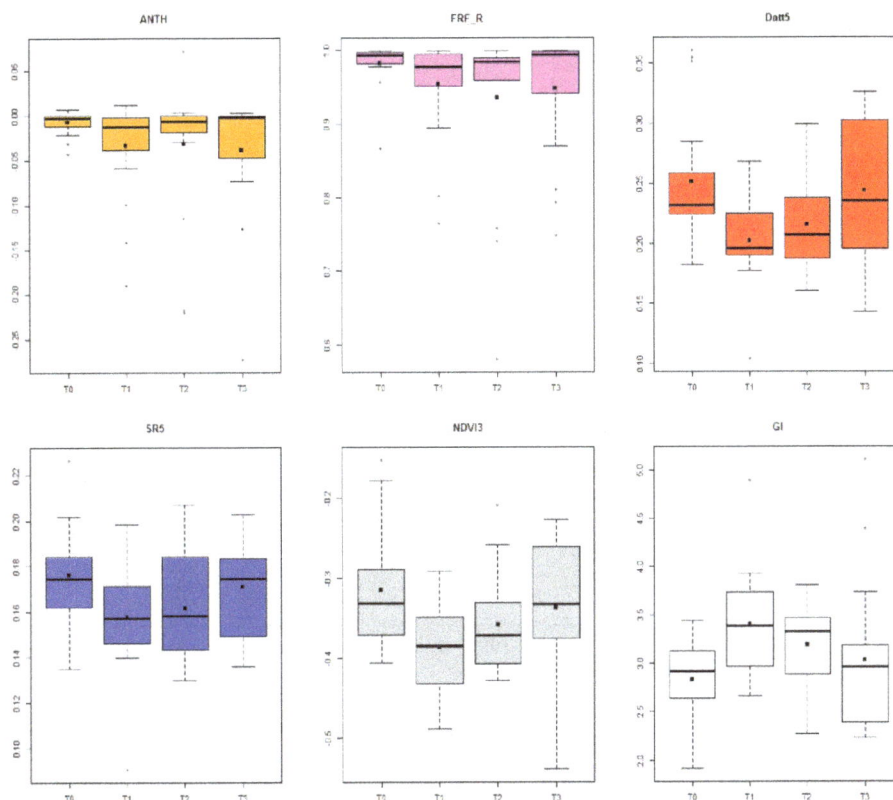

Figure 5. Boxplots of vegetation indices (ANTH, FRF_R, Datt5, SR5, NDVI3 and GI) for each treatment group for Orishele variety. Means are shown by fat dots; boxes indicate median and interquartile range, with vertical lines depicting the range. Dots outside vertical range are outliers' data

In Orishele variety, improved effects of mycorrhizae on chlorophyll content and photosynthetic activity appear and are respectively shown by higher value of red-edge spectral feature (RatiodRE_703) and double peak indice (DPI) in T1 and T2 mycorrhized treatments (Figure 6). Experience conducted by Zarco-Tejada et al (2003) have demonstrated that natural fluorescence emission is observable on the first derivative reflectance spectra as a double-peak feature in the 690–710 nm spectral region. DPI is capable for tracking natural steady-state fluorescence and is closely related to photosynthetic activity so that DPI is low when photosynthetic rates are high. Nethertheless, NBI_R boxplot means show that plants with inoculated mycorrhizae have low nitrogen content even if we can observe highest value of this index (1.3-1.51) for T1 treatment. NBI_R which is expressed by the ratio between epidermal phenolic compounds and chlorophyll is positively well correlated to the leaf N content (Agati, Foshi, Grossi & Volterrani, 2015). Figure 6 illustrates distribution of vegetation indices (NBI_R, RatiodRE_703 and DPI) according to mycorrhizal treatments for Orishele variety.

Step-down Dunnett test helps us identifying the most effective mycorrhizal treatments in comparison to control. From multiple comparison analysis, only significant vegetation index with four groups in ANOVA are concerned. Results in Table 6 show clearly that for Fhia 21 cultivar treatment T1 is the best and then treatment T2 follows concerning with improvement of water content performance. Reflectance indexes in Table 7 indicate that treatment T1 and T2 are efficient with few precision about the first one and the second one according to step-down Dunnett statistical test. In both varieties, following indexes namely Carter5, GI, RatiodRE_703 and NWI5 were not able to discriminate mycorrhizal treatments with this test. In addition, treatment T3 is not at all significant.

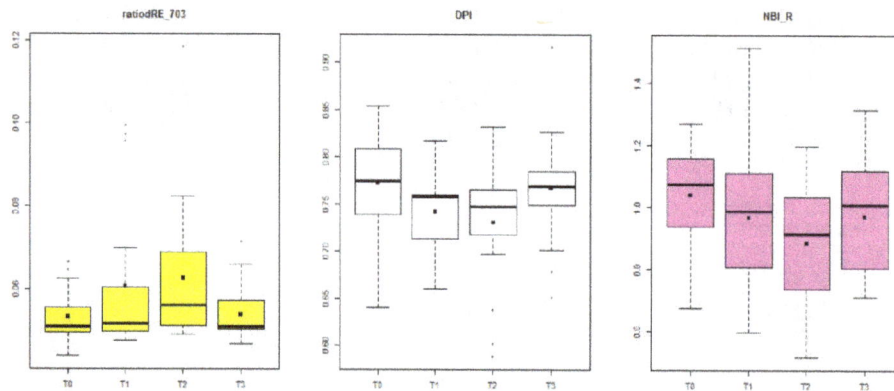

Figure 6. Boxplots of vegetation indices for each treatment group for Orishele variety (RatiodRE_703, NBI_R, and DPI). Means are shown by fat dots; boxes indicate median and interquartile range, with vertical lines depicting the range. Dots outside vertical range are outliers' data

Table 6. Multiple comparisons of means to a control of plantain cultivar Fhia 21 under greenhouse condition

Mycorrhizal Treatment vs control	Step-down Dunnett statistical significance								
	Carter5	Datt5	SR5	GI	NDVI3	NWI	NWI4	NWI5	min950_970
T1-T0	NS	***	**	NS	***	*	**	NS	**
T2-T0	NS	**	*	NS	*	NS	NS	NS	*
T3-T0	NS	NS	NS	NS	NS	NS	NS	NS	NS

Significance code: * Significant at 0.05, ** significant at 0.01, *** significant at 0.001, NS Not significant.

Table 7 Multiple comparisons of means to a control of plantain cultivar Orishele under greenhouse conditions

Mycorrhizal Treatment vs control	Step-down Dunnett statistical significance					
	DPI	RatiodRE_703	NWI	NWI5	dNIRmin920_980	min950_970
T1-T0	*	NS	**	NS	**	**
T2-T0	**	NS	*	NS	**	**
T3-T0	NS	NS	NS	NS	NS	NS

Significance code: * Significant at 0.05, ** significant at 0.01, NS Not significant.

Analysis of mycorrhizal effects on Fhia 21 and Orishele plantain cultivars leads us discovering a new water index as well as others relevant and varied fluorescence and reflectance indices. Spectral responses analysis of plants show essentially improved effects of mycorrhizae in water content for both varieties. Good effects are also revealed in anthocyanin, chlorophyll, photosynthetic activity and nitrogen content whose treatments T1 and T2 are mostly responsible for.

3.3 LogitBoost Classification Evaluation

In addition to finding most significant vegetation indexes and efficient mycorrhization treatment, additive logistic regression models are constructed to automatically predict specific class of a of mycorrhized plantain. Principal components used as variables input of models explains 99.47% of spectral features variability for Fhia 21 variety and 94.163 % in Orishele one. The PC1 and PC2 extracted allow this as shown in Table 8.

Cross-validation gives us a first insight in classification performance. On total instances of 72 samples, 90.2778 % are correctly classified while 9.7222 % are misclassified for both varieties (Table 10). Detailed performance metrics are given in Table 9. Precisely, classification accuracy is higher for treatment without any mycorrhizae T0 (94.4 %) than those with (88.9 %) in overall variety. In the mycorrhized classes treatment T3 has the best precision parameter (94.1 %) because of its low FP Rate. In addition, we can explain higher value of recall in control class in such way that the FN samples are low. It is also notable to precise that mycorrhized treatment T1 and T2 present the same performance metrics.

Table 8. Principal components results

Variety	Principal component	Variance explained (%)
Fhia 21	PC1	89.44
	PC1, PC2	99.47
Orishele	PC1	80.04
	PC1, PC2	94.163

Table 9. Classification performance metrics of the LogitBoost classifier by class

Variety	Evaluation method	Class	TP Rate	FP Rate	Precision	Recall
Fhia 21	Stratified cross-validation	T0	94.4	3.70	89.5	94.4
		T1	88.9	3.70	88.9	88.9
		T2	88.9	3.70	88.9	88.9
		T3	88.9	1.90	94.1	88.9
	Holdout	T0	100	11.1	75.0	100
		T1	60.0	0.00	100	60.0
		T2	100	5.30	83.3	100
		T3	87.5	0.00	100	87.5
Orishele	Stratified cross-validation	T0	94.4	3.70	89.5	94.4
		T1	88.9	3.70	88.9	88.9
		T2	88.9	3.70	88.9	88.9
		T3	88.9	1.90	94.1	88.9
	Holdout	T0	100	11.1	75.0	100
		T1	60.0	0.00	100	60.0
		T2	100	5.30	83.3	100
		T3	87.5	0.00	100	87.5

Table 10. General performance of the LogitBoost classifier

Variety	Evaluation method	Correctly classified (%)	Incorrectly classified (%)
Fhia 21	Stratified cross-validation	90.2778	9.7222
	Holdout	87.5	12.5
Orishele	Stratified cross-validation	90.2778	9.7222
	Holdout	87.5	12.5

Predictions by treatment with holdout method display two tendencies for both varieties (Table 10). Treatment T0 and T2 are highly accurate with no FN samples (100 %) while T1 and T3 treatments accuracy are low respectively 60 % and 87.5 % with no FP rate.

From models analysis, we can assess that general predictive classification in cross-validation results are good and reflect prediction on each treatment but in testing the models, treatment T1 is overestimated. Anyway, models developed for each cultivar appear acceptable and greatly encouraging even if 10 times tenfold cross-validation should be performed for reliable results and accurate error estimation.

4. Conclusion

In this article, we focus on feasibility for remotely detecting at early stage two plantain varieties inoculated with different Arbuscular Mycorrhizal Fungi. Our results firstly emphasize a new water content indice NWI5 highly significant with Datt5 index at P <0.001. New findings also concern 3 fluorescence features (ANTH, FRF_R, and NBI_R) related to anthocyanin, foliar density and nitrogen content, and 10 others in reflectance spectra (SR5, NDVI3, ratiodRE_703, NWI, NWI4, dNIRmin920_980, min950_970, Carter5, DPI, GI) linked to chlorophyll content, water content, photosynthetic activity and leaf area index. Secondly, we derive predictive multiclass classification models for each cultivar with around 90.3 % and 87.5 % accuracy respectively in stratified cross-validation and holdout evaluation. This demonstrates capability of our fluorescence and reflectance-based instrumentation for quantitatively investigating crop growth under mycorrhizae fertilization. Effective vegetation indices identification in crops specialty as well as rapid pattern recognition of mycorrhizal treatment represent advanced key for developing non-destructive and cost-effective crops monitoring detectors which will more contribute to sustainability in agriculture. Furthermore, we might expect better results in this study using improved version of the LogitBoost classifier. Elaborating a unique model for both plantain varieties could also represent another challenge.

Acknowledgments

We would like to thank ISP (the International Science Program) for their financial support. We would also like to express our gratitude to all research colleagues from agricultural science and engineering, division of Vegetable and Microbial Biotechnology in National Polytechnic Institute Houphouet-Boigny who provided biological sample and biotechnology expertise that greatly allowed this research.

References

Agati, G., Foschi, L., Grossi, N., & Volterrani, M. (2015). In field non-invasive sensing of the nitrogen status in hybrid bermudagrass (Cynodon dactylon × C. transvaalensis Burtt Davy) by a fluorescence-based method. *European Journal of Agronomy, 63*, 89-96. http://doi.org/10.1016/j.eja.2014.11.007

Agati, G., Pinelli, P., Cortés Ebner, S., Romani, A., Cartelat, A., & Cerovic, Z. G. (2005). Nondestructive evaluation of anthocyanins in olive (Olea europaea) fruits by in situ chlorophyll fluorescence spectroscopy. *Journal of agricultural and food chemistry, 53*(5), 1354–1363. http://doi.org/10.1021/048381d

Arduino - Home. (n.d.). Retrieved November 26, 2015, from https://www.arduino.cc/

Babar, M. A., Reynolds, M. P., van Ginkel, M., Klatt, A. R., Raun, W. R., & Stone, M. L. (2006). Spectral Reflectance Indices as a Potential Indirect Selection Criteria for Wheat Yield under Irrigation. *Crop Science, 46*(2), 578. http://doi.org/10.2135/cropsci2005.0059

Baret, F., Guyot, G., & Major, D. (1988). Coupled Fluorescence and Reflectance Measurements to Improve Crop Productivity Evaluation. In H. K. Lichtenthaler (Eds.), *Applications of Chlorophyll Fluorescene in Photosynthesis Research, Stress Physiology, Hydrobiology and Remote Sensing* (pp. 319-324). Dordrecht: Springer Netherlands.

Barnes, R., Dhanoa, M., & Lister, S. (1993). Letter: Correction to the description of Standard Normal Variate (SNV) and De- Trend (DT) ransformations in Practical Spectroscopy with Applications in Food and everage Analysis–2nd Edition. *Journal of Near Infrared Spectroscopy, 1*(1), 185. http://doi.org/10.1255/jnirs.21

Barton, C. V. ., & North, P. R. (2001). Remote sensing of canopy light use efficiency using the photochemical reflectance index. *Remote Sensing of Environment, 78*(3), 264–273. http://doi.org/10.1016/S0034-4257 (01)00224-3

Belanger, M. C., Viau, A. A., Samson, G., & Chamberland, M. (2005). Determination of a multivariate indicator of nitrogen imbalance (MINI) in potato using reflectance and fluorescence spectroscopy. *Agronomy journal, 97*(6), 1515–1523. http://doi.org/10.2134/agronj2005.0040

Bewick, V., Cheek, L., & Ball, J. (2004). Statistics review 9: One-way analysis of variance. *Critical Care, 8*(2), 130-136. http://doi.org/10.1186/cc2836

Bramley, R. G. V., Le Moigne, M., Evain, S., Ouzman, J., Florin, L., Fadaili, E. M., … Cerovic, Z. G. (2011). On-the-go sensing of grape berry anthocyanins during commercial harvest: development and prospects: On-the-go sensing of grape anthocyanins. *Australian Journal of Grape and Wine Research, 17*(3), 316–326. http://doi.org/10.1111/j.1755-0238.2011.00158.x

Bretz, F., Hothorn, T., & Westfall, P. (2010). *Multiple Comparisons Using R*. CRC Press.

Buschmann, C., Langsdorf, G., Lichtenthaler, H.K. (2008) Blue, green, red, and far-red fluorescence signatures of plant tissues, their multicolor fluorescence imaging, and application for agrofood assessment, in: Manuela, Z., (Ed.), *Optical Monitoring of Fresh and Processed Agricultural Crops* (pp. 272–319). CRC Press.

Cameron, K. C., Di, H. J., & Moir, J. L. (2013). Nitrogen losses from the soil/plant system: a review. *Annals of Applied Biology, 162*(2), 145-173. http://doi.org/10.1111/aab.12014

Cartelat, A., Cerovic, Z. G., Goulas, Y., Meyer, S., Lelarge, C., Prioul, J.-L., … Moya, I. (2005). Optically assessed contents of leaf polyphenolics and chlorophyll as indicators of nitrogen deficiency in wheat (*Triticum aestivum* L.). *Field Crops Research, 91*, 35-49. http://doi.org/10.1016/j.fcr.2004.05.002

Carter, G. A. (1994). Ratios of leaf reflectance in narrow wavebands as indicators of plant stress. *International Journal of Remote Sensing, 15*(3), 697-703. http://doi.org/10.1080/01431169408954109

Cerovic, Z. G., Ounis, A., Cartelat, A., Latouche, G., Goulas, Y., Meyer, S., & Moya, I. (2002). The use of chlorophyll fluorescence excitation spectra for the non-destructive in situ assessment of UV-absorbing compounds in leaves. *Plant, Cell & Environment, 25*(12), 1663–1676. http://doi.org/10.1046/j.1365-3040.2002.00942.x

Chappelle, E. W., Kim, M. S., & McMurtrey, J. E. I. I. I. (1992). Ratio analysis of reflectance spectra (RARS): an algorithm for the remote estimation of the concentrations of chlorophyll a, chlorophyll b and carotenoids in soybean leaves. *Remote Sensing of Environment, 39* (3), 239–247. http://doi.org/10.1016/0034-4257(92)90089-3

Cvetković, S., Stojanović, M. B., & Nikolić, S. V. (2015). Multi-channel descriptors and ensemble of Extreme Learning Machines for classification of remote sensing images. *Signal Processing: Image Communication, 39*, 111-120. http://doi.org/10.1016/j.image.2015.09.004.

Datt, B. (1998). Remote sensing of chlorophyll a, chlorophyll b, chlorophyll a + b and total carotenoid content in Eucalyptus leaves. *Remote Sensing of Environment, 66*(2), 111–121. http://doi.org/10.1016/Soo34-4257(98)00046-7

Declerck, S., Cevos, B., Delvaux, B., & Plenchette, C. (1994). Growth response of micropropagated banana plants to VAM inoculations. *Fruits, 49*, 103–109.

Filella, & Penuelas, J. (1994). The red edge position and shape as indicators of plant chlorophyll content, biomass and hydric status. *International Journal of Remote Sensing, 15*(7), 1459-1470. http://doi.org/10.1080/01431169408954177

Friedman, J., Hastie, T., & Tibshirani, R. (2000). Special Invited Paper. Additive Logistic Regression: A Statistical View of Boosting. *The Annals of Statistics, 28*(2), 337-374. http://doi.org/10.1214/aos/1016218223

Gaidashova, S. V., Nsabimana, A., Asten, P. J. A. van, Delvaux, B., Elsen, A., & Declerck, S. (2013). Impact of arbuscular mycorrhizal fungi on growth of banana genotypes in three different, pasteurized and non-pasteurized soils of Rwanda. In G. Blomme, P. van Asten, & B. Vanlauwe (Eds.), *Banana systems in the humid highlands of sub-Saharan Africa: enhancing resilience and productivity* (p. 72-82). Wallingford: CABI.

Gandia, S., Fernández, G., García, J. C., & Moreno, J. (2004). Retrieval of vegetation biophysical variables from CHRIS/PROBA data in the SPARC campaign. *Proceedings of the 2nd CHRISProba Workshop ESAESRIN Frascati Italy 2830*. http://earth.esa.int/workshops/chris_proba_04/papers/12_Gandia.pdf

Garbaye, J. (2013). La symbiose mycorhizienne: une association entre les plantes et les champignons. Editions Quae.

Gerdemann, J. W., & Nicolson, T. H. (1963). Spores of mycorrhizal Endogone species extracted from soil by wet sieving and decanting. *Transactions of the British Mycological Society, 46*(2), 235–244. http://doi.org/10.1016/S0007-1536(63)80079-0

Gianinazzi, S. (1981). L'endomycorhizafion contrôlée en agriculture, en horticulture et en arboriculture : problèmes et progrès. In S. Gianinazzi, V. Gianinazzi-Pearson, & A. Trouvelot (Eds), *Les Mycorhizes, Partie Intégrante de la Plante : Biologie et Perspectives d'Utilisation* (Les colloques de l'INRA, Vol. 13, pp. 231-241). Paris : INRA.

Hall, M., Frank, E., Holmes, G., Pfahringer, B., Reutemann, P., & Witten, I. H. (2009). The WEKA Data Mining Software: An Update; *ACM SIGKDD Explorations, 11*(1), 10. http://doi.org/10.1145/1656274.1656278

J. Baluja, M. P. D. (2012). Assessment of the spatial variability of grape anthocyanins using a fluorescence sensor. Relationships with vine vigour and yield. *Precision Agriculture, 13*(4). http://doi.org/10.1007/s11119-012 -9261-x

Jefwa, J. M., Rurangwa, E., Gaidashova, S. V., Kavoo, A. M., Mwashasha, M., Robinson, J., & Vanlauwe, B. (2013). Indigenous arbuscular mycorrhizal fungi and growth of tissue-cultured banana plantlets under nursery and field conditions in Rwanda. In G. Blomme, P. van Asten, & B. Vanlauwe (Eds.), *Banana systems in the humid highlands of sub-Saharan Africa: enhancing resilience and productivity* (pp. 83-92). Wallingford: CABI.

Josep Peñuelas, I. F. (1998). Visible and near-infrared reflectance techniques for diagnosing plant physiological status. *Trends in Plant Science, 3*(4), 151-156. http://doi.org/10.1016/S1360-1385(98)01213-8

Kanamori, T., & Takenouchi, T. (2013). Improving Logitboost with prior knowledge. *Information Fusion, 14*(2), 208-219. http://doi.org/10.1016/j.inffus.2011.11.004

Kassir, L. N. (2014). Environmental Impact of Phosphate Fertilizers and By-Products on Agricultural Soils, in: López-valdez, F. and Fernández-luqueño, F. (Eds.), *Fertilizers: Components, Uses in Agriculture and Environmental Impacts* (pp. 45-66). Nova Science Publishers, Inc., New York.

Krishnaraj, Y., & Reddy, C. (2009). Protein Structure Classification Using Machine Learning Methods. In S. Lonardi & J. Chen (Eds.), *Biological Data Mining* (Vol. 20093941, pp. 69-87). Chapman and Hall/CRC.

Lichtenthaler, H. K., Wenzel, O., Buschmann, C., & Gitelson, A. (2006). Plant Stress Detection by Reflectance and Fluorescencea. *Annals of the New York Academy of Sciences, 851*(1), 271-285. http://doi.org/10.1111/j.1749-6632.1998.tb09002.x

Misra, A. N., & Singh, M. M. and R. (2012). *Chlorophyll Fluorescence in Plant Biology.* http://doi.org/10.5772/35111

Moreno, R., Corona, F., Lendasse, A., Graña, M., & Galvão, L. S. (2014). Extreme learning machines for soybean classification in remote sensing hyperspectral images. *Neurocomputing, 128*, 207-216. http://doi.org/10.1016/j.neucom.2013.03.057

Moshou, D., Pantazi, X.-E., Kateris, D., & Gravalos, I. (2014). Water stress detection based on optical multisensor fusion with a least squares support vector machine classifier. *Biosystems Engineering, 117*, 15-22. http://doi.org/10.1016/j.biosystemseng.2013.07.008

Mountrakis, G., Im, J., & Ogole, C. (2011). Support vector machines in remote sensing: A review. *ISPRS Journal of Photogrammetry and Remote Sensing, 66*(3), 247-259. http://doi.org/10.1016/j.isprsjprs.2010.11.001

Mulla, D. J. (2013). Twenty five years of remote sensing in precision agriculture: Key advances and remaining knowledge gaps. *Biosystems Engineering, 114*(4), 358-371. http://doi.org/10.1016/j.biosystemseng.2012.08.009

Murmu, S., & Biswas, S. (2015). Application of Fuzzy Logic and Neural Network in Crop Classification: A Review. *Aquatic Procedia, 4*, 1203-1210. http://doi.org/10.1016/j.aqpro.2015.02.153

Nwaga, D., Tenkouano, A., Tomekpe, K., Fogain, R., Kinfack, D. M., Tsané, G., & Yombo, O. (2011). Multi-functional Properties of Mycorrhizal Fungi for Crop Production: The Case Study of Banana Development and Drought Tolerance. In A. Bationo, B. Waswa, J. M. Okeyo, F. Maina, & J. M. Kihara (Eds.), *Innovations as Key to the Green Revolution in Africa* (pp. 523-531). Dordrecht: Springer Netherlands.

Peñuelas, J., Filella, I., Biel, C., Serrano, L., & Savé, R. (1993). The reflectance at the 950–970 nm region as an indicator of plant water status. *International Journal of Remote Sensing, 14*(10), 1887-1905. http://doi.org/10.1080/01431169308954010

Pinheiro, F., & Gusmo dos Anjos, W. de P. (2014). Optical Sensors Applied in Agricultural Crops. In M. Yasin (Ed.), *Optical Sensors - New Developments and Practical Applications*. InTech.

R Core Team. (2014). R: A language and environment for statistical computing. R Foundation for Statistical Computing, Vienna, Austria. Retrieved November 26, 2015, from https://www.R-project.org/.

Sahraoui, A. L. (2013). La Mycorhize à arbuscules : quels bénéfices pour l'homme et son environnement dans un contexte de développement durable ? *Synthèse: Revue des Sciences et de la Technologie, 26*(1), 06-19.

Smith, R. C.G., Adams, J., Stephens, D. J., & Hick, P. T. (1995). Forecasting wheat yield in a Mediterranean-type environment from the NOAA satellite. *Australian Journal of Agricultural Research, 46*(1), 113–125. http://doi.org/10.1071/AR9950113

Sun, P., Reid, M. D., & Zhou, J. (2014). An improved multiclass LogitBoost using adaptive-one-vs-one. *Machine Learning, 97*(3), 295-326. http://doi.org/10.1007/s10994-014-5434-3

Tropical Soil Biology and Fertility Institute. (2007). Exploration of Integrated Soil Fertility Management for Banana Production and Marketing in Uganda and Kenya: Arbuscular Mycorrhizal Fungi help Establishment and Production of Tissue Culture Banana. *Report to Rockefeller Foundation grant. 2003–2007.* TSBF-CIAT, Nairobi.

Weber, O. B. (2014). Biofertilizers with Arbuscular Mycorrhizal Fungi in Agriculture. In Z. M. Solaiman, L. K. Abbott, & A. Varma (Eds.), *Mycorrhizal Fungi: Use in Sustainable Agriculture and Land Restoration* (Vol. 41, pp. 45–66). Berlin, Heidelberg: Springer Berlin Heidelberg.

Witten, I. H., Frank, E., & Hall, M. A. (2011). Data Mining: Practical machine learning tools and techniques, (3rd ed.) Morgan Kaufmann, Publishers Inc. San Fransisco, CA, USA.

Zarco-Tejada, P. J., Pushnik, J. C., Dobrowski, S., Ustin, S. L. (2003). Steady-state chlorophyll a fluorescence detection from canopy derivative reflectance and double-peak red-edge effects. *Remote Sensing of Environment, 84*(2), 283–294. http://doi.org/10.1016/S0034-4257(02)00113-X

12

The Connection between Gravity and Electricity According to Twin Physics and a Survey of the Results So Far, Including Neutron Decay

Anna C.M. Backerra[1]

[1] Gualtherus Sylvanusstraat 2, 7412 DM Deventer, The Netherlands

Correspondence: Anna C.M. Backerra, Gualtherus Sylvanusstraat 2, 7412 DM Deventer, The Netherlands. E-mail: annabackerra@gmail.com

Abstract

Unlike electricity, gravity is active in only one direction, namely attraction. If both phenomena are considered using the mathematical description called complementary language, the range of gravity turns out to be larger than the range of electricity. This seems to be the reason for the apparent one-sidedness of gravity. Beyond the reach of electromagnetic effects, repulsion of physical bodies is possible and this can be identified with expansion of the universe. With this last piece of the puzzle, our dualistic way of considering the universe, called twin physics, seems in principle to be completed. A coherent survey of the results so far is given, including a possible description of neutron decay. The results are compared with the features of the Standard Model.

Keywords: gravity, electricity, expanding universe, complementarity, twin physics, neutron decay, neutrino

1. Introduction

This is the fifth paper in a series exploring the power of twin physics (Backerra, 2010, 2012, 2014, 2015), a way of considering the universe based upon complementary mathematical language, describing determinate as well as indeterminate aspects of phenomena. Twin physics unites three new principles deviating from classical physics, one philosophical, one mathematical and one physical.

The *philosophical principle* can be simply expressed as *'every thing has its own nothing'*, saying that a physical object is not only characterized by its presence in terms of mass, geometry and electromagnetic features, but also by its absence, which is reflected by a finite space around the object, being an independent physical item to which energy is ascribed. The concept of space developed from the classical Greek notion of ether as an extremely thin substance filling the universe, to the assumption of Newton (1687) that space is empty, invisible, unmoving and having an infinite extension. Subsequent scientists changed their minds several times about this notion, until Michelson and Morley (1887), who were unable to find an experimental difference between the velocity of light moving perpendicular to the track of the earth and light moving parallel to it. Since then, space has been considered to be empty and unbounded. By contrast, twin physics considers space as a finite physical object, containing energy and occurring free as well as bounded to a particle.

The *mathematical principle* underlying complementary language is the joint description of determinate and indeterminate mathematical items in sets. These sets contain not only algebraic but geometric expressions as well, such as spheres, which is unusual. According to Steven Weinberg (2015), geometry played the main role in the brilliant Greek mathematics of the third century BC, whereas algebra was far less developed. By contrast, the modern physicist tends to derive geometric facts algebraically even when geometry is important. Geometry and algebra can be combined by using set theory (Kahn, 1967).

The *physical principle* is a renewed conception of potential energy. This term was introduced in the 19th century (Rankine, 1853), but it has links to Greek philosopher Aristotle's concept of potentiality. It is conceived as the energy possessed by a body by virtue of its position relative to others, the stresses within itself, its electrical charge, and other factors. Potential energy is associated with the forces that act on a body in a way that depends only on the body's position in space. These forces can be represented by a vector at every point in space forming a vector field of forces. However, potential energy can never be observed in reality because if it is observed, it is no longer potential energy, but actual energy. Twin physics is restricted to the description of physical observations and for that reason we consider potential energy as a mathematical item to deal with the transformation of energy from one type into another. To express this basic difference between a mathematical and physical concept of energy, we have introduced the term 'actual energy'. Consequently, the law of conservation of energy is expressed in twin physics as the law of conservation of *actual energy*.

These three principles deviate considerably from current concepts in physics, resulting in a completely different theoretical structure. Instead of considering elementary particles as the building blocks of the universe, the *Heisenberg unit* (H-unit), being a unit of potential energy, is introduced as fundamental. Thus, a mathematical item is taken as the starting point instead of a physical phenomenon. The H-unit manifests itself by interaction with another H-unit, as a result of which potential energy is converted into actual energy. Because not necessarily all potential energy is used by a specific interaction, the potential energy of a H-unit acts as an upper boundary for the generation of actual energy. This leads to an upper boundary for the mass of elementary particles instead of a lower boundary, the inverse of the traditional approach, which was to search for the smallest particles in existence.

At the outset of twin theory we constructed a set of four elements called the *zipper*, a formula collecting the descriptions of all possible physical observations, but it was not clear yet how these results could be identified. We started out rather intuitively, suggesting as a first trial sometimes identifications for specific cases which later turned out to be impossible. One example is the electron, which we had previously thought could be described by the first element of the zipper (Backerra, 2012, section VII.A); later it became clear that only heavy particles like neutrons and protons can be described by this zip. A second recent example is that we are only now certain that a zipper of the first order, containing eight zips, does not describe more observations than a zipper of the second order, containing only four zips, so from now on we will use only zippers of the second order.

It was only after years of laborious jigsaw-puzzling that the relationship between the set elements of the zipper and the manifold of possible observations became comprehensible and coherent. We were encouraged by our success in describing gravity (Backerra, 2014 section IV) and deriving the radius of a proton using its mass, the equivalence of mass and energy and the energy balance between particle and waves (Backerra, 2014 section III.A), as first proof that twin theory describes real physics. It was even more encouraging to be able to describe a photon (Backerra, 2015, section 8), including the quantization of its energy and its constant velocity, without using a postulate. Because the deduction of the photon was based upon both large- and small-scale physics, this description represented a bridge between the two.

It seems appropriate to compare our results with those of the Standard Model, to find out whether it might be an addition or perhaps even a viable alternative. Before investigating this, however, we would like to clear up one intriguing detail about gravity. In a previous paper (Backerra, 2014), we explained how gravity can be described by twin physics, but without considering electricity. At that point, we didn't answer the question why gravity leads only to accelerated attraction between masses, whilst electricity leads to accelerated attraction or *repulsion* between charged masses. What could be the reason for the restriction of gravity to only one of these two apparently obvious possibilities? This question has occupied many scientists since the discovery of electricity, but we could not find any theories explaining this typical difference between gravity and electricity. Using our results from 2014, we now will take the next step by considering both phenomena together.

After that, we will survey all of the theoretical results obtained so far using twin physics. In the previous four papers we have improved our definitions step by step, resulting in changes of interpretations, some of them drastic. Each time we have presented rather long-winded repetitions of our theory in order to explain these adaptations. This time, only a slight adaptation in the transformation formulas for space attributes in section 2.4 was necessary, to improve the description of the electron. We will now repeat our final definitions and formulas once more, this time as briefly as possible, as a 'dry theory', just to have our theoretical instruments at hand. A thorough consideration of the energy of the zipper is added and based upon it, a description of the neutron decay. More information about the underlying ideas can be found in previous publications, indicated when appropriate.

2. A Manual for the Use of Twin Physics

Potential energy is defined as the power to interact. An ***H-unit*** is an ***elementary unity of potential energy***, expressed in complementary terms; it can manifest itself only by interacting with other H-units. We suppose that an H-unit in an isolated system starts to interact as soon as another H-unit exists in this system. Then the potential energy will partly or fully be converted into actual energy, resulting in the appearance of one or more ***Heisenberg-events***, written as H-events. H-events are physical realities like a particle or a space. We suppose that the conversion of potential energy into phenomena occurs in a complementary way, as a contraction or as an expansion. In our first publication (Backerra, 2010) you can find more about the basics of twin physics.

The law of conservation of energy is supposed to be valid for interacting H-units. Considering the universe as an isolated system, built up by interacting H-units, this means that the total amount of actual energy, being generated by all H-units together, is constant. Thus if the universe has a certain starting amount of actual energy, an increasing conversion from potential into actual energy by some H-units has to be balanced by the opposite by others.

Because no independent frame of reference is used, any movement of an H-event is only possible with respect to another H-unit, so relativity is incorporated from scratch. To be able to describe physical items with the zipper, we introduced *three mathematical qualities: time, space and mark* (the last one being the mathematical precursor of electromagnetic features). For each quality, an H-unit is supplied with pairs of complementary mathematical attributes, each pair containing a determinate and an indeterminate attribute.

If we express the Heisenberg's principle as "each observation of certainty implies an amount of uncertainty", an extended Heisenberg's principle can be created by adding the complementary sentence "each observation of uncertainty implies an amount of certainty". This fundamental feature of nature is anchored in our theory by supposing that all attributes add to any observation in pairs, which are restricted by the following *axiom*: Attributes contribute to an observation in pairs, which are such that one member is of major and the other of minor importance. Therefore an H-unit H_i is, for each quality, supplied with a set h_i of mathematical attributes in two types: Determinate attributes are represented by D_i or d_i, and indeterminate ones by U^i or u^i; attributes of major importance are indicated in capitals and that of minor importance in lower case. They are collected in the general *set of attributes*:

$$h_i = \left\{ D_i, U^i, d_i, u^i \right\} \tag{1}$$

Each H-unit has a set of four attributes for each quality, so all together it has twelve attributes. The interaction between two H-units H_i and H_j is written as $H_i * H_j$. *Any interaction is based upon the exchange of attributes*, according to mathematical rules which may differ for each quality. By transforming exchanged attributes from mathematical into physical items, potential energy is converted into actual energy, describing phenomena. An H-unit may have no mark attributes; then it is called a *neutral H-unit*, written as H_0. A neutral H-unit has the same potential energy as a marked H-unit and so its spatial dimensions are larger.

We defined two operators to describe exchange: the join operator $\bowtie$, pronounced as 'joined with', and the link operator $\propto$, pronounced as 'linked to'. The *join operator* connects one major with one minor attribute to a joined pair. The members of one pair may be both determinate, both indeterminate or mixed. After transformation into a physical space, joined pairs occur combined in any phenomenon and so are necessarily observed together. The rules according to which the operation 'joining' is carried out for each quality, will be given after defining time, space and mark attributes. The *link operator* couples joined pairs of attributes to chains. If joined pairs are linked and transformed, their intersection may represent a physical observation. The link $X \propto Y$ of mathematical objects X and Y is defined as their intersection. If $X \propto Y$, then $Y \propto X$; if $X = Y$, then $X \propto Y = X$; $X \propto \varnothing = \varnothing$ in which $\varnothing$ is an empty set.

With these two operators, the interaction between H-units can be defined by collecting all possible combinations of their attributes of one and the same quality, two by two joined to pairs and these pairs linked to each other in chains. There exist 40 distinct chains for each of the three qualities time, space and mark, but this amount reduces drastically by realizing that *complementary observations* cannot be observed simultaneously, as we have learned from quantum mechanics. This is expressed as *the exclusion principle*, saying that a joined pair containing a *determinate major attribute* cannot link with a joined pair containing an *indeterminate major attribute of the same quality and the same H-unit*.

The two consequences of the exclusion principle are, that chains of more than four joined pairs cannot be transformed into actual H-events, and that only four distinct chains of four joined pairs exist. Smaller chains do not add more information, but they are less practical to work with.

These four chains are not yet enough to define the zipper, because the operators are defined such that, after carrying out the operations joining and linking, the small-scale elements contain only minor attributes, which will lead to only small-scale observations. To be able to describe as well large-scale observations, each chain is stripped down to the participating two major attributes, which will be used along with the original chains. In doing so, each phenomenon will be characterized by a large- as well as a small-scale description.

Each of the four original chains is, together with its belonging stripped chain, collected in a set called a **zip**. In this set, the large-scale element (containing only major attributes) is placed at the left hand and the small-scale element (containing both major and minor attributes) at the right hand.

Then the **zipper in general,** describing the complete interaction $H_i * H_j$, can be written as the set of four zips in which each zip is a set of two elements:

$$Z\left(h_i * h_j\right) = \left\{z_1, z_2, z_3, z_4\right\} = \begin{cases} \left[\left\{D_i \propto D_j\right\}, \left\{\left(D_i \bowtie u^i\right) \propto \left(D_j \bowtie u^j\right) \propto \left(D_i \bowtie u^j\right) \propto \left(D_j \bowtie u^i\right)\right\}\right] \\ \left[\left\{U^i \propto U^j\right\}, \left\{\left(U^i \bowtie d_i\right) \propto \left(U^j \bowtie d_j\right) \propto \left(U^i \bowtie d_j\right) \propto \left(U^j \bowtie d_i\right)\right\}\right] \\ \left[\left\{D_i \propto U^j\right\}, \left\{\left(D_i \bowtie u^i\right) \propto \left(U^j \bowtie d_j\right) \propto \left(D_i \bowtie d_j\right) \propto \left(U^j \bowtie u^i\right)\right\}\right] \\ \left[\left\{D_j \propto U^i\right\}, \left\{\left(D_j \bowtie u^j\right) \propto \left(U^i \bowtie d_i\right) \propto \left(D_j \bowtie d_i\right) \propto \left(U^i \bowtie u^j\right)\right\}\right] \end{cases} \quad (2)$$

The square brackets around each zip indicate the transformation of mathematical objects into physical items by representing them in a physical space. The indication $h_i * h_j$ will be dropped later.

After transformation, each zip is a set of two elements, of which the first is the large-scale description and the second the small-scale one. Usually only two of them can be transformed into an H-event. Because of the principle of exclusion, *distinct zips cannot be observed simultaneously.*

Before transformation, the potential energy of the zipper is equal to the sum of potential energy of the participating H-units, which is twice the potential energy of one H-unit. The potential energy of an H-unit is by definition a unity, which will be indicated by the constant V, as a reminiscence to the original definition by Rankine (1853). So after transformation, the *actual energy E of the zipper*, and thus the actual energy of the interaction can be expressed as:

$$E\left(H_i * H_j\right) \le 2 \times V \quad (3)$$

The actual energy of the zipper is also supposed to be the sum of the generated actual energies of the four zips:

$$E\left(H_i * H_j\right) = \sum_{n=1}^{n=4} E\left(z_n\right). \quad (4)$$

The *potential energies of the zips* are supposed to be equal. By transforming a zip, its potential energy will fully, partly or not at all be converted into actual energy, depending on the type of interaction, so *actual energies of the zips* are not necessarily equal. The potential energy of the complete zipper may be fully, partly or not at all be converted into actual energy. Equations (3) and (4) suggest that the potential energy of one zip is maximum equal to ½ x V; the minimum of actual energy of the complete zipper is by definition zero (if H_i and H_j have no interaction at all). We suppose that the maximum occurs if H_i and H_j have coinciding time and space attributes. As we know from previously deduced zippers for two coinciding H-units (for instance Backerra, 2015, sections 3.1 and 4.1), then zips z_3 and z_4 are empty. Because the sum of actual energy of the four zips is equal to $2 \times V$ and the zips are supposed to have equal potential energies, this implies that for coinciding H-units the actual energy of one zip is equal to V. Thus before interacting, the *potential energy of one zip also must be equal to V*. Added to the experience that we never found more than two actual zips in a complete zipper, we suppose that the conversion of potential into actual energy for one zip in general is restricted to this amount, which can be written as:

$$E(z_i) \le V. \quad (5)$$

All attributes have to be inserted in general zipper (2) for each quality separately. Only after having reduced all elements according to the considered case, the time zipper will be combined with the space zipper and eventually with the mark zipper. A tilde indicates that the attribute describes a mathematical object. After transformation into a physical space, this indication has to be dropped.

The set of attributes h_i of H-unit H_i for the quality time $\tilde{t}$ is defined as:

120

$$h_i\left(\tilde{t}\right)=\left\{\tilde{T}_i,\tilde{F}^i\backslash\tilde{T}_i,\tilde{\tau}_i,\tilde{f}^i\right\}\ ,\tag{6}$$

for the quality space **x** as:

$$h_i\left(\tilde{\mathbf{x}}\right)=\left\{\tilde{P}_i,\tilde{S}^i\backslash\tilde{P}_i,\tilde{p}_i,\tilde{s}^i\right\}\tag{7}$$

and for the quality mark $\tilde{q}$ as:

$$h_i\left(\tilde{q}\right)=\left\{\left\{\tilde{\mathbf{E}}_i,\tilde{Q}_i\right\},\left\{\tilde{\mathbf{B}}^i,\tilde{Q}_i\times i\right\},\left\{\tilde{\nabla},1\right\},\left\{\partial/\partial\tilde{t},i\right\}\right\}.\tag{8}$$

Each set of attributes is defined by subsequently two major and two minor attributes, each duo containing one determinate and one indeterminate member. Both major attributes together constitute a closed mathematical system; the same is valid for both minor attributes. The origin of these definitions can be found in previous work (time: see Backerra, 2012, section IV; space: 2012, section III; mark: 2010, section IV and 2015, section 6). Previously the minor attributes of mark were erroneously not included in the set of attributes.

Note that the attributes of time and space are similarly defined, with the only difference that time is defined in one dimension and space in three. It is necessary to split up time and space because by using four-dimensional spacetime, the description of a part of the H-events gets lost. Time is taken as the first quality, because after transformation it does not contain energy, but acts as a label, informing us about the status of the energy of the described H-event.

2.1 Time Attributes

Our way of dealing with time is different from classical physics. According to twin physics, time does not run self-evident. If H-units in a closed system have no interaction, which is the case if their major spaces have an empty intersection, then the time in this system stands still. Only if they overlap each other and have interaction, time is existing. Time runs only *regularly* if the interaction between H-units generates a cyclic phenomenon, providing a clock; in that case, the major point of time will travel along a one-dimensional axis, taking the remaining three attributes of time with it. The past is not incorporated, because it is not observable. In our second publication (Backerra, 2012, section IV) more can be found about the choice of time attributes.

Another very unusual feature of time is, that we do not ascribe energy to time attributes. In twin physics, a time observation acts as a **_label_**, attached to an observation, to inform us about the status of the energy, being potential, actual or a combination of both, and if the generated H-event is static or dynamic. An empty time label indicates a potential physical existence, being timeless. A non-empty label indicates the actual existence of an H-event, expressed by exchanged and transformed time attributes. A phenomenon is considered as static if its velocity is zero or constant and it is considered as dynamic if it moves accelerated; this is indicated by the time zipper.

The time attributes are defined as follows. The major determinate attribute is **major point of time** $\tilde{T}_i$, being the first point of interval $\tilde{F}^i$, which is an arbitrary large span of time. The major indeterminate attribute is **future**

$$\tilde{F}^i\backslash\tilde{T}_i=\left\{\tilde{t}\,|\,\tilde{T}_i<\tilde{t}<\tilde{T}_{ei}\right\},\tag{9}$$

in which $\tilde{t}$ is a mathematical point of time. So the future is an interval of time with $\tilde{T}_{ei}$ at the end. The major time attributes exclude each other, are complementary and constitute together the major time system of H_i, being $\tilde{F}^i$. The minor indeterminate attribute is the **flying time**

$$\tilde{f}^i=\left\{\tilde{t}\,|\,\tilde{T}_i\le\tilde{t}<\tilde{t}_i\right\},\tag{10}$$

being an *extended present*. The minor determinate attribute is the **flash of time**

$$\tilde{\tau}_i=\left\{\tilde{t}\,|\,\tilde{t}_i\le\tilde{t}\le\tilde{t}_i+d\tilde{t}_i\right\},\tag{11}$$

being an infinitesimal small time interval describing a *change*. The minor time attributes exclude each other, are complementary and constitute together the minor time system, being $\tilde{f}^i\cup\tilde{\tau}_i$.

The **_time zipper_** $Z(t)$ is decided by the distance $\tilde{T}_{ij}$ between $\tilde{T}_i$ and $\tilde{T}_j$ with respect to the flying times and the flashes of time. Previously we distinguished four distinct time cases, but it became apparent that time case 2 had to be split in two separate possibilities and so now we define five time cases.

Time cases for all interactions: Time case 1 is defined by: $\widetilde{T}_i$ and $\widetilde{T}_j$ are coinciding, or by: $\widetilde{T}_{ij}=0$. **Time case 2** is defined (for $T_j > T_j$) by: $\widetilde{T}_i$ exists in the flying time of H_j and the flashes of time are partly overlapping, or by: $0 < \widetilde{T}_{ij} \leq dt_i$. **Time case 3** is defined (for $T_j > T_j$) by: $\widetilde{T}_i$ exists in the flying time of H_j and the flashes of time are not overlapping, or by: $dt_i < \widetilde{T}_{ij} < t_i$. **Time case 4** is defined (for $T_i > T_j$) by: $\widetilde{T}_i$ exists in the flash of time of H_j, or by $t_i \leq \widetilde{T}_{ij} \leq t_i + dt_i$. **Time case 5** is defined by: The overlapping futures do not contain minor attributes, or by: $\widetilde{T}_{ij} > t_i + dt_i$.

2.2 Space Attributes

Our conception of space is different from the classical one. Classical space is infinite; in twin physics, on the contrary, spaces are finite. Classical space has no energy; in twin physics, spaces contain energy. As with time, space exists as a possible result of the interaction of H-units. In classical physics, a space contains an infinite amount of points, each being in principle observable with a certain accuracy; in twin physics one point of an independent space is not observable separately, because the space acts as one organism. In our second publication (Backerra, 2012, section III) you can find a description of the choices of space attributes. They are defined in a three-dimensional mathematical space as follows.

The major determinate attribute is **major point of space** $\widetilde{P}_i$, being a point. The major indeterminate attribute is **major space** $\widetilde{S}^i \setminus \widetilde{P}_i$, being a finite sphere with its central point $\widetilde{P}_i$ excluded, having a large but not infinite radius $\widetilde{R}$ which might have an astronomic magnitude, the border excluded. The major space attributes exclude each other, are complementary and constitute together the major space system of H_i, being $\widetilde{S}^i$.

The minor indeterminate attribute is **minor space** $\widetilde{s}^i$, being a small sphere with central point $\widetilde{P}_i$, having radius $\widetilde{r} << \widetilde{R}$, its border excluded (but $\widetilde{P}_i$ included). The minor determinate attribute is **pellicle** $\widetilde{p}_i$, being the border of this minor space, having an infinitely small thickness $d\widetilde{r}$. The minor space attributes exclude each other, are complementary and constitute together the minor space system, being $\widetilde{s}^i \cup \widetilde{p}_i$.

The **space zipper** $Z(\mathbf{x})$ is decided by the distance $\widetilde{P}_{ij}$ of $\widetilde{P}_i$ and $\widetilde{P}_j$ with respect to the radii of the major spaces R_i and R_j and those of the minor spaces r_i and r_j. If the major spaces do not overlap each other at all, so if $\widetilde{P}_{ij} \geq 2 \times \widetilde{R}$, then there is no interaction between the H-units.

For a quick comprehension, we advise to use two-dimensional models of H-units, circles cut out of plastic sheets; take the radii for the neutral H-units about five times larger than the marked ones.

Space cases for equally sized H-units (both neutral or both with marking):

For H-units having equal sizes, indicated by $H_i * H_j$ or by $H_{0i} * H_{0j}$ (both marked or both neutral), we define seven space cases, which are given in a geometric as well as an algebraic form.

Case 1 is defined by: $\widetilde{P}_i$ and $\widetilde{P}_j$ coincide, or by: $\widetilde{P}_{ij}=0$. **Case 2** is defined by: $\widetilde{P}_i$ and $\widetilde{P}_j$ exist inside of each others minor spaces, but they do not coincide, or by: $0 < \widetilde{P}_{ij} < r$. **Case 3** is defined by: $\widetilde{P}_i$ and $\widetilde{P}_j$ exist inside each others pellicles, or by: $r \leq \widetilde{P}_{ij} \leq r + dr$. **Case 4** is defined by: The small spaces are partly overlapping and do not contain $\widetilde{P}_i$ and $\widetilde{P}_j$, or by: $r + dr < \widetilde{P}_{ij} < 2 \times r$. **Case 5** is defined by: The pellicles touch each other, or by: $2 \times r \leq \widetilde{P}_{ij} \leq 2 \times (r + dr)$. **Case 6** is defined by: The major spaces overlap each other such, that the intersection contains all minor attributes, but the pellicles have an empty intersection, or by: $2 \times (\widetilde{r} + d\widetilde{r}) < \widetilde{P}_{ij} < \widetilde{R}$. **Case 7** is defined by: The major spaces overlap each other partly, but they overlap neither each others major points, nor each others minor attributes, or by: $\widetilde{R} + \widetilde{r} \leq \widetilde{P}_{ij} < 2 \times \widetilde{R}$.

Space cases for mixed sized H-units:

Interactions of H-units having different sizes (one marked and one neutral) are indicated by

$H_i * H_{0i}$. Marked H-units are smaller than neutral ones; we assume that $R_0 > R_i$ and $r_0 > r_i$. For mixed interactions, only case 7 can be used; below six more cases are defined. The distance of $\widetilde{P}_i$ and $\widetilde{P}_{0i}$ is indicated by $\widetilde{P}_{i0i}$. The algebraic description of these cases will not be given if this is superfluous complicated. Mind that previously another numbering of mixed cases is used (Backerra, 2010).

Case 8 is defined by: $\widetilde{P}_i$ and $\widetilde{P}_{0i}$ coincide, or by: $\widetilde{P}_i = \widetilde{P}_{0i}$. **Case 9** is defined by: The charged minor space exists inside the neutral one, overlapping the neutral point of space, but the major points do not coincide, or by: $0 < P_{i0i} < r_i$. **Case 10** is defined by: The major neutral point exists inside the charged pellicle, or by: $r_i \leq P_{i0i} < r_i + dr$. **Case 11** is defined by: The marked minor space exists completely inside the neutral one, not overlapping the neutral major point of space, or by: $r_i + dr < P_{i0i} < r_{0i} - (r_i + dr)$. **Case 12** is defined by: The marked major point of space exists inside the neutral pellicle, or by: $r_{0i} \leq P_{i0i} \leq r_{0i} + dr$. **Case 13** is defined by: the major points of space exist in each others major spaces, but their minor spaces and pellicles have empty intersections, or by:

$r_i + r_{0i} + 2 \times dr < P_{i0i} < R_i$. **Case 14** is defined by: The marked minor space exists completely inside the neutral major space, but the marked major space is not overlapping the neutral major point of space, or by: $R_i \leq P_{i0i} \leq R_{0i} - r_i$.

2.3 Mark Attributes

Mark attributes are defined in a complementary way, like we did with space and time attributes, and such that an entire H-unit is distinguished from another one by having a positive or negative charge. If an H-unit is not marked, it is called neutral. More about the quality mark can be found in our first and fourth paper (Backerra, 2010, the introduction of section IV, and 2015, section 6). The mark attributes are as follows defined in mathematical time and three-dimensional mathematical space.

First we consider the major mark attributes. In the major determinate mark set $\left\{\widetilde{\mathbf{E}}_i, \widetilde{Q}_i\right\}$, *electric field* $\widetilde{\mathbf{E}}_i\left(\tilde{t}\right)$ is defined as a 3-dimensional, radial oriented, time dependent vector field with $\widetilde{\mathbf{E}}_i\left(\widetilde{P}_i\right) = \mathbf{0}$, and its source is *real charge* $\widetilde{Q}_i$, defined as a real number. In major indeterminate mark set $\left\{\widetilde{\mathbf{B}}^i, \widetilde{Q}_i \times i\right\}$, *magnetic field* $\widetilde{\mathbf{B}}^i\left(\tilde{t}\right)$ is a 3-dimensional, time dependent vector field with $\widetilde{\mathbf{B}}^i\left(\widetilde{P}_i\right) = \mathbf{0}$, and its source is *imaginary charge* $\widetilde{Q}_i \times i$ (with $i = \sqrt{-1}$), defined as an imaginary number. The magnetic field has by definition one vector in each point, having a direction tangent to a spherical surface with $\widetilde{P}_i$ as the central point. Adjacent vectors may point in an infinite variety of directions and there is no information about the absolute value of each vector. To create an absolute indeterminate magnetic field, the *principle of uniqueness* is added to the definition of $\widetilde{\mathbf{B}}^i$, saying that $\widetilde{\mathbf{B}}^i \neq \widetilde{\mathbf{B}}^j$ for any two H-units H_i and H_j.

Secondly we define the minor mark attributes. In the minor determinate mark set $\left\{\widetilde{\nabla}, 1\right\}$, $\widetilde{\nabla}$ is defined as the *nabla operator* $\left(\partial/\partial \tilde{x}, \partial/\partial \tilde{y}, \partial/\partial \tilde{z}\right)$ and the number 1 is the unity of real numbers. In the minor indeterminate mark set $\left\{\partial/\partial \tilde{t}, i\right\}$, $\partial/\partial \tilde{t}$ is defined as the mathematical time derivative and i is defined by $i = \sqrt{-1}$. Mark zipper $Z(q)$ is decided by the charges $\widetilde{Q}_i$ and $\widetilde{Q}_j$.

Mark cases for all interactions:

Mark case 1 is defined by $\widetilde{Q}_i = \widetilde{Q}_j$. **Mark case 2** is defined by $\widetilde{Q}_i = -\widetilde{Q}_j$. **Mark case 3** is defined by $\widetilde{Q}_{0i} = 0$. In the third case, a marked H-unit H_i interacts with a neutral one H_{0i}; mind that the size of the marked H-unit is smaller than that of the neutral one.

Previously we called $\widetilde{\mathbf{E}}_i\left(\tilde{t}\right)$ and $\widetilde{\mathbf{B}}^i\left(\tilde{t}\right)$ 'determinate and indeterminate vector fields' and $\widetilde{Q}_i$ and $\widetilde{Q}_i \times i$ 'real and imaginary numbers'; only after transformation we called them 'electric and magnetic fields' and 'charges'. Later we switched over to the same indication for both conditions, remembering that a tilde indicates a mathematical existence, and the absence of a tilde a physical existence; they may occur mixed in cases that we do not know yet all transformations.

2.4 Operators

To obtain rules for interacting major and minor objects, we will return to the extended Heisenberg's uncertainty principle, saying that each 'certain observation' contains a small influence of 'uncertainty', and the other way round, each 'uncertain observation' contains a small influence of 'certainty'. The operator 'joining' is defined such that these conditions are met, and in a similar way for time and space. Mind that the difference in dimensions of time and space in some cases may result in zippers which are not at all similar.

Joining of time attributes, if major *points of time T* are involved, is defined as:

$$T_i \bowtie f^j = f^j \text{ if } T_i \in f^j \, ; T_i \bowtie f^j = \varnothing \text{ if } T_i \notin f^j \, ;$$
$$T_i \bowtie \tau_j = \tau_j \text{ if } T_i \in \tau_j \, ; T_i \bowtie \tau_j = \varnothing \text{ if } T_i \notin \tau_j \tag{12}$$

and if major *futures F* are involved, as:

$$F^i \bowtie \tau_j = \tau_j \text{ if } \tau_j \subset F^i \, ; F^i \bowtie \tau_j = \varnothing \text{ if } \tau_j \not\subset F^i \, ;$$
$$F^i \bowtie f^j = f^j \text{ if } f^j \subset F^i \, ; F^i \bowtie f^j = \varnothing \text{ if } f^j \not\subset F^i \tag{13}$$

in which $\varnothing$ indicates an empty set.

Linking of joined pairs of time is defined as taking their intersection.

Joining of space attributes, if major *points of space P* are involved, is defined as:

$$P_i \bowtie s^j = s^j \text{ if } P_i \in s^j \; ; \; P_i \bowtie s^j = \varnothing \text{ if } P_i \notin s^j \; ;$$
$$P_i \bowtie p_j = p_j \text{ if } P_i \in p_j \; ; \; P_i \bowtie p_j = \varnothing \text{ if } P_i \notin p_j \; . \tag{14}$$

and if major *spaces S* are involved:

$$S^i \bowtie p_j = p_j \text{ if } p_j \subset S^i \; ; \; S^i \bowtie p_j = \varnothing \text{ if } p_j \not\subset S^i \; ;$$
$$S^i \bowtie s^j = s^j \text{ if } s^j \subset F^i \; ; \; S^i \bowtie s^j = \varnothing \text{ if } s^j \not\subset S^i \; . \tag{15}$$

Linking of joined pairs of space is defined as taking their intersection.

Joining of mark attributes is defined for fields and charges separately.

For fields, joining is defined as:

$$\widetilde{\mathbf{A}} \bowtie \widetilde{\nabla} = \widetilde{\nabla} \times \widetilde{\mathbf{A}} \quad \text{and} \quad \widetilde{\mathbf{A}} \bowtie \partial / \partial \widetilde{t} = \partial \widetilde{\mathbf{A}} / \partial \widetilde{t} \; , \tag{16}$$

in which $\widetilde{\nabla} \times \widetilde{\mathbf{A}}$ is a vector-product.

For charges, joining is defined as multiplication of major attributes with minor attributes 1 and i, subsequently. This results in the possible pairs of mark attributes $\widetilde{Q}_i$, $\widetilde{Q}_i \times i$ and $-\widetilde{Q}_i$.

Linking of joined pairs of mark attributes is as well defined for fields and charges separately.

For fields, linking is defined as vector addition of the joined pairs; it is important to carry out this operation after joining. According to the definition of linking, two equal vector fields link to a single vector field, so if $\widetilde{\mathbf{E}}_i = \widetilde{\mathbf{E}}_j$ than $\widetilde{\mathbf{E}}_i \propto \widetilde{\mathbf{E}}_j = \widetilde{\mathbf{E}}_i$. Note that electric fields are only equal in case they have two equal charges *and* two coinciding major points of space. Magnetic fields cannot be equal, because of the principle of uniqueness.

For charges, linking is defined as numerical addition. Two equal charges link to one single charge, so if $\widetilde{Q}_i = \widetilde{Q}_j$ than $\widetilde{Q}_i \propto \widetilde{Q}_j = \widetilde{Q}_i$.

The *transformation* of a mathematical element $\widetilde{X}$ into a physical item X is written as $\left[\widetilde{X}\right] = X$; by carrying out this transformation, tildes above attributes disappear. The *transformations of time elements* of a zip are defined by

$$\left[\widetilde{T}_i\right] = T_i \; ; \left[\widetilde{F}^i \backslash \widetilde{T}_i\right] = F^i \backslash T_i \; ; \left[\widetilde{\tau}_i\right] = \tau_i \; ; \left[\widetilde{f}^i\right] = f^i \; , \tag{17}$$

Transformed time elements do not contain energy.

The *transformations of space elements* of a zip are defined by

$$\left[\widetilde{P}\right] = P \; ; \left[\widetilde{S} \backslash \widetilde{P}\right] = S \backslash P \; ; \left[\widetilde{s}^i\right] = s^i \; ; \left[\widetilde{p}_i\right] = \pi(p_i) \; ; \left[\widetilde{p}_i \cap \widetilde{p}_j\right] = \pi(p_i \cap p_j) \; . \tag{18}$$

The last two objects are minor spherical spheres, occupying the full width of the pellicle.

A new definition is introduced for the transformation of the small-scale geometric element $s^i \cap p_j$ in zip z_3 (and mirrored in z_4):

$$\left[\widetilde{s}^i \cap \widetilde{p}_j\right] = e_{s^i \cap p_j} \; , \tag{19}$$

in which e is a tiny sphere. This definition affiliates better with the two previous definitions in which pellicles are involved and spherical minor particles are generated. This geometric object will be considered further in section 2.5. Transformed *space elements contain energy* proportional to the magnitude of the space, and so a point of space contains no energy.

The *transformations of mark elements* of a zip are defined by:

$$\left[\widetilde{\mathbf{E}}_i\right] = \mathbf{E}_i \; ; \left[\widetilde{\mathbf{B}}^i\right] = \mathbf{B}^i \; ; \left[\widetilde{\nabla} \times \widetilde{\mathbf{E}}_i\right] = \nabla \times \mathbf{E}_i \; ; \left[\widetilde{\nabla} \times \widetilde{\mathbf{B}}^i\right] = \nabla \times \mathbf{B}^i \; ;$$
$$\left[\widetilde{Q}_i\right] = Q_i \; ; \left[\widetilde{Q}_i \times i\right] = 0 \; ; \left[-\widetilde{Q}_i\right] = -Q_i \tag{20}$$

Because the electric field is decided by its charge, and because three distinct transformations of charge exist, we distinguish *three types of H-units*: positively marked, negatively marked, and neutral. In general marked H-units are written as H_i and neutral ones as H_0.

All transformed mark elements add actual energy to geometric items, on the understanding that the mathematical charges and infinite vector fields after transformation are restricted to the described time and space elements of the zip. The actual energy of charge Q_i is defined as a constant.

After having transformed the zipper completely, and thus having removed all square brackets, the potential energy of each zip is partly or fully converted into actual energy. As we saw in equation (5), the actual energy of a zip (so the actual energy of an H-event) cannot exceed the unity of potential energy, which is V. The final interpretation in physical reality of zip $z_i(t,\mathbf{x},q)$, is called the ***appearance*** $A_i(z_i)$, describing an H-event, which is a phenomenon having actual energy. The set of appearances of the zipper is defined as a set containing one interpretation of each zip, which can be written as:

$$A\big(Z(t,\mathbf{x},q)\big) = \big\{A_1(t,\mathbf{x},q), A_2(t,\mathbf{x},q), A_3(t,\mathbf{x},q), A_4(t,\mathbf{x},q)\big\}. \tag{21}$$

As far as we have found, for each case maximum two appearances of a zipper are non-empty.

If the transformed zip contains actual energy but the appearance is not observable, the appearance will be written between round brackets, so as $\big(A_i(t,\mathbf{x},q)\big)$. If desirable, the qualities t, $\mathbf{x}$ and q may be dropped in the formulation and replaced by more relevant indications.

2.5 Defining Appearances of Actual Energetic Items

It is important to understand the difference between the notion of energy by common consent and that in twin physics. Because both mass and space are generated by interacting H-units, both are considered as energetic spatial objects, called ***energetic space***. The difference between mass and space is supposed to be only in the concentration of energy. An actual object with a high energetic density is called a ***compact space***, an actual object with low energetic density is called an ***extended space***. For convenience 'compact space' and 'extended or free space' may be expressed in the classical terms 'mass' and 'space'. We suppose that these two energy densities are constants. Moreover, we suppose that the energy is homogeneously distributed over the occupied volume.

The quality time carries no energy. ***Time is a label of energy***, informing us about the status of a physical item, existing potentially or actual, in the present or in the future, in a static or a dynamic way. Note that the past is not represented.

By marking an H-unit, it is supplied with ***potential mark energy***. Because the potential energy of an H-unit is constant, this implies that by marking, the potential space energy will be reduced and thus ***the sizes of the space attributes are reduced***. This implies that the total energy density of a marked energetic space is higher than that of a neutral energetic space. Supposing that for a marked H-unit the proportion of potential space and mark energy is a constant, and that the two possible densities of mark energy also are constants, we obtain four distinct energy densities. The highest one is related to charged mass, followed by neutral mass, marked space, and neutral space.

Below we will list the chosen names for appearances of actual energetic objects. They are only related to the geometric appearances, without considering the quality 'mark'.

First we consider extended spaces. The appearance of (a part of) a major space is called ***major extended space*** Θ. The appearance of (a part of) a minor space is called ***minor extended space*** θ. A space without an accompanying particle in another zip, is called a ***free space***; its two types are ***major free space*** Θ^0 and ***minor free space*** Θ^0.

Second we consider compact spaces, which may appear as major or minor mass-carrying particles or as a point particle. The appearance of $s^i \cap s^j$ or s^i is called a ***major particle*** σ; it may be identified with a proton or a neutron. The appearance of $p_i \cap p_j$ is called a ***minor particle*** π, being a relatively small sphere existing inside the intersecting pellicles, such that the width is maximum used, having no charge (because the second zip, in which it appears, cannot describe charge); its actual energy is equal to the potential energy of the complete mathematical object $p_i \cap p_j$. It may be identified with a gluon or a neutrino. The appearance of $s^i \cap p_j$ is called a ***dot particle*** δ, being a relatively very small sphere, possibly charged (because the third or fourth zip, in which it appears, may describe charge), existing inside the indicated geometric object, having an actual energy equal to the potential energy of the complete mathematical object; it may be identified with an electron (previously the transformation to a small sphere was not yet defined, as we supposed that the disc-like form could be actual).

Third we consider items having neither an extended, nor a compact space. The appearance of point of space P_i is called a **point particle** Π; this carries no spatial energy, but it might carry mark energy in the form of a charge or an electromagnetic field; it may be identified with a photon.

2.6 Reducing the Zipper to Identify H-events

With the theoretical tools as given above, we are ready to deduce time, space and mark zippers in specific cases, based upon the general zipper (2). First we carry out the operation joining, and second linking (see section 2.4). After that, the zippers will be combined into one time-space-mark zipper. Each non-zero zip of the zipper is a set, containing two elements; the first is a stripped chain and the second is a chain of exchanged attributes. Each zip has to be transformed into a physical space. For some elements this is obvious; however, often it is not clear how to transform. Below we give some indications how to combine the large- and small-scale elements of each zip into a phenomenon.

The most easy indication is that, if both corresponding elements of a zip contain *empty time elements*, no potential energy can be converted into actual energy. The same is valid for corresponding *empty space elements*. In that case, the zip will be replaced by an empty set. If only one time element is empty, the belonging geometric element might exist independent of time.

Corresponding large- and small-scale elements have to be *compatible* for each quality. If, for instance, a large-scale geometric point occurs in combination with a small-scale geometric object, then we require that the point is an element of the object. If this is not possible, the two elements are not compatible and the zip will be replaced by an empty set. This requirement can be compared with an every-day zipper, having a series of left and right "teeth" which closes properly only by bringing them in the right contact with each other.

If a *large-scale element is empty* in all three qualities, the small-scale element still may be observable and so an H-event might exists. On the contrary, if the *small-scale element is empty* in all three qualities, the observation is incompatible with the extended Heisenberg's principle, implying that a large scale attribute cannot be observed without a small scale addition. In that case there might be an H-event, so potential energy might be converted into actual energy, but the resulting H-event is *not observable*. Previously we replaced the zip by an empty set in that case, but in describing the photon, it appeared to be useful to describe unobservable actual items as well; then the resulting physical item will be placed between round brackets.

After these first checks, we are ready for the fine-tuning of interpretations. The complementary starting-point of twin theory is reflected in each zip in a different way, so we have to consider each zip separately.

2.6.1 The first zip mainly describes the determinate aspects of nature, so concentrated items, which is expressed by large-scale element $D_i \propto D_j$. The combination of the large- and small-scale element may describe a mass-carrying particle, geometric characterized by the small-scale element and localized by the large-scale element. Thus the small-scale element softens the extreme determinate character of the large-scale element.

2.6.2 The second zip mainly describes the indeterminate aspect of nature, so divergent items like space, which is expressed by large-scale element $U^i \propto U^j$. The large-scale element of the second zip is non-empty for each interaction, in whatever case. Previously we supposed that space was generated in all cases, the small-scale element adding specific details. This was an erroneous reflex, originating in having considered space in the classically way, so as 'always and everywhere existing'. If in each interaction actual space would be generated, having a constant energy density, then a larger number of interacting H-units, generating many overlapping spaces, would cause an energetic problem, because space is supposed to have a constant energy density. We discovered that, if the small-scale geometry describes a minor particle, then the large-scale space has to be reduced to the spatial track of this particle. Then the small-scale element softens the extreme indeterminate character of the large-scale space. Below we consider three distinct possibilities for the second zip, the first two of them depending on the first zip.

If zip z_1 *describes a heavy mass-carrying particle* σ (see section 2.5), then $\left[\tilde{p}_i \cap \tilde{p}_j \right]$ in z_2 is defined as a light-weight minor particle π appearing in the intersection of pellicles around the mass. Because there is no boundary between them, σ and π may be considered together as one single H-event, in which π is identified with *a gluon*, providing the **spin** of the particle (see Backerra, 2012, VII.E). The large-scale element of z_2, being $\tilde{s}^i \cap \tilde{s}^j$, is reduced to the track of the gluon.

If zip z_1 *is empty*, then the transformation of $\left[\tilde{p}_i \cap \tilde{p}_j \right]$ in z_2 can be defined as a lightweight minor particle π, existing independently of another particle. This particle is identified with a ***neutrino***, its radius depending on the type of involved pellicles. Thus the large-scale element of z_2, being $\tilde{S}^i \cap \tilde{S}^j$, is reduced to the track of the neutrino.

If the small-scale space element of z_2 *is empty*, then the generated space $\tilde{S}^i \cap \tilde{S}^j$, containing actual energy, is unobservable and the appearance (see section 2.5) will be written between round brackets as (Θ). If two neutral H-units are interacting, then the space is written as $S^{0i} \cap S^{0j}$ and the appearance is an unobservable **major extended space** (Θ^0); if a small-scale element of time is indicated (τ_{0i} or $\tau_{0i} \cap \tau_{0j}$) without a small-scale geometry, then this element will be assigned to the major space, so the space is moving. If two marked H-units are interacting, than the generated space is smaller and magnetized; the appearance is $(\Theta^{\mathbf{B}})$, being an actual, unobservable, magnetized extended space.

2.6.3 The third and fourth zip are symmetrical with respect to the indices. Mind that the resulting H-events in general are *not symmetrical*, which will become clear by considering mixed cases (the interaction of a marked and a neutral H-unit). These two zips mainly describe the exchange between determinate and indeterminate aspects in nature, like in electromagnetic waves, which is expressed by their large-scale elements $D_i \propto U^j$ and $D_j \propto U^i$. If *one time or one geometric element is empty*, the mark element exists independent of time or space and will be ascribed to the corresponding element of the same zip.

2.7 Time Zippers, Space Zippers and Mark Zippers

The zippers for separate qualities are derived by inserting the sets of attributes (6), (7) and (8) in the general zipper (2) and adapt them to the cases in consideration (sections 2.1, 2.2 and 2.3). A tilde above a symbol says that it is a mathematical symbol; square brackets around an element say that the element has to be transformed to a real physical space, possibly even when the attributes are already transformed. Different as in sections 2.1 and 2.2, the cases are not described in a geometric or functional algebraic way any more, but in set algebra according to Kahn (1967).

For time zippers, all elements are transformed, and so all square brackets and tildes are removed. All space attributes are, as a first step, transformed to a physical space, but if an element involves pellicles, it is still written with square brackets, because it may be a non-singular object; only after having considered the remaining zips, the final appearance can be obtained. For mark zippers, the operations joining, linking and transforming are carried out as far as possible, but in general they can be carried out only after the time and space parts are completely transformed.

Time zippers:

Cases 2 through 5 are defined for $T_i > T_j$; for $T_i < T_j$ the zipper is found by interchanging indices i by j, and the same for zips z_3 and z_4. All elements are transformed to physical time items, so no tildes have to be written any more.

Case 1 (so $T_i \cap T_j = \varnothing$):

$$Z(t) = \left\{ \left\{ T_i, f^i \right\}, \left\{ F^i \setminus T_i, \tau_i \right\}, \varnothing, \varnothing \right\}. \tag{22}$$

Case 2 (so $T_i \in f^j$ and $\tau_i \cap \tau_j \neq \varnothing$):

$$Z(t) = \left\{ \varnothing, \left\{ F^i \setminus T_i \cap F^j, \tau_i \cap \tau_j \right\}, \left\{ T_i, \varnothing \right\}, \varnothing \right\}. \tag{23}$$

Case 3 (so $T_i \in f^j$ and $\tau_i \cap \tau_j = \varnothing$):

$$Z(t) = \left\{ \varnothing, \left\{ F^i \setminus T_i \cap F^j, \varnothing \right\}, \left\{ T_i, \varnothing \right\}, \varnothing \right\}. \tag{24}$$

Case 4 (so $T_i \in \tau_j$):

$$Z(t) = \left\{ \varnothing, \left\{ F^i \setminus T_i \cap F^j, \varnothing \right\}, \left\{ T_i, f^i \cap \tau_j \right\}, \varnothing \right\}. \tag{25}$$

Case 5 (so $T_i \in F^j$ and $F^i \cap \tau_j = \varnothing$):

$$Z(t) = \left\{ \varnothing, \left\{ F^i \setminus T_i \cap F^j, \varnothing \right\}, \left\{ T_i, \varnothing \right\}, \varnothing \right\}. \tag{26}$$

Space zippers:

First the zippers for equally sized H-units are given; after that, the zippers for mixed H-units. All attributes are transformed to physical space items, so no tildes any more. However, if an element contains a pellicle, it is still written between square brackets because the transformation cannot yet be completed.

Case 1 for H-units of equal size, with $P_i \cap P_j = P_i$:

$$Z(\mathbf{x}) = \left\{ \left\{ P_i, s^i \right\}, \left[S^i \setminus P_i, p_i \right], \varnothing, \varnothing \right\}. \tag{27}$$

Case 2 for H-units of equal size, with $P_i \in s^i \cap s^j$ and $P_i \cap P_j = \varnothing$:

$$Z(\mathbf{x}) = \left\{ \left\{ \varnothing, s^i \cap s^j \right\}, \left[S^i \setminus P_i \cap S^j \setminus P_j, p_i \cap p_j \right], \left\{ P_i, \varnothing \right\}, \left\{ P_j, \varnothing \right\} \right\}. \tag{28}$$

(Note that $z_1(\mathbf{x})$ contains $s^i \cap s^j$, whilst $z_1(t) = \varnothing$ because time is one-dimensional).

Case 3 for H-units of equal size, with $P_i \in p_j$ and thus $P_j \in p_i$:

$$Z(\mathbf{x}) = \left\{ \varnothing, \left[S^i \setminus P_i \cap S^j \setminus P_j, p_i \cap p_j \right], \left\{ P_i, s^i \cap p_j \right\}, \left\{ P_j, s^j \cap p_i \right\} \right\}. \tag{29}$$

Case 4 for H-units of equal size, with $s^i \cap s^j \neq \varnothing$ and $P_i \notin s^i \cap s^j$:

$$Z(\mathbf{x}) = \left\{ \varnothing, \left[S^i \setminus P_i \cap S^j \setminus P_j, p_i \cap p_j \right], \left\{ P_i, \varnothing \right\}, \left\{ P_j, \varnothing \right\} \right\}. \tag{30}$$

Case 5 for H-units of equal size, with $p_i \cap p_j = p_{ij}$ in which p_{ij} is a singular object:

$$Z(\mathbf{x}) = \left\{ \varnothing, \left[S^i \setminus P_i \cap S^j \setminus P_j, p_{ij} \right], \left\{ P_i, \varnothing \right\}, \left\{ P_j, \varnothing \right\} \right\}. \tag{31}$$

Case 6 for H-units of equal size, with $P_i \in S^i \cap S^j$ and $P_j \in S^i \cap S^j$, but $p_i \cap p_j = \varnothing$:

$$Z(\mathbf{x}) = \left\{ \varnothing, \left\{ S^i \setminus P_i \cap S^j \setminus P_j, \varnothing \right\}, \left\{ P_i, \varnothing \right\}, \left\{ P_j, \varnothing \right\} \right\}. \tag{32}$$

Case 7 for two H-units in general, so $S^i \cap S^j \neq \varnothing$, with $p_i \notin S^j$ and $p_j \notin S^i$:

$$Z(\mathbf{x}) = \left\{ \varnothing, \left\{ S^i \cap S^j, \varnothing \right\}, \varnothing, \varnothing \right\}. \tag{33}$$

Case 8 for H-units of different size, with $P_i \cap P_{0i} = P_i$:

$$Z(\mathbf{x}) = \left\{ \left\{ P_i, s^i \right\}, \left\{ S^i \setminus P_i, \varnothing \right\}, \varnothing, \varnothing \right\}. \tag{34}$$

Case 9 for H-units of different size, with $P_{0i} \in s^i$ and $P_i \cap P_{0i} = \varnothing$:

$$Z(\mathbf{x}) = \left\{ \left\{ \varnothing, s^i \right\}, \left\{ S^i \setminus P_i, \varnothing \right\}, \left\{ P_i, \varnothing \right\}, \left\{ P_{0i}, \varnothing \right\} \right\}. \tag{35}$$

Case 10 for H-units of different size, with $P_{0i} \in p_i$:

$$Z(\mathbf{x}) = \left\{ \varnothing, \left\{ S^i \setminus P_i, \varnothing \right\}, \left\{ P_i, \varnothing \right\}, \left[P_{0i}, p_i \right] \right\} \tag{36}$$

Case 11 for H-units of different size, with $s^i \subset s^{0i}$ and $P_{0i} \notin s$:

$$Z(\mathbf{x}) = \left\{ \varnothing, \left\{ S^i \setminus P_i, \varnothing \right\}, \left\{ P_i, \varnothing \right\}, \left\{ P_{0i}, \varnothing \right\} \right\}. \tag{37}$$

Case 12 for H-units of different size, with $P_i \in p_{0i}$:

$$Z(\mathbf{x}) = \left\{ \varnothing, \left[S^i \setminus P_i \cap S^{0i} \setminus P_{0i}, p_i \cap p_{0i} \right], \left\{ P_i, s^i \cap p_{0i} \right\}, \left\{ P_{0i}, \varnothing \right\} \right\}. \tag{38}$$

Case 13 for H-units of different size, with $P_i \cap S^{0i} \neq \varnothing$ and $P_{0i} \cap S^i \neq \varnothing$, but $p_i \cap p_{0i} = \varnothing$:

$$Z(\mathbf{x}) = \left\{ \varnothing, \left[S^i \setminus P_i \cap S^{0i} \setminus P_{0i}, \varnothing \right], \left\{ P_i, \varnothing \right\}, \left\{ P_{0i}, \varnothing \right\} \right\}. \tag{39}$$

Case 14 for H-units of different size, with $s^i \subset S^{0i}$ and $P_{0i} \notin S^i$:

$$Z(\mathbf{x}) = \left\{ \varnothing, \left\{ S^i \setminus P_i \cap S^{0i} \setminus P_{0i}, \varnothing \right\}, \left\{ P_i, \varnothing \right\}, \varnothing \right\}. \tag{40}$$

Mark zippers:

The mark zipper **for two marked H-units** H_i and H_j in case 1 ($Q_i = Q_j$) and in case 2 ($Q_i = -Q_j$) is:

$$Z(q) = \begin{cases} \left[\left\{\tilde{Q}_i \propto \tilde{Q}_j , \tilde{\mathbf{E}}_i \propto \tilde{\mathbf{E}}_j\right\} , \partial\tilde{\mathbf{E}}_i / \partial\tilde{t} \propto \partial\tilde{\mathbf{E}}_j / \partial\tilde{t}\right] \\ \left[\left\{\left\{0 , \mathbf{B}^i + \mathbf{B}^j\right\}\right\} , \left\{\tilde{\nabla}\times\mathbf{B}^i \propto \tilde{\nabla}\times\tilde{\mathbf{B}}^j\right\}\right] \\ \left[\left\{\tilde{Q}_i , \tilde{\mathbf{E}}_i \propto \tilde{\mathbf{B}}^j\right\} , \partial\tilde{\mathbf{E}}_i / \partial\tilde{t} \propto \tilde{\nabla}\times\tilde{\mathbf{B}}^j \propto \tilde{\nabla}\times\tilde{\mathbf{E}}_i \propto \partial\tilde{\mathbf{B}}^j / \partial\tilde{t}\right] \\ \left[\left\{\tilde{Q}_j , \tilde{\mathbf{E}}_j \propto \tilde{\mathbf{B}}^i\right\} , \partial\tilde{\mathbf{E}}_j / \partial\tilde{t} \propto \tilde{\nabla}\times\tilde{\mathbf{B}}^i \propto \tilde{\nabla}\times\tilde{\mathbf{E}}_j \propto \partial\tilde{\mathbf{B}}^i / \partial\tilde{t}\right] \end{cases}. \tag{41}$$

Because the fields are time and space dependent, the mark zipper can only be transformed further after combining with the time and the space zippers. In case 1: $\tilde{Q}_i \propto \tilde{Q}_j = \tilde{Q}_i$. In case 2: $\tilde{Q}_i \propto \tilde{Q}_j = 0$.

In case 3 ($\tilde{Q}_{0i} = 0$), the non-existing fields and charges of H_{0i} in (41) are dropped and the zipper is:

$$Z(q) = \begin{cases} \left\{\left\{Q_i , \mathbf{E}_i\right\} , \partial\mathbf{E}_i / \partial t\right\} \\ \left\{\left\{0 , \mathbf{B}^i\right\} , \nabla\times\mathbf{B}^i\right\} \\ \left\{\left\{Q_i , \mathbf{E}_i\right\} , \left\{\partial\mathbf{E}_i / \partial t + \nabla\times\mathbf{E}_i\right\}\right\} \\ \left\{\left\{0 , \mathbf{B}^i\right\} , \left\{\nabla\times\mathbf{B}^i + \partial\mathbf{B}^i / \partial t\right\}\right\} \end{cases}. \tag{42}$$

The *limiting conditions* (for the derivation: see Backerra, 2015), are in general:

$$\tilde{\nabla}\cdot\tilde{\mathbf{E}}_i = \tilde{Q}_i \,; \tilde{\nabla}\cdot\tilde{\mathbf{B}}^i = 0 \,; \partial\tilde{\mathbf{E}}_i / \partial\tilde{t} = \tilde{\nabla}\times\tilde{\mathbf{B}}^i \,; \partial\tilde{\mathbf{B}}^i / \partial\tilde{t} = -\tilde{\nabla}\times\tilde{\mathbf{E}}_i. \tag{43}$$

and similar for transformed fields (without tildes). Note that two identically marked H-units are in general not identical in geometrical sense. Bear in mind that identical magnetic fields cannot exist, due to the principle of uniqueness.

3. Gravity and Electricity

In our previous paper about gravity (Backerra, 2014), we considered only zippers of the first order, in the erroneous conviction that zips of the second order could not be identified with gravity. Because there turned out to be no difference between the results of first and second order zippers, we will consider this subject now with zippers of only the second order.

Essential for the existence of gravity is, that **spatial attributes of neutral H-units are larger than those of marked H-units,** because they do not have to spend potential energy to the quality mark. For convenience we assume that $S^{0i} \gg S^i$ and $s^{0i} \gg s^i$, which means that the difference in size is large enough for our considerations (and the same for index j). Like in the previous paper, we will consider **two charged particles** σ_i and σ_j, each generated by the interaction of one marked and one neutral H-unit. The marked H-units are indicated by H_i and H_j, the neutral ones by H_{0i} and H_{0j}. There are six combinations of two H-units out of four, so in principle six interactions. In this paper, we consider time cases as distinct ways of labeling separate interactions and so these six interactions, participating in the gravitational process, may be described by distinct time cases. In this paper we do not consider the possible physical meanings of differences in time cases, as we want to consider only the energetic process and time does not contain energy.

3.1 Generation of Two Stable, Charged Particles

Previously we did not pay much attention to the exact description of the particles, involved in the gravitational process. Only now we found a mistake (Backerra, 2014, section IV.A, equation (55)), describing them in time case 2, but then the relevant element is empty; these particles can only be described in time case 1. However, it makes no difference for the considerations about gravity. This time we will go more into the details of the particles and deduce the zippers properly in time case 1.

Two stable, charged particles, existing isolated from each other, are described by the two interactions $H_i * H_{0i}$ and $H_j * H_{0j}$. The zipper is considered in **time case 1, space case 8 and mark case 3**, which means that we have mixed H-units with $T_i \cap T_{0i} = T_i$, $P_i \cap P_{0i} = P_i$ and $Q_{0i} = 0$. Then the zipper for $H_i * H_{0i}$ can be written as (see (22), (34) and (42)):

$$Z(t, \mathbf{x}, q) = \{z_1, z_2\} = \left\{ \begin{array}{c} \left[\{T_i, P_i, \{Q_i, \mathbf{E}_i\}\}, \{f^i, s^i, \partial \mathbf{E}_i / \partial t\} \right] \\ \left[\{F^i \setminus T_i \cap F^j \setminus T_j, S^i \setminus P_i, \mathbf{B}^i\}, \{\tau_i, \varnothing, \nabla \times \mathbf{B}^i\} \right] \end{array} \right\}. \tag{44}$$

The mark elements can not completely be transformed yet and so the square brackets are still present.

In the *first zip* z_1, the small-scale mark element $\partial \mathbf{E}_i / \partial t$ must be a constant field of derivatives, because time element f^i cannot describe a change, so it is called $\mathbf{C}$. This field exists inside s^i, so the particle moves with a constant velocity with respect to another particle, but because it is an isolated particle, the velocity is zero. The small-scale mark element of the *second zip* z_2 is $\nabla \times \mathbf{B}^i = \partial \mathbf{E}_i / \partial t$ (see limiting conditions (43)), so this term is also empty. Thus the zipper can be written as:

$$Z(H_i * H_{0i}) = \{z_1, z_2\} = \left\{ \begin{array}{c} \left[\{T_i, P_i, Q_i\}, \{f^i, s^i, \mathbf{O}\} \right] \\ \left[\{F^i \setminus T_i \cap F^j \setminus T_j, S^i \setminus P_i, \mathbf{B}^i\}, \{\tau_i, \varnothing, \mathbf{O}\} \right] \end{array} \right\}, \tag{45}$$

in which the qualities in Z are replaced by the interaction in consideration. Small-scale time element τ_i is assigned to the large-scale element $S^i \setminus P_i$, a space which cannot move (because the particle is isolated) and not shrink (because the radius of S^i is by definition constant). So z_2 has no appearance and the appearance of the zipper can be written as:

$$A(H_i * H_{0i}) = \sigma_{s^i}. \tag{46}$$

Thus H_i and H_{0i} generate an observable **major particle** σ_{s^i}, having geometry s^i and charge Q_i; in the following we will write this shortly as σ_i. In this isolated situation, no space appears. In similar way particle σ_j is generated.

As soon as neutral H-unit H_{0i} starts to interact with the second neutral H-unit H_{0j}, they will generate actual space. As we pointed out previously (Backerra, 2014), which will be repeated shortly in the next sections, these H-units move accelerated towards each other and the particles will be taken in this movement. As soon as this happens, $\mathbf{E}_i$ changes accelerated, so in equation (44), in the first zip $\partial \mathbf{E}_i / \partial t$ changes linear and so in the second zip $\nabla \times \mathbf{B}^i$ changes also linear. So the vector fields in the zipper are not zero any more and zipper (45) is no longer valid. Consequently, the small-scale time element τ_i is not assigned to the large-scale element, but to the magnetic term (being independent of space). Then appearance (46) contains not only a particle, but also a space:

$$A(H_i * H_{0i}) = \left\{ \sigma_i, \left(\Theta_{S^i \setminus P_i}^{\mathbf{B}^i} \right) \right\}, \tag{47}$$

being particle σ_i and actual unobservable, **magnetized, extended space** $\Theta_{S^i \setminus P_i}^{\mathbf{B}^i}$, and similar for $H_j * H_{0j}$, generating the appearances σ_j and $\Theta_{S^j \setminus P_j}^{\mathbf{B}^j}$. Thus the interactions $H_i * H_{0i}$ and $H_j * H_{0j}$, each generating a particle, start to generate as well magnetized spaces as soon as the neutral H-units start to overlap each other. This is a third interaction, being $H_{0i} * H_{0j}$, which will be considered in the next section.

3.2 Gravity between the Two Particles

The accelerated movement is described by a third zipper (after (45) and a similar zipper for indices j and $0j$), concerning the interaction between the two neutral H-units. At the start, the distance of their major points is such, that they have a relatively very small overlapping area, which is not overlapping the two magnetic spaces around the particles. The positions of H_{0i} and H_{0j} with respect to each other corresponds with **space case 7**. Because we want to describe an accelerated movement, we take **time case 2**. Then the zipper for $H_{0i} * H_{0j}$ contains only one zip, which can be written as:

$$Z(H_{0i} * H_{0j}) = \{z_2\} = \left\{ \{F^{0i} \setminus T_{0i} \cap F^{0j} \setminus T_{0j}, S^{0i} \cap S^{0j}\}, \{\tau_{0i} \cap \tau_{0j}, \varnothing\} \right\}. \tag{48}$$

Small-scale time element $\tau_{0i} \cap \tau_{0j}$ is independent of space, so it will be assigned to the large-scale element and thus describes an accelerated change of the major space intersection $S^{0i} \cap S^{0j}$. This space is not observable because the corresponding small-scale space element is empty. Then the appearance (see section 2.5) can be written as:

$$A\left(H_{0i} * H_{0j} \right) = \left(\Theta^0_{S^{0i} \cap S^{0j}} \right). \tag{49}$$

This is an actual, unobservable, neutral, **_major free space_**, having the disc-like shape of two spherical segments. The indicated change $\tau_{0i} \cap \tau_{0j}$ can only be an accelerated change of the radii of S^{0i} and S^{0j}, or a change of the distance between P_{0i} and P_{0j}. A change of the radii is not possible because the size of an H-unit is constant, and so the H-units move from or towards each other. If moving from each other, the case would be finished in a short time, so we consider the second possibility: P_{0i} and P_{0j} move in an accelerated way towards each other. Consequently, the distance between the particles σ_i and σ_j, existing in connection with P_{0i} and P_{0j}, also decreases in an accelerated way and thus they move accelerated towards each other. This is identified with **_gravity_**.

Major free space $\Theta^0_{S^{0i} \cap S^{0j}}$ exists in between of the particles, not covering them, nor overlapping a part of their

magnetic spaces. As P_{0i} and P_{0j} move accelerated towards each other, the neutral space grows larger, so more and more potential energy of the neutral H-units is converted into actual energy.

This process might be circumscribed as **_gravity produces actual space_**.

3.3 Involvement of Magnetism

As soon as $\Theta^0_{S^{0i} \cap S^{0j}}$ is large enough to overlap the magnetic spaces $\Theta^{\mathbf{B}^i}_{S^i \backslash P_i}$ and $\Theta^{\mathbf{B}^j}_{S^j \backslash P_j}$, each on one side,

electromagnetism will be involved in the attracting process. From that moment, the interaction occurs between *all four H-units*. Because our zipper is suited only for the interaction of two H-units, we will consider each of the possible interactions between the four H-units separately; after that, we will consider them as a *cluster of interactions*, under the restriction that the energy of the complete phenomenon has to be equal to the sum of the isolated H-events.

We will start with marked H-unit H_i and neutral one H_{0j} (note that the index of the neutral H-unit now is j). The zipper for $H_i * H_{0j}$ in **_time case 2_** (for $T_{0i} < T_{0j}$) and **_space case 7_** is:

$$Z\left(H_i * H_{0j} \right) = \{ z_2 \} = \left\{ \left\{ F^{0i} \cap F^{0j} \backslash T_{0j}, S^i \cap S^{0j}, \mathbf{B}^i \right\}, \left[\tau_i \cap \tau_{0j}, \varnothing, \nabla \times \mathbf{B}^i \right] \right\}. \tag{50}$$

Compared with Equation (48), a magnetic field and a magnetic derivations field are added. Small-scale mark element $\nabla \times \mathbf{B}^i$ is independent of space and, according to the limiting conditions (43), equal to $\partial \mathbf{E}_i / \partial t$. Because σ_i moves accelerated, $\mathbf{E}_i$ changes with the second derivative of time, so $\partial \mathbf{E}_i / \partial t$ with the first derivative, which cannot be described by $\tau_i \cap \tau_{0j}$. So the magnetic field derivative will be replaced by a zero vector field and the small-scale time element will be assigned to the large-scale element, describing an accelerated change of the major space. The appearance can be written as:

$$A\left(H_i * H_{0j} \right) = \left(\Theta^{\mathbf{B}^i}_{S^i \cap S^{0j}} \right). \tag{51}$$

The same is valid for the mirrored interaction $H_j * H_{0i}$, generating the appearance:

$$A\left(H_j * H_{0i} \right) = \left(\Theta^{\mathbf{B}^j}_{S^j \cap S^{0i}} \right). \tag{52}$$

Thus the actual energy of the neutral space $\Theta^0_{S^{0i} \cap S^{0j}}$ is enriched with the actual energy of the two magnetized

spaces $\Theta^{\mathbf{B}^i}_{S^i \cap S^{0j}}$ and $\Theta^{\mathbf{B}^j}_{S^j \cap S^{0i}}$. This implies that the energy of the gravitational process increases as soon as the

magnetized spaces start to play a role in the interaction.

3.4 Involvement of Electricity

There will be a moment that the two particles have come so close to each other, that their magnetized major spaces $\Theta^{\mathbf{B}^i}_{S^i\backslash P_i}$ and $\Theta^{\mathbf{B}^j}_{S^j\backslash P_j}$ start to overlap each other. From that moment, interaction $H_i * H_j$ will be involved, the last of the six possible interactions between the four H-units. We will consider this interaction for *time case 2* and *space case 7* (so $S^i \cap S^j \neq \varnothing$, $p_i \notin S^j$ and $p_j \notin S^i$). Then the zipper contains only zip z_2:

$$Z(t,\mathbf{x},q)=\{z_2\}=\left[\left\{F^i\backslash T_i \cap F^j\backslash T_j, S^i \cap S^j, \mathbf{B}^i+\mathbf{B}^j\right\},\left\{\tau_i\cap\tau_j,\varnothing,\widetilde{\nabla}\times\widetilde{\mathbf{B}}^i\propto\widetilde{\nabla}\times\widetilde{\mathbf{B}}^j\right\}\right]. \tag{53}$$

Using a field condition (see (43)), the small-scale mark element can be written as $\partial\widetilde{\mathbf{E}}_i/\partial\widetilde{t}\propto\partial\widetilde{\mathbf{E}}_j/\partial\widetilde{t}$ and this term transforms into an actual energetic H-event only if it is accelerated changing. Because the particle moves accelerated, $\mathbf{E}_i$ changes with the second derivative of time, so $\partial\mathbf{E}_i/\partial t$ with the first derivative, which cannot be described by $\tau_i\cap\tau_{0j}$. Thus the small-scale magnetic term will be replaced by an empty set. Then the small-scale time element $\tau_i\cap\tau_{0j}$ will be assigned to the large-scale element, so describing an accelerated change of the major space intersection. This change may be a shrinking or growing of $S^i\cap S^j$. It is not observable, because the small-scale space element is empty. Then the appearance can be written as

$$A(t,\mathbf{x},q)=\left(\Theta^{\mathbf{B}^i+\mathbf{B}^j}_{S^i\cap S^j}\right), \tag{54}$$

being and an actual, unobservable, magnetized, major extended space, changing in an accelerated way.

If $Q_i = Q_j$, the field lines of $\mathbf{E}_i$ and $\mathbf{E}_j$ inside space $\Theta^{\mathbf{B}^i+\mathbf{B}^j}_{S^i\cap S^j}$ have opposite components in the direction of

the line segment P_iP_j, and so the actual energy density of the marked space will be smaller than in case only one

electric field exists. However, the actual energy density of marked spaces is a constant, so to raise this decreasing

density, $\Theta^{\mathbf{B}^i+\mathbf{B}^j}_{S^i\cap S^j}$ will shrink, implying that P_i and P_j move away from each other. If $Q_i = -Q_j$, the field lines

have equal components in the direction of P_iP_j, and so the actual energy of the marked space will be larger than

in case only one electric field exists. However, the actual energy density of marked spaces is a constant, so to

reduce this increasing density, $\Theta^{\mathbf{B}^i+\mathbf{B}^j}_{S^i\cap S^j}$ will grow, implying that P_i and P_j move towards each other. Thus

equal charges are repulsing, and opposite charges are attracting each other.

As soon as the major points P_i and P_j approach each other far enough to enter each others major spaces, *space case 6* is valid; then the zipper contains one zip more, being z_3. Because z_2 is the same as in (53), we write only the new one:

$$z_3=\left[\left\{T_i,P_i,Q_i\right\},\left\{\varnothing,\varnothing,\partial\widetilde{\mathbf{E}}_i/\partial\widetilde{t}\propto\widetilde{\nabla}\times\widetilde{\mathbf{B}}^j\propto\widetilde{\nabla}\times\widetilde{\mathbf{E}}_i\propto\partial\widetilde{\mathbf{B}}^j/\partial\widetilde{t}\right\}\right]. \tag{55}$$

The small-scale field term is independent of time and space, so it will be assigned to the large-scale element, where it appears as an electromagnetic vector in major point P_i. This vector does not add energy because there is no space. The point is storing the actual energy of the charge, being unobservable because the small-scale geometry is empty. The appearance (54) is enriched with an unobservable charged point, so:

$$A(H_i * H_j)=\left\{\left(\Pi_{P_i,Q_i}\right),\left(\Theta^{\mathbf{B}^i+\mathbf{B}^j}_{S^i\cap S^j}\right)\right\}. \tag{56}$$

Thus, as soon as P_i and P_j enter each others major spaces, a charge enters the energetic scene, together with the already existing magnetized space.

3.5 The Involvement of the Neutral Pellicles

As the last detail of the interaction between the four H-units, we will consider interaction $H_{0i} * H_{0j}$ once more. The space cases, having started with space case 7 at the start, pass subsequently through cases 6, 5 and 4. For these three space cases, time case 2 is used, so the time zipper:

$$Z(t) = \left\{ \varnothing, \left\{ F^i \setminus T_i \cap F^j \setminus T_j, \tau_i \cap \tau_j \right\}, \left\{ T_i, \varnothing \right\}, \varnothing \right\}. \tag{57}$$

Time zippers z_1 and z_4 are empty and so the appearances as well. In space cases 6, 5 and 4, $z_3 = \left\{ P_i, \varnothing \right\}$, which can generate no energy in combination with time zip $\left\{ T_i, \varnothing \right\}$. Thus the subsequent zippers contain only zip z_2. Written for increasing overlapping of S^{0i} and S^{0j}, the zipper in space case 6 is:

$$Z_6(t, \mathbf{x}) = \left\{ z_2 \right\} = \left[\left\{ F^i \setminus T_i \cap F^j \setminus T_j, S^i \setminus P_i \cap S^j \setminus P_j \right\}, \left\{ \tau_i \cap \tau_j, \varnothing \right\} \right], \tag{58}$$

in which the index of Z indicates the number of the space case. The appearance is $\left(\Theta^0{}_{S^{0i} \cap S^{0j}} \right)$, being the same unobservable, accelerating space as in (49), so nothing new. The zipper in space case 5 is:

$$Z_5(t, \mathbf{x}) = \left\{ z_2 \right\} = \left[\left\{ F^i \setminus T_i \cap F^j \setminus T_j, S^i \setminus P_i \cap S^j \setminus P_j \right\}, \left\{ \tau_i \cap \tau_j, p_{ij} \right\} \right], \tag{59}$$

and in space case 4:

$$Z_4(t, \mathbf{x}) = \left\{ z_2 \right\} = \left[\left\{ F^i \setminus T_i \cap F^j \setminus T_j, S^i \setminus P_i \cap S^j \setminus P_j \right\}, \left\{ \tau_i \cap \tau_j, p_i \cap p_j \right\} \right]. \tag{60}$$

Now a change is described: in both cases minor particle $\pi_{p_i \cap p_j}$ appears. The minor particle pops up in equation (59) and starts to travel through the pellicle intersection in equation (60). As a consequence, no extended space appears any more in interaction $H_{0i} * H_{0j}$, because space $\left(\Theta^0{}_{S^{0i} \cap S^{0j}} \right)$ reduces to the track of the minor particle. Thus, from the moment that space case 5 is reached, **the gravitational process has ended**. Particle $\pi_{p_i \cap p_j}$ is neutral and has a small mass; its existence ends as soon as space case 4 is reached. We identify this H-event with a **Higgs particle**, suddenly appearing in the last phase of the gravitational process and existing only during the small interval of time $\tau_i \cap \tau_j$.

Summarizing, the introduction of electromagnetic features adds a second type of accelerated movement to the gravitational process, away or towards each other, depending on the signs of the charges with respect to each other. Because magnetized major spaces are smaller then neutral ones, this addition can appear *only after the gravitational process has started*. In its first phase, *magnetic energy* is added to the total process; in the second phase, also *charge energy* is added. The Higgs particle is generated immediately after the gravitational process has ended and exists only during an extreme small period of time.

The results as presented in this section, can explain the **instability in the experimentally observed constants of gravity**. If we take the radius of S^{0i} as a first estimation as $R_{0i} = 100 \times R_i$, then the volume of S^{0i} is 10^6 larger than that of S^i, and, taking the energy density of a marked space as 100 times that of a neutral space, then the total additional actual energy of the marked space (counting with completely overlapping spaces) is about 10^{-4} times the total actual energy of the neutral space. The official value of G is 6.673889×10^{-11} N·(m/kg)^2, but the 13 measured values analyzed in a study of Lisa Zyga (2015) range from approximately 6.672×10^{-11} N·(m/kg)^2 to 6.675×10^{-11} N·(m/kg)^2, which is a variation of about 10^{-4}.

3.6 Expansion of the Universe

If the four H-units, as considered in the previous sections, would exist in an isolated system, their interactions would be impossible, because an increase of the total actual energy of the system is not allowed by the law of conservation of actual energy. If the system contains much more H-units, the increase of actual energy by the gravitational process can be compensated by another process. The possibility of repulsing of the same type of interaction is suited for this. To consider the possibility of an accelerated moving of P_{0i} and P_{0j} away from each other, we repeat (48):

$$z_2(t, \mathbf{x}) = \left\{ \left\{ F^{0i} \cap F^{0j} \setminus T_{0j}, S^{0i} \cap S^{0j} \right\}, \left\{ \tau_{0i} \cap \tau_{0j}, \varnothing \right\} \right\}. \tag{61}$$

The appearance $\left(\Theta^0_{S^{0i} \cap S^{0j}} \right)$, shortly written as Θ^0, is an actual, unobservable neutral space, moving in an accelerated way such that its size and so its actual energy decreases. Adding two marked H-units to these two neutral H-units in a similar way as we did to describe gravity, we have a second cluster of four H-units, describing two charged particles σ_i and σ_j. In this repulsing interaction, the marked spaces of these particles are supposed to exist outside the reach of Θ^0, but on the other hand, space $S^{0i} \cap S^{0j}$ is supposed to be large enough to contain P_{0i} and P_{0j}. Then the particles move accelerated away from each other. Space Θ^0 shrinks until it is empty and all actual space energy is converted back into potential energy. This could be expressed by saying that *actual space is swallowed up* by the expansion.

During the interaction, each of the H-units has an accelerated growing velocity with respect to the other one, as if they are launched with respect to each other. After the interaction has finished by losing contact between their major spaces, the velocity is not accelerating any more, but constant. So the H-units will continue to move away from each other, with the same velocity as they have reached. The particles σ_i and σ_j inside, move away from each other without ever being able to interact, and so their electromagnetic features stay out of the appearance. Thus the first phase of expansion is an accelerated period and, after the neutral spaces lost contact, switches over to a moving away from each other with a constant velocity.

We identify this H-event with the *expanding universe*, which in classical physics is considered as a separate phenomenon. According to twin physics, gravity and the expanding universe could be considered as two distinct aspects of one type of interaction.

3.7 The Cluster of Interactions Considered Together

Summarizing section 3, gravity is generated by space energy, whilst electric attraction and repulsion are generated by mark energy. **The difference of gravity and electricity is in their range**, resulting from the different sizes of neutral and marked major spaces. Because the range of electric attraction or repulsion is smaller than the range of gravity, spatial effects are allowed to act separately from marking effects.

Although attraction and expansion have the same origin, namely an accelerated change of a space, they show a striking difference. Gravity will be combined with electric repulsion or attraction as soon as the magnetic spaces around the particles are reached. Expansion, on the contrary, never combines with electromagnetic effects. At an astronomic scale, these two phenomena are adding to conserve the total actual energy of the universe by balancing their production and swallowing of space. If this happens with a conglomeration of H-units, then clusters of gravitating H-units may develop, drifting apart from each other due to expansion.

4. Overview of the Results of Twin Physics

Until now, we obtained for each investigated case one or two zips describing H-events, so having actual energy; the remaining elements of the zipper were empty. If two zips appeared, we considered them in combination with each other, to get a view on complementarity. However, in this section we want to prepare our results for a comparison with those of the standard model, and therefore a series of zips is considered separately, keeping in mind that the story might be not complete without considering the complete zipper. By far not all possible cases are considered. With these rather brief examples, not going into the details, we try to evoke a feeling for the basic character of each zip. After that, in section 4.5, we will use these results to describe the decay of a neutron.

Interactions of two marked H-units (called marked interactions) are indicated by $H_i * H_j$; interactions involving one neutral H-unit (mixed interactions) are indicated by $H_i * H_{0i}$, and if both H-units are neutral (neutral interactions) by $H_{0i} * H_{0j}$. Bear in mind that *neutral* major as well as minor spaces (indicated by a zero index) are larger than *marked* spaces. In general the time and space cases are chosen such, that the zips show as much as possible, or that they are useful for describing the neutron decay. All time attributes are transformed to a physical space, so the tildes and square brackets are removed; for shortness, in general time elements will only be explained if indispensable. Most of the space attributes are transformed to a physical space; elements containing pellicle intersections are still written with square brackets, because usually this is a non-singular object. Possibly only in the last step of transformation the appearance of these objects is clear. The operations joining, linking and transforming in the mark zippers (41) and (42) are carried out as far as possible; as long as the transformation cannot be fully decided, the operators will remain in the zipper.

4.1 The First Zip

Zip z_1 describes particles, but only in time case 1 and if the major spaces overlap each other such that the major points of space are inside the overlapping region. Then the appearance may be a spherical or slightly flattened major particle, having mass and eventually charge. It may be identified with black matter, a neutron, a proton, or an antiproton. This zip is essential in the decay process of the neutron (see section 4.5). Moreover this zip is the basis of the so-called 'strong force'.

4.1.1 For two neutral H-units in time case 1 ($T_{0i} \cap T_{0j} = T_{0i}$) and space case 1 ($P_{0i} \cap P_{0j} = P_{0i}$):

$$z_1\left(H_{0i} * H_{0j}\right) = \left\{\left\{T_{0i}, P_{0i}\right\}, \left\{f^{0i}, s^{0i}\right\}\right\}. \tag{62}$$

Appearance $A(z_1)$ is $\sigma_{s\,0i}$, being a major particle without electromagnetic features and having geometry s^{0i}. Its radius is much larger than that of a neutron. It is identified with **black matter**.

4.1.2 For two marked H-units in time case 1 ($T_i \cap T_j = T_i$) and space case 1 ($P_i \cap P_j = P_i$):

$$z_1\left(H_i * H_j\right) = \left\{\left\{T_i, P_i, \left[\tilde{Q}_i \propto \tilde{Q}_j\right]\right\}, \left\{f^i, s^i, \left[\partial\tilde{\mathbf{E}}_i / \partial\tilde{t} \propto \partial\tilde{\mathbf{E}}_j / \partial\tilde{t}\right]\right\}\right\}. \tag{63}$$

If in the considered system no other H-units exists to move to or from, then $\partial\tilde{\mathbf{E}}_i / \partial\tilde{t} = \partial\tilde{\mathbf{E}}_j / \partial\tilde{t} = \mathbf{O}$. Appearance $A(z_1)$ is σ_{s^i}, being a spherical major particle. If $Q_i = Q_j$, the particle is neutral and identified with a **neutron**. If $Q_i = Q_j$, the particle is charged and identified with a **proton** or an **anti-proton**. Mind that $\left[\tilde{Q}_i \propto \tilde{Q}_i\right] = \left[\tilde{Q}_i\right] = Q_i$.

4.1.3 For one marked and one neutral H-unit in time case 1 ($T_i \cap T_{0j} = T_i$) and space case 8

($P_i \cap P_{0j} = P_i$):

$$z_1\left(H_i * H_{0i}\right) = \left\{\left\{T_i, P_i, Q_i\right\}, \left\{f^i, s^i, \partial\mathbf{E}_i / \partial t\right\}\right\}. \tag{64}$$

If in the considered system no other H-units exists to move to or from, then $\partial\mathbf{E}_i / \partial t = \mathbf{O}$. Appearance $A(z_1)$ is σ_{s^i}, being a major particle having geometry s^i. For positive Q_i, it is again identified with a **proton** and for negative charge with an **antiproton**. This zip is the basis of the **strong force**, because the neutral minor space s^{0i} supplies the connecting background of an atom, without generating attracting or repulsing electric field derivatives.

4.1.4 For two marked H-units in time case 1 ($T_i \cap T_j = T_i$) and space case 2 ($P_i \in s^i \cap s^j$ and $P_i \cap P_j = \emptyset$):

$$z_1\left(H_i * H_j\right) = \left\{\left\{T_i, \emptyset, \left[\tilde{Q}_i \propto \tilde{Q}_j\right]\right\}, \left\{f^i, s^i \cap s^j, \left[\partial\tilde{\mathbf{E}}_i / \partial\tilde{t} \propto \partial\tilde{\mathbf{E}}_j / \partial\tilde{t}\right]\right\}\right\}. \tag{65}$$

If in the considered system no other H-units exists to move to or from, then $\partial\tilde{\mathbf{E}}_i / \partial\tilde{t} \propto \partial\tilde{\mathbf{E}}_j / \partial\tilde{t} = \mathbf{O}$.

The difference with (63) is, that here the large-scale space element is empty. The appearance is the same, so a **neutron, a proton or an anti-proton**, without a large-scale major point of space.

4.1.5 For one marked and one neutral H-unit in time case 2 (so $T_i \in f^{0i}$ and $\tau_i \cap \tau_{0i} \neq \emptyset$) and space case 9 ($P_{0i} \in s^i$ and $P_i \cap P_j = \emptyset$):

$$z_1\left(H_i * H_{0i}\right) = \left[\left\{\emptyset, \emptyset, Q_i\right\}, \left\{\emptyset, s^i, \partial\mathbf{E}_i / \partial t\right\}\right]. \tag{66}$$

This is a demonstration that the first zip in another time case than time case 1 has **no appearance**. The interaction generates no actual energy at all in this zip and there is no H-event.

4.2 The Second Zip

Zip z_2 generates actual energy for all interactions, if only the major spaces have a non-empty intersection and so can interact in general. The described H-events all are related to 'a large space', or to 'space to travel through' and in general they are changing with time. If the small-scale geometry is empty, the large-scale one appears as an actual, *unobservable* major extended space, which might be magnetized; it is active by changing its size or by turning around. If the small-scale geometry is not empty, the zip describes a particle, existing in the intersection of pellicles, having no charge and being considerably smaller than a major particle. In these cases, the major space is restricted to the track of the particle. Two types of gluons or a neutrino may appear. This zip is the basis for gravity, expansion, electric attraction and repulsion.

4.2.1 For two neutral H-units in time case 1 (so $T_{0i} \cap T_{0j} = T_{0i}$) and space case 1 ($P_{0i} \cap P_{0j} = P_{0i}$):

$$z_2\left(H_{0i} * H_{0j}\right) = \left[\left\{F^{0i} \backslash T_{0i}, S^{0i} \backslash P_{0i}\right\}, \left\{\tau_{0i}, p_{0i}\right\}\right]. \tag{67}$$

Appearance $A(z_2)$ is $\pi_{p_{0i}}$, being a minor particle, having a small mass (compared with a major particle), no charge, and traveling with a constant velocity over the surface of s^{0i} (see section 4.1.1). It is identified with a ***gluon***.

4.2.2 For two marked H-units in time case 1 (so $T_i \cap T_j = T_i$) and space case 1 (so $P_i \cap P_j = P_i$):

$$z_2\left(H_i * H_j\right) = \left[\left\{F^i \backslash T_i, S^i \backslash P_i, \mathbf{B}^i + \mathbf{B}^j\right\}, \left\{\tau_i, p_i, \left[\widetilde{\nabla} \times \widetilde{\mathbf{B}}^i \propto \widetilde{\nabla} \times \widetilde{\mathbf{B}}^j\right]\right\}\right]. \tag{68}$$

Mind that $\mathbf{B}^i + \mathbf{B}^j$ cannot be reduced to zero, because each major magnetic field is by definition unique (see section 2.3). Appearance $A(z_2)$ is π_{p_i}, being a minor particle, having no charge, traveling with a constant velocity through p_i over the surface of s^i (see section 4.1.2). Like in the previous example, this is identified with a ***gluon.***

4.2.3 For one marked and one neutral H-unit in time case 1 (so $T_i \cap T_{0i} = T_i$) and space case 8 (so $P_i \cap P_{0j} = P_i$):

$$z_2\left(H_i * H_{0i}\right) = \left[\left\{F^i \backslash T_i, S^i \backslash P_i, \mathbf{B}^i\right\}, \left\{\tau_i, \varnothing, \nabla \times \mathbf{B}^i\right\}\right]. \tag{69}$$

If in the considered system no other H-units exists, a movement is not possible and so $\nabla \times \mathbf{B}^i = \mathbf{O}$ (because of conditions (43)), so the small-scale time element τ_i will be assigned to the large-scale element of z_2. But because the major space cannot move, nor grow or shrink, there is ***no appearance***. If at least one H-unit more exists in the considered system, a movement is possible, the magnetic field term is not a zero field and τ_i indicates a change in the small-scale element. Then the appearance $A(z_2)$ is $\left(\Theta_{S^i \backslash P_i}^{\left(\mathbf{B}^i\right)}\right)$, in which the brackets around $\mathbf{B}^i$ indicate a changing field. This can be identified with an actual, unobservable, ***major extended space,*** carrying a changing magnetic field.

4.2.4 For one marked and one neutral H-unit in time case 2 (so $T_i \in f^{0i}$ and $\tau_i \cap \tau_{0i} \neq \varnothing$) and space case 9 ($P_{0i} \in s^i$ and $P_i \cap P_{0i} = \varnothing$), and also for space case 11 ($s^i \subset s^{0i}$ and $P_{0i} \notin s^i$):

$$z_2\left(H_i * H_{0i}\right) = \left[\left\{F^i \backslash T_i \cap F^{0i} \backslash T_{0i}, S^i \backslash P_i, \mathbf{B}^i\right\}, \left\{\tau_i \cap \tau_{0i}, \varnothing, \nabla \times \mathbf{B}^i\right\}\right]. \tag{70}$$

Small-scale time element $\tau_i \cap \tau_{0i}$ indicates an accelerated change; this is possible if $S^i \backslash P_i$ turns around P_{0i}.

Then appearance $A(z_2)$ is $\left[\Theta_{S^i \backslash P_i}^{\left(\mathbf{B}^i\right)}\right]$, being an actual, unobservable, major extended space, carrying a quadratic changing magnetic field. But because of field conditions (43), this would be possible only if also $\partial \mathbf{E}_i / \partial t$ would change quadratic and so if $\mathbf{E}_i$ would change in the third order. We cannot identify this with experimental results, so we suppose that the small-scale field is not appearing.

4.2.5 For two marked H-units in time case 2 (so $T_i \in f^j$ and $\tau_i \cap \tau_j \neq \varnothing$) and space case 2 (so $P_i \in s^i \cap s^j$ and $P_i \cap P_j = \varnothing$):

$$z_2\left(H_i * H_j\right) = \left[\left\{F^i \backslash T_i \cap F^j \backslash T_j, S^i \backslash P_i \cap S^j \backslash P_j, \mathbf{B}^i + \mathbf{B}^j\right\}, \left\{\tau_i \cap \tau_j, p_i \cap p_j, \nabla \times \mathbf{B}^i \propto \nabla \times \mathbf{B}^j\right\}\right], \tag{71}$$

with the large-scale element written above the small-scale one. Appearance $A(z_2)$ is $\pi_{p_i \cap p_j}$, being a minor particle, having a small mass and no charge, and traveling accelerated through $p_i \cap p_j$ along the border of $s^i \cap s^j$. The H-event could again be identified with a gluon, but there is a difference with the previous two: This one accelerates, as if being launched. If it continues its existence, it has a constant velocity close to the velocity of light (see Backerra 2012). This will be identified with a ***neutrino***.

4.2.6 For two neutral H-units in time case 2 (so $T_{0i} \in f^{0j}$ and $\tau_{0i} \cap \tau_{0j} \neq \varnothing$) and space case 7 (so $S^i \cap S^j \neq \varnothing$, $p_i \notin S^j$ and $p_j \notin S^i$):

$$z_2\left(H_{0i} * H_{0j}\right) = \left\{\left\{F^{0i} \backslash T_{0i} \cap F^{0j} \backslash T_{0j}, S^{0i} \cap S^{0j}\right\}, \left\{\tau_{0i} \cap \tau_{0j}, \varnothing\right\}\right\}. \tag{72}$$

The small-scale time element $\tau_{0i} \cap \tau_{0j}$ is assigned to the large-scale one. The intersection of major spaces has

the possibility to change, by growing or shrinking. Appearance $A(z_2)$ is $\left(\Theta^0_{S^{0i} \cap S^{0j}}\right)$, being an actual,

unobservable, ***major extended space***, changing accelerated. Thus the major points of space P_{0i} and P_{0j} move accelerated towards or from each other. The first possibility is identified with ***gravity***, the second with ***expansion***. Note that both phenomena are observable only if major particles exist in the centre of the neutral H-units, so if two marked H-units are involved (see section 3).

4.2.7 For two marked H-units in time case 2 (so $T_i \in f^j$ and $\tau_i \cap \tau_j \neq \varnothing$) and space case 7 (so $S^i \cap S^j \neq \varnothing$, $p_i \notin S^j$ and $p_j \notin S^i$):

$$z_2\left(H_i * H_j\right) = \left\{\left\{F^i \backslash T_i \cap F^j, S^i \cap S^j, \mathbf{B}^i + \mathbf{B}^j\right\}, \left\{\tau_i \cap \tau_j, \varnothing, \left[\tilde{\nabla} \times \tilde{\mathbf{B}}^i \propto \tilde{\nabla} \times \tilde{\mathbf{B}}^j\right]\right\}\right\}. \tag{73}$$

Appearance $A(z_2)$ is $\left(\Theta^{\mathbf{B}^i + \mathbf{B}^j}_{S^i \cap S^j}\right)$, being an actual, unobservable, magnetized ***major extended space***, changing

in an accelerated way. Because of field conditions (43), the small-scale mark element appears only if

$\partial \tilde{\mathbf{E}}_i / \partial \tilde{t} \propto \partial \tilde{\mathbf{E}}_j / \partial \tilde{t}$ would change quadratic, and thus $\mathbf{E}_i \propto \mathbf{E}_j$ in the third order, so we

suppose that this field term does not appear. As pointed out in section (3.4), if $Q_i = Q_j$, the H-units repel each other; if $Q_i = Q_j$, they attract each other. This is identified with ***electric repulsion and attraction***, respectively. Like in the previous example, this is only observable if two more (neutral) H-units are involved.

4.2.8 For one marked and one neutral H-unit in time case 4 (so $T_i \in \tau_{0i}$ for $T_{0i} > T_0$, and space case 12 (so $P_i \in p_{0i}$):

$$z_2\left(H_i * H_{0i}\right) = \left[\left\{F^i \backslash T_i \cap F^{0i} \backslash T_{0i}, S^i \backslash P_i \cap S^{0i} \backslash P_{0i}, \mathbf{B}^i\right\}, \left\{\varnothing, p_i \cap p_{0i}, \nabla \times \mathbf{B}^i\right\}\right]. \tag{74}$$

Appearance $A(z_2)$ is $\pi_{p_i \cap p_{0i}}$, being a minor particle, traveling with a constant velocity through $p_i \cap p_0$. The field term $\nabla \times \mathbf{B}^i$ is independent of space. Depending on the remaining part of the zipper, it might be a ***gluon*** or a ***neutrino***.

4.2.9 For one marked and one neutral H-unit in time case 1 (so $T_i \cap T_{0i} = T_i$) and space case 13 (so $s^i \subset S^{0i}$ and $P_{0i} \notin S^i$), is:

$$Z\left(H_i * H_{0i}\right) = \{z_2\} = \left\{\left\{F^i \backslash T_i, S^i \backslash P_i, \mathbf{B}^i\right\}, \left\{\tau_i, \varnothing, \nabla \times \mathbf{B}^i\right\}\right\}. \tag{75}$$

The appearance is non-empty if $\nabla \times \mathbf{B}^i$ changes linear, so if $\partial \mathbf{E}_i / \partial t$ changes linear, then $\mathbf{E}_i$ changes with the second order and thus the generated H-event moves accelerated. The H-units have eccentric positions with respect of each other, so P_i and P_{0i} may move accelerated towards or from each other, or turn around each

other. Appearance $A(z_2)$ is $\left(\Theta^{\mathbf{B}^i}_{S^i \backslash P_i}\right)$, being an unobservable, magnetized major extended space. We will

meet this case again in section 4.5, considering the ***neutron decay***.

4.3 The Third Zip

Zip z_3 is in general empty if the major points of the two H-units are coinciding. Here we will consider only mixed interactions, generating an electron.

4.3.1 For one marked and one neutral H-unit in time case 5 (so $T_i \in F^j$ and $F^i \cap \tau_j = \varnothing$) and space case 10 (so $P_{0i} \in p_i$):

$$z_3\left(H_i * H_{0i}\right)=\left[\left\{T_i, P_i, Q_i\right\}, \left\{\varnothing, \varnothing, \partial \mathbf{E}_i / \partial t + \nabla \times \mathbf{E}_i\right\}\right]. \tag{76}$$

In this case, the small-scale time and space elements are empty, so small-scale mark element $\partial \mathbf{E}_i / \partial t + \nabla \times \mathbf{E}_i$ is independent of time and space. Appearance $A\left(z_3\right)$ is $\left(\Pi_{Q_i}\right)$, being an unobservable, charged point particle, storing the actual energy of Q_i and carrying a vector of electromagnetic field $\partial \mathbf{E}_i / \partial t + \nabla \times \mathbf{E}_i$. Previously we identified this particle with an **anti-photon** (see Backerra, 2015), because in that context it acted as a counterbalance for the photon.

For $T_i < T_{0i}$, the zip is empty.

4.3.2 For one marked and one neutral H-unit in time case 4 (so $T_i \in \tau_{0j}$ for $T_i > T_{0i}$) and space case 12 ($P_i \in p_{0i}$):

$$z_3\left(H_i * H_{0i}\right)=\left[\left\{T_i, P_i, Q_i\right\}, \left\{f^i \cap \tau_{0i}, s^i \cap p_{0i}, \partial \mathbf{E}_i / \partial t + \nabla \times \mathbf{E}_i\right\}\right]. \tag{77}$$

Major point P_i exists inside the small-scale geometry $s^i \cap p_{0i}$, turning together with a constant velocity around P_{0i} over the neutral pellicle p_{0j}. Appearance $A\left(z_3\right)$ is $\delta_{s^i \cap p_{0i}}$, being a dot particle inside the pellicle, having a small mass, a charge Q_i, and by definition shaped as a small sphere (see section 2.5); inside this particle, an electromagnetic field exists. It is identified with an **electron**.

For $T_i < T_{0i}$, the zip is empty.

4.4 The Fourth Zip

Zip z_4 is empty in cases with coinciding major points. Here we will consider only a photon.

Then we need the time zipper in case 5 (so $T_i \in F^j$ and $\tau_j \cap F^i = \varnothing$, for $T_i < T_{0i}$), which is $\left\{\varnothing, \left\{F^i \setminus T_i \cap F^j, \varnothing\right\}, \varnothing, \left\{T_{0i}, \varnothing\right\}\right\}$. The space case is similar, so $P_i \cap S^{0i} \neq \varnothing$ and $P_{0i} \cap S^i \neq \varnothing$, but $P_i \cap P_{0i} = \varnothing$. For one marked and one neutral H-unit in time case 5 and space case 13:

$$z_4\left(H_i * H_{0i}\right)=\left[\left\{T_{0i}, P_{0i}, \mathbf{b}^i\right\}, \left\{\varnothing, \varnothing, \nabla \times \mathbf{B}^i + \partial \mathbf{B}^i / \partial t\right\}\right]. \tag{78}$$

Mind that in our previous paper (Backerra 2015, section 7, equation (63)) the photon is described by the third zip in equation (64) instead of the fourth, because the major points of time were taken oppositely upon the time axis (so $T_i > T_{0i}$). The small-scale mark element is independent of time and space. Appearance $A\left(z_4\right)$ is $\left(\Pi_{\mathbf{b}^i}\right)$, being an unobservable, neutral, massless point particle, having a magnetic vector $\mathbf{b}^i$ attached, and an electromagnetic vector. The generated particle is identified with a **photon**, the elementary particle without mass and charge, having an unobservable major location and carrying observable electromagnetic energy through space, with a constant velocity (see Backerra, 2015, section 7.1). The photon travels with a constant velocity around P_i, so it acts as a clock.

For $T_i > T_{0i}$, the zip is empty.

4.5 Decay of the Neutron

A free neutron is known to decay into a proton, an electron and a neutrino. We will try to obtain a proper description of this process by considering three H-units, one being neutral and two oppositely charged, in subsequently four steps. Each of the three possible interactions between two H-units will be considered separately and we will connect them by using the law of conservation of actual energy. In principle all combinations of time cases with space cases are allowed, if only they are compatible with each other and if the generated actual energy matches with the conservation law.

The neutral H-unit is H_0; the marked ones are H_i and H_j, having charges Q_i and Q_j, respectively, with $Q_i = -Q_j$ and Q_i positive. The three possible interactions are written as $H_i * H_j$, $H_i * H_0$ and $H_j * H_0$. Their sets of appearances are indicated by A_{ij}, A_{i0} and A_{j0}, respectively.

Step 1: At the start we have a closed mathematical system containing these three H-units.

The two marked ones, H_i and H_j, are considered as coinciding in **time case 1** and **space case 1** (so $T_i \cap T_j = T_i$ and $P_i \cap P_j = P_i$). Together they exist inside S^0, which is the major space of H_0, close to the border. Then the two non-zero zips of our *first considered interaction* $H_i * H_j$ appear together as one H-event (see (63) and (68)):

$$A_{ij}(z_1, z_2) = \left\{ \sigma_{s^i}, \pi_{p_i} \right\}. \tag{79}$$

This is a **neutron**, written as σ_{ij}^0 (indices ij indicate the involved H-units and 0 the charge), with a **gluon** π_{ij} traveling over its surface with a constant velocity.

The appearance of our *second considered interaction*, $H_i * H_0$, in **time case 1** (so $T_i \cap T_{0i} = T_i$) and **space case 13** (so $s^i \subset S^{0i}$ and $P_{0i} \notin S^i$) is (see (75)):

$$A_{i0}(H_i * H_0) = \left(\Theta_{S^i \setminus P_i}^{\mathbf{B}^i} \right). \tag{80}$$

This is an unobservable, **magnetized major extended space**, written as $\Theta_{i0}^{\mathbf{B}^i}$, **moving with a constant velocity** towards or from P_0. If the H-units move away from each other, the case ends if the contact is lost and nothing will happen. So we consider them as moving towards each other.

Our *third considered interaction*, $H_j * H_0$, could in principle appear likewise, but the last step of transforming the zipper is not allowed by the law of conservation of actual energy, because then the actual energy of $\Theta_{i0}^{\mathbf{B}^i}$ would be doubled, and by definition the energy density of space is a constant.

Then the actual energy of the total system can be written as:

$$E_1\left(H_i * H_j * H_0 \right) = E\left(H_i * H_j \right) + E\left(H_i * H_0 \right), \tag{81}$$

in which the index 1 of E indicates "before the interaction", and so with (79) and (80) as:

$$E_1\left(H_i * H_j * H_0 \right) = E\left(\left(\sigma_{ij}^0 + \pi_{ij} \right) + \Theta_{i0}^{\mathbf{B}^i} \right). \tag{82}$$

Step 2: As the neutron moves linear towards P_0, there will be a moment that P_j (still coinciding with P_i) reaches neutral pellicle p_0. Then $H_j * H_0$ can generate actual energy in **space case 12** (so $P_j \in p_0$). We choose the time case similar to the space case: $T_j \in \tau_0$ (for $T_j > T_0$), so **time case 4**.

The zipper is (see (74) and (77)):

$$Z\left(H_j * H_0 \right) = \{z_2, z_3\} = \left\{ \begin{array}{l} \left[\left\{ F^j \setminus T_j \cap F^0, S^j \setminus P_j \cap S^0 \setminus P_0, \mathbf{B}^j \right\}, \left\{ \varnothing, p_j \cap p_0, \nabla \times \mathbf{B}^j \right\} \right] \\ \left[\left\{ T_j, P_j, Q_j \right\}, \left\{ f^j \cap \tau_0, s^j \cap p_0, \nabla \times \mathbf{E}_j \right\} \right] \end{array} \right\} \tag{83}$$

and the appearance is:

$$A_{j0}(z_2, z_3) = \left\{ \pi_{p_j \cap p_0}, \delta_{s^j \cap p_0} \right\}. \tag{84}$$

The first element is a minor particle, traveling with a constant velocity through $p_j \cap p_0$, being a small circular track inside the relatively large neutral pellicle p_0. This track is the border of minor space $s^j \cap p_0$, containing P_j in its centre; the transformation of this object is a dot particle, being identified with an **electron** e_{j0}^- (see section 4.3.2). Its small-scale time element indicates that it moves with a constant velocity and this can only be a movement around P_0. The electron exists in the middle of the circulating minor particle, so the second particle is identified with a **neutrino** v_{j0}, having a small mass, no charge and being distinct from a gluon (because it does not glues to the surface of a major particle). Then the actual energy of (only this) interaction can be written as:

$$E\left(H_j * H_0 \right) = E\left(e_{j0}^- + v_{j0} \right). \tag{85}$$

Step 3: Because H_j is involved in generating the electron and the neutrino, interaction $H_i * H_j$ has less potential energy available to generate the neutron, but this requires all its potential energy. The only possibility to reduce the generated actual energy is a shift of P_i and P_j, leading to **space case 2**. Supposing that the time shifts in the same way, we reach **time case 2**. In time case 2, no particle can be generated, so the actual energy of σ_{ij}^0 transforms back to potential energy and the **neutron annihilates**; it is not clear yet if the gluon annihilates also, or continues its existence in another way. The relatively large decrease of the actual energy in z_1 is not immediately violating the energy law, because only after the flying time f^i (which is an extended present) this change has to be described in zip z_1, giving the opportunity to rearrange the remaining interactions.

As a consequence of the loss of the neutron energy, interaction $H_i * H_0$, generating space $\Theta_{i0}^{\mathbf{B}^i}$ until the pellicle is reached, has to compensate the loss of actual energy (by the annihilation of the neutron). This is possible by **generating a new major particle**. Major point P_i goes further inside s^0, until it coincides with P_0. Then **space case 8** is reached, which for **time case 1** (so $T_i \cap T_0 = T_i$) generates a major particle (see equations (64) and (69)):

$$Z\left(H_i * H_0\right) = \left\{z_1, z_2\right\} = \left\{ \begin{matrix} \left\{\left\{T_i, P_i, Q_i\right\}, \left\{f^i, s^i, \partial \mathbf{E}_i / \partial t\right\}\right\} \\ \left[\left\{F^i \setminus T_i, S^i \setminus P_i, \mathbf{B}^i\right\}, \left\{\tau_i, \varnothing, \nabla \times \mathbf{B}^i\right\}\right] \end{matrix} \right\}. \tag{86}$$

The mark term $\nabla \times \mathbf{B}^i$ is independent of space, so $\partial \mathbf{E}_i / \partial t$ as well; both may be replaced by a constant field $\mathbf{C}$. The appearance is:

$$A\left(z_1, z_2\right) = \left\{\sigma_{s^i, Q_i}, \left(\Theta_{S^i \setminus P_i}^{\mathbf{B}^i}\right)\right\}. \tag{87}$$

This is a **proton** σ_{i0}^+ existing in the coinciding points P_i and P_0, and an actual, unobservable, **major extended space**, written as $\Theta_{i0}^{\mathbf{B}}$. Considering this together with step 2, the proton circles around the electron (and the neutrino still accompanies the electron).

Step 4: During the flying time, the three H-units have rearranged such, that $P_j \in p_0$ and $P_i = P_0$, so after the flying time, the total energy of the system can be written as:

$$E_2\left(H_i * H_j * H_0\right) = E\left(\left(e_{j0}^- + \nu_{j0}\right) + \left(\sigma_{i0}^+ + \Theta_{i0}^{\mathbf{B}^i}\right)\right), \tag{88}$$

in which the index 2 of E indicates "after the interaction". A proton, an electron and a neutrino are generated; the magnetized space, existing at the start, appears again. The decay energy is still missing; in classical physics, this is supposed to be delivered by the kinetic energy of the electron. Because kinetic energy is related to velocity, and velocity is only possible in a space, we assume that according to twin physics, this energy can be found in a second zip (because this is the only one which may describe an extended space), in one of the three joint interactions.

To find this part of the decay process, we consider interaction $H_i * H_j$ again. In the beginning this generated the neutron with a gluon. After meeting the neutral pellicle, the neutron and the gluon annihilated and each of the marked H-units started a new career in combination with H_0, generating an electron with a neutrino and a proton, respectively. At the end of the decay story, the major points of space of H_i and H_j have a constant distance, equal to the radius of s^0, and they circle around each other. Because s^i and s^j are much smaller than s^0, their pellicles are supposed not to intersect.

Then, with $P_i \in S^j$ and $P_j \in S^i$, **space case 6** is valid. If we take **time case 2**, for $T_i > T_j$, zipper $Z\left(H_i * H_j\right)$ contains three zips, being z_2, z_3 and z_4.

$$Z\left(H_i * H_j\right) = \left\{ \begin{matrix} \left[\left\{F^i \setminus T_i \cap F^j, S^i \setminus P_i \cap S^j \setminus P_j, \mathbf{B}^i + \mathbf{B}^j\right\}, \left\{\tau_i \cap \tau_j, \varnothing, \tilde{\nabla} \times \tilde{\mathbf{B}}^i \propto \tilde{\nabla} \times \tilde{\mathbf{B}}^j\right\}\right] \\ \left[\left\{T_i, P_i, Q_i\right\}, \left\{\varnothing, \varnothing, \partial \tilde{\mathbf{E}}_i / \partial \tilde{t} \propto \tilde{\nabla} \times \tilde{\mathbf{B}}^j \propto \tilde{\nabla} \times \tilde{\mathbf{E}}_i \propto \partial \tilde{\mathbf{B}}^j / \partial \tilde{t}\right\}\right] \\ \left[\left\{\varnothing, P_j, Q_j\right\}, \left\{\varnothing, \varnothing, \partial \tilde{\mathbf{E}}_j / \partial \tilde{t} \propto \tilde{\nabla} \times \tilde{\mathbf{B}}^i \propto \tilde{\nabla} \times \tilde{\mathbf{E}}_j \propto \partial \tilde{\mathbf{B}}^i / \partial \tilde{t}\right\}\right] \end{matrix} \right\}. \tag{89}$$

The appearance of the middle zip (z_3) is a **charged point particle** $\Pi\left(P_i, Q_i\right)$ in P_i; its actual energy is equal to the actual energy of the charge. The last zip (z_4) has no appearance, because both time elements are empty, but for $T_i < T_j$ the situation is mirrored and the appearance of this zip is a **second charged point particle** $\Pi\left(P_j, Q_j\right)$ in P_j. We suppose that these energies are already included in equations (83) and (87), so these appearances would double the amount of actual charge energy in equation (88) and thus they are not allowed. Possibly they play a role at the very moment of the annihilation of the neutron, which we did not consider in all its details.

The first zip (z_2) describes a new item. Appearance $A\left(z_2\right)$ is an **accelerated extended space**, describing the turning around of the overlapping major spaces and thus the turning of P_i and P_j around each other; this will be written as $\Theta_{ij}^{(\mathbf{B})}$ (the upper index with round brackets indicates the changing magnetic fields). We identify this **extended space energy** with the kinetic energy of the electron.

Besides of this, it is experimentally known that in about one in 1000 decays an extra particle is produced, being a photon. Because a photon can only be generated by a mixed interaction, we will consider the mixed interaction products once again. If we consider $H_j * H_0$ not only in time case 4, as we did above, but also in time case 5, then z_2 in eq. (83) is the same, zip z_3 is empty and zip z_4 describes something new. For $T_j < T_0$ in **time case 5** and **space case 12**, zip z_4 is (see (78)):

$$z_4\left(H_j * H_0\right) = \left[\left\{T_0, P_0, \mathbf{b}^j\right\}, \left\{\varnothing, \varnothing, \nabla \times \mathbf{B}^j + \partial \mathbf{B}^j / \partial t\right\}\right]. \tag{90}$$

Appearance $A\left(z_4\right)$ is $\left(\Pi_{j0}^0\right)$, being an unobservable, neutral, massless point particle, having a magnetic vector $\mathbf{b}^i$ attached and carrying an electromagnetic field vector of $\nabla \times \mathbf{B}^j + \partial \mathbf{B}^j / \partial t$.

This can be identified with a **photon** γ_{j0}, traveling with a constant velocity around P_i and carrying a part of the **decay energy**. Adding the energy of the photon and that of the extended space to equation (88), the resulting energy after the decay of the neutron can be written as:

$$E_2\left(H_i * H_j * H_0\right) = E\left(\left(\Theta_{ij}^{(\mathbf{B})}\right) + \left(e_{j0}^- + \nu_{j0} + \gamma_{jo}\right) + \left(\sigma_{i0}^+ + \Theta_{i0}^{\mathbf{B}^i}\right)\right). \tag{91}$$

This is supposed to be the complete description after the decay of the neutron, in case a photon is generated. Inserting the energies for the neutron 939.56 MeV, for the proton 938.27 MeV and for the electron 0.51 MeV, with the decay energy of 0,78 MeV, the energy before the decay can be written as:

$$E_1\left(H_i * H_j * H_0\right) = 939.56 \text{ MeV} + E\left(\pi_{ij}\right) + E\left(\Theta_{i0}^{\mathbf{B}^i}\right). \tag{92}$$

and after the decay as:

$$E_2\left(H_i * H_j * H_0\right) = E\left(\left(\Theta_{ij}^{(\mathbf{B})}\right) + \left(0.51 + \nu_{j0} + \gamma_{jo}\right) + \left(938.27 + \Theta_{i0}^{\mathbf{B}^i}\right)\right). \tag{93}$$

Actual energies (92) and (93) have to be equal. Supposing that the decay energy is equal to the actual energy of the extended space plus the eventually generated photon, we insert in equation (93):

$$E\left(\Theta_{ij}^{(\mathbf{B})}\right) + E\left(\gamma_{jo}\right) = 0.78 \text{ MeV}. \tag{94}$$

Then the energies of the gluon (before the decay) and the neutrino (after the decay) are related as:

$$E\left(\pi_{ij}\right) = E\left(\nu_{j0}\right). \tag{95}$$

Because the gluon (before the decay) and the neutrino (after the decay) have equal energies, the gluon is supposed not to annihilate together with the neutron, but to change its existence from 'bound to the neutron', into a 'free existence' in the neutral pellicle of the neutral H-unit, without being glued to a mass carrying particle.

Summarizing, a neutron generated by two oppositely marked H-units, having a gluon upon its surface, becomes unstable under the influence of the major space of a neutral H-unit. First it moves with a constant velocity towards the centre of this H-unit. As soon as the neutron touches the neutral pellicle, it annihilates; the generation of its gluon is taken over from interaction $H_i * H_j$ to interaction $H_j * H_0$, and so the track of the gluon changes. Major point P_j, not fixed any more inside the neutron, sticks to the neutral pellicle; then, in one and the same new interaction, **an electron is generated** and **the identity of the gluon is changed into a neutrino**. Because a neutrino has a velocity close to that of light, we suppose that during the annihilation of the neutron, which is maximal the time necessary to cross the pellicle, the gluon accelerates to this velocity. After that, the *positively* marked H-unit, having crossed the neutral pellicle without sticking to it, enters the neutral minor space and goes further to the centre of the neutral minor space, to form **the proton**. The decay energy might be stored in an electromagnetic extended space representing the electron kinetic energy, or in an emitted photon, or both. To describe this phenomenon, we need three H-units and all four zips, to derive three zippers in several time and space cases.

It would be interesting to investigate if a hydrogen atom can be described by the interaction of four H-units.

5. Comparison with the Standard Model

The Standard Model is a theory of particle physics, classifying all subatomic particles known, as well as concerning the electromagnetic, weak, and strong nuclear interactions. It was developed throughout the latter half of the 20th century by collaborating scientists around the world, theoretical and experimental particle physicists alike. The current formulation was finalized in the mid-1970s upon experimental confirmation of the existence of quarks. The discoveries of the top quark (1995), the tau neutrino (2000), and the Higgs boson (2012) have given further credence to the Standard Model. For theorists, the Standard Model is a paradigm of a quantum field theory, which exhibits a wide range of physics including spontaneous symmetry breaking, anomalies and non-perturbative behavior.

However, the Standard Model leaves some phenomena unexplained and it falls short of being a complete theory of fundamental interactions; it does not incorporate the full theory of gravitation as described by general relativity and it does not account for the accelerating expansion of the universe.

The most fundamental concept of the Standard Model is, that the universe is built up in a deterministic way. Determinism is still strongly defended by many physicists, for instance by Rietdijk (2003), who used this concept to construct a theory about hidden variables. However, there are signs of indeterminism pressing onward, like in the recent paper of Langley (2016, section 8.1), who needs 'true randomness' in his theory about gravity; this can be conceived as a notion of indeterminism. Possibly even the incompatibility of Einstein's relativity theory with quantum mechanics is related to its deterministic basis.

By contrast, twin physics is based upon the concept that the world is dualistic; determinism and indeterminism are considered as intimately connected. To be able to describe indeterminate objects independently of deterministic ones, a mathematical language called 'complementary language', based upon set theory, has been developed. Spaces are considered as limited, energetic items; also the Heisenberg's uncertainty relations are translated into a dual description. An important difference is, that the concept of 'force' is not represented in twin physics; energetic space supplies a substitution for forces.

The basic element in twin physics is the H-unit, being a mathematical unity of potential energy; this is strongly related to the original conception of quarks, introduced by Gell-Mann (1964), in which he distinguished mathematical and physical quarks, the former being mathematical expedients. Quarks might be the deterministic precursors of H-units. An important difference between the Standard Model and twin physics is in conceiving potential energy; in twin physics, potential energy is not a real energy which can be assigned to a body, but an intermediate mathematical item.

In this paper, we described with one set of four elements (the zipper), containing entangled attributes of time, space and charge, all basic phenomena in physics, like the neutron, the proton and its anti-particle, the electron and its anti-particle, the gluon, the neutrino, the photon, the Higgs particle, including spin (Backerra, 2012). Moreover, the laws of Maxwell emerge from the definitions; gravity can be connected to electric attraction and repulsion; the phenomena expansion of the universe, the weak force and the strong force can be described in principle. The constant velocity of light and the quantisation of Planck can be deduced (Backerra, 2015), instead of taking them as postulates.

6. Conclusion

According to twin physics, the range of gravity is larger than that of electricity. This is related to the restriction of spaces to finite sizes, combined with an essential difference in size between neutral and marked H-units. Marked H-units have to spend a part of their potential energy to charges and fields, so not all potential energy is available for space. This difference in ranges might explain the instability in the experimentally observed constants of gravity (Zyga, 2015). The type of interaction between H-units which is identified with gravity not only leads to attraction but may also generate repulsion. Because this latter phenomenon occurs only beyond the reach of electromagnetic influences, expansion of the universe is unjustly considered to be an independent phenomenon.

The decay of a free neutron can be described using three H-units. Although the zipper in general describes only the interactions between two H-units, it seems possible to describe this phenomenon properly by considering the three interaction possibilities of the H-units separately and using the law of conservation of actual energy to connect them. The decay of the neutron, described by two marked H-units, is initialized if a third, neutral H-unit crosses its space. At that moment the marked H-units can no longer spend all their potential energy on the previous type of interaction, so the H-units will rearrange themselves to generate a proton, an electron, a neutrino and decay energy. This explains in principle why the moment of decay for a single neutron is unpredictable; the average time of decay might give insight into the proportions of the universe. The description predicts that upon a free neutron a gluon exists, which during the decay transforms into a neutrino.

An important feature of twin physics is the formulation of one zipper for all phenomena. Heisenberg uncertainty relationships, relativity, and small-scale as well as large-scale observations are incorporated within the zipper. The use of four-dimensional space-time descriptions, being difficult to imagine, is not necessary and is even undesirable. As in the Standard Model, the elementary particles – neutron, proton, electron, gluon, neutrino, Higgs particle and photon – can be described. In twin physics, dark matter can also be described and the existence of an anti-photon is predicted. Not all experimentally found variants of elementary particles have been considered yet. The quark is considered to be the deterministic precursor of the complementary H-unit.

In contrast to the Standard Model, twin physics makes no use of the concept of forces. This is made possible by considering finite spaces as independent energetic items, connecting particles. The strong force is supplied by the minor space of neutral H-units. The weak force is represented by a rearrangement of H-units resulting from their constant potential energy in combination with the law of conservation of actual energy. The electromagnetic force is supplied by coupling appropriate features to geometric features of H-units. The laws of Maxwell are not plugged in, but emerge from these definitions. The gravitational field has its analogue in twin physics by taking interacting neutral H-units as a background for electromagnetic sensible mass; the possibility of expansion as a second feature of the same interaction pops up.

While quantum mechanics and astronomy are not compatible in the Standard Model, in twin physics small- and large-scale physics are reconciled from the start. The constancy of the velocity of light and the quantization of the photon energy can be formally deduced in twin physics. This strengthens the supposition that twin physics may cover the complete field in an undivided way.

For the time being, the disadvantage of twin theory lies in the restriction of the system used to only a few H-units. The zipper describes the interaction of two H-units only and it seems appropriate to take a few more into consideration by incorporating the energetic balance. The description of systems containing a large number of H-units needs to be developed.

The major disadvantage of the Standard Model is the incompatibility of its constituent parts. The view of Einstein's relativity theory as an unimpeachable theory is at least open to question. The relatively large attention to the deterministic aspects of physics may be unjustified.

Acknowledgments

I owe the conceptual idea for this paper to Willem van Erk, who also assisted me in correcting the paper. I am very grateful to Wim Graef, who challenged me to defend the principles of twin theory in a series of fundamental discussions. Special thanks to the Taurida International Symphony Orchestra in Saint Petersburg (Russia), under the guidance of Mikhail Golikov, for allowing me to work on this subject while attending their rehearsals and concerts. Their music provided me with inspiration that I could make full use of thanks to their warm hearted reception, which turned Russia into my second homeland.

References

Backerra, A. C. M. (2010). Uncertainty as a principle. *Physics Essays, 23*(3), 419-441.

Backerra, A. C. M. (2012). The unification of elementary particles. *Physics Essays, 25*(4), 601-619.

Backerra, A. C. M. (2014). The quantum-mechanical foundations of gravity. *Physics Essays, 27*(3), 380-397.

Backerra, A. C. M. (2015). A bridge between quantum mechanics and astronomy. *Applied Physics Research, 8*(1).

Gell-Mann, M. (1964). *A Schematic Model of Baryons and Mesons. Physics Letters, 8* (3): 214-215

Hudson, J. J., Kara, D. M., Smallman, I. J., Sauer, B. E., Tarbutt, M. R., & Hinds, E. A. (2011). Improved measurement of the shape of the electron. *Nature, 473*(7348), 493-496.

Kahn, P. J. (1967). *Introduction to linear algebra.* London: Harper & Row, Ltd.

Langley, R. (2016). Quantum Gravity as Higher Dimensional Perspective. *Applied Physics Research, 8*(4).

Michelson, A. A., & Morley, E. W. (1887). *On the Relative Motion of the Earth and the Luminiferous Ether. American Journal of Science, 34*, 333–345.

Newton, I. (1687). The mathematical principles of natural philosophy (or: Principia), page 77. Retrieved from http://docs.lib.noaa.gov/rescue/Rarebook_treasures/QA803A451846.PDF

Rankine, W. J. (1853). On the general law of the transformation of energy. *Philosophical Magazine, 4*(5), 106-117.

Rietdijk, C. W. (2003). How do hidden variables fit in natural law? *Physics Essays, 16*(1), 42-62.

Weinberg, S. (2015). *To explain the world*. Great Brittain: Allen Lane.

Zyga, L. (2015). *Why do measurements of the gravitational constant vary so much?* Retrieved from http://phys.org/news/2015-04-gravitational-constant-vary.html#jCp

13

The Path of the Shortest Time

Masoud Asadi-Zeydabadi[1] & Clyde Zaidins[1]

[1] Department of physics, University of Colorado Denver, Denver, CO 80217, USA

Correspondence: Masoud Asadi-Zeydabadi, Department of physics, University of Colorado Denver, Denver, CO 80217, USA. E-mail: Masoud.Asadi-Zeydabadi@ucdenver.edu

Abstract

The principal of least action is one of the fundamental ideas in physics. The path of the shortest time of a particle in the presence of gravity is an example of this principal. In this paper some methods are introduced to teach the optimal path in introductory physics courses. The optimal path (path of the shortest time) is calculated for a few families of paths. Finally a numerical method according to Snell's law in a discrete medium is used to find the general optimal path and is compared with the brachistochrone path.

Keywords: optimal path, path of the shortest time, brachistochrone, the principal of least action, Fermat's principal, Senll's law

1. Introduction

The principle of least action is one of the fundamental ideas in physics. The calculus of variations is a powerful mathematical tool used to understand and identify the path for the shortest time (Boas, 2006; Taylor, 2005; Thornton & Marion, 2003). This is an advanced mathematical method that is taught usually at the junior level. Some knowledge about the optimal problems such as brachistochrone, tautochrone and catenary problems are useful for broad range of students (Erlichson, 1999; Babb & Currie, 2008; Aravind, 1981; Gomez-Aiza, Gomez, & Marquina, 2006; McKinley, 1979). Here we will introduce several ways to teach this fundamental concept to those students at any level who are not familiar with calculus of variations. This is not only a fundamental idea in physics but also since optimization is a very important concept in any field this knowledge would be useful even for non-science majors. In this paper we will study several families of curves. For each family a variable parameter is used and the optimal value of the parameter corresponding to the shortest time is calculated. In two cases a combination of inclined surfaces is used. For those cases the students can calculate time analytically as function of the height of the inclined surface. This is the parameter that should be found for the optimal path. In order to find the minimum time we need to calculate the derivative of time with respect to the variable parameter. This derivative even for the simple case is complicated and we calculate it numerically. The next example that is considered in this paper is a parabolic family of curves. We will find the corresponding time as a function of a parameter that will be defined for this case. Then the shortest time and the corresponding parabolic path will be found. At the end a numerical technique according to Snell's law for a series of discrete media is used to find the general optimal path. In the numerical method the shooting method is used in order to find the path that passes through the end points. This path numerically starts from the initial position and by adjusting the initial angle we can calculate a path that passes through the end point. The brachistochrone path is also introduced in this paper and all optimal paths, including the numerical Snell's path are compared to the brachistochrone curve. The numerical Snell's path in the limit as the numerical intervals goes to zero approaches the brachistochrone curve. These are basically the same curve but there is a numerical error for the second method. All students have access to computers and they are familiar with some software packages such as Excel, Python, or MATLAB. In general we strongly believe students should be exposed to challenging problems by using the advantage of computational techniques. We use numerical methods to explore a variety of challenging problems for the students (Asadi-Zeydabadi, 2014; Asadi-Zeydabadi & Sadun, 2013; Asadi-Zeydabadi & Sadun, 2014).

We have introduced this problem to a broad range of audiences, high school teachers, and students at different levels. We found out that the majority of them miss the main idea that there is an optimal path. We think that it needs to be discussed with more detail in in different ways. This is one reason for writing this paper.

2. Simple Linear (Polygon) Paths

In this section two simple models are introduced. These models can be introduced to any student who has an algebra background at the high school or college level. The students do not need to have calculus background and they do not need to take derivatives to find the optimal values. There are plenty of linear paths similar to these that can be used. These are common examples that are used for a variety of purposes in any introductory physics course as sample problems in a lecture, homework problems, and demonstrations in a class or as part of an experiment in a lab. Because these are familiar examples and students know their relevant kinematic equations, we will use them to introduce the idea of the optimal path (the path for the shortest time).

Figure (1) shows two different paths from A to B: AB and AMB. In this paper we call the AMB track a triangular path. Suppose that a point mass starts from rest at point A. In this example the second path is a combination of two inclined surfaces. Point M is at the middle of the horizontal distance between A and B, $x_m = x_f/2$. The height of the surfaces changes by y and is the variable parameter for this case. This is one of the common demonstrations that we use in class. One question that we ask students is to guess (or determine) the path corresponding to the shortest time. An interesting question is how time changes as y increases. In all examples in this paper we use positive y downward.

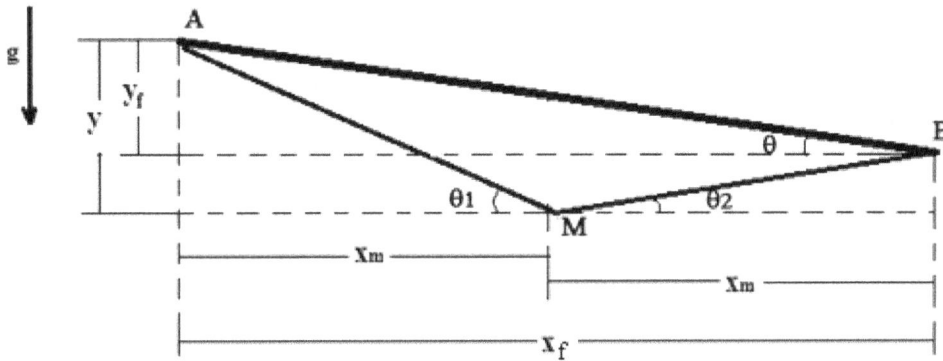

Figure 1. Two frictionless paths between the same points. A simple inclined surface, AB and a combination of two inclined surfaces, triangular AMB path

It is obvious that (according to the triangular inequality theorem) path AB is shorter than path AMB. A common mistake is that most students think the path of the shortest length is the same as the path of shortest time. This is true if the motion was uniform (speed was constant and the same for both paths).

The time for AB and AMB paths can be found by using the simple kinematic equations. We assume both paths are frictionless. The final velocities for both paths are the same and one can use conservation of mechanical energy to find the final speed at point B.

$$v_B = \sqrt{2gy_f} \tag{1}$$

Where g is the gravitational acceleration and y_f is the vertical position of point B relative to A as shown in Fig 1. The corresponding time for path AB is given by

$$T = \frac{v_B}{g \sin \theta} = \frac{1}{\sin \theta} \sqrt{\frac{2y_f}{g}} \tag{2}$$

In a similar way we can find the time for AM:

$$t_{AM} = \frac{v_M}{g \sin \theta_1} = \frac{1}{\sin \theta_1} \sqrt{\frac{2y}{g}} \tag{3}$$

where $\theta_1 = \tan^{-1}\left(\frac{y}{x_m}\right) = \tan^{-1}\left(\frac{2y}{x_f}\right)$ and $x_f = \frac{y_f}{\tan \theta}$.

The time for path MB is given in terms of the speed at points M and B, $v_M = \sqrt{2gy}$ and $v_B = \sqrt{2gy_f}$:

$$t_{MB} = \frac{v_M - v_B}{g \sin \theta_2} = \frac{1}{\sin \theta_2} \sqrt{\frac{2}{g}} \left(\sqrt{y} - \sqrt{y_f} \right) \tag{4}$$

where $\theta_2 = \tan^{-1} \left(\frac{y - y_f}{x_m} \right) = \tan^{-1} \left(\frac{2(y - y_f)}{x_f} \right)$.

The time for path AMB is the sum of t_{AM} and t_{MB}:

$$t = t_{AM} + t_{MB}. \tag{5}$$

Time t depends on y but it is not a monotonic function. Now two questions can be asked of the students. First how does t change as a function of y and secondly how does its value change compare with the time for path AB, T. One of the interesting questions is does $t(y)$ have an optimal (minimum) value as y changes. In order to find the minimum $t(y)$, we need to find $\frac{dt}{dy} = 0$. Time is minimized if $\frac{d^2 t}{dy^2} > 0$ at the optimal point. Even for this simple problem finding the optimal value of y corresponding to the minimum value of t needs a numerical solution. Student can either solve $\frac{dt}{dy} = 0$ numerically and test $\frac{d^2 t}{dy^2} > 0$ at the optimal value or plot $t(y)$ directly and observe the minimum value. The plan is not to use any sophisticated mathematical proofs. We want to demonstrate existence of the path of the shortest time to the students, for example, at high school level.

We can demonstrate with two paths similar to Figure 1 that the time for path AMB was shorter than for AB. After the students find out for this particular case that the time for AMB is shorter than for AB, then we can ask what does happen if y increases (or show them some additional paths with different y). Most of them think that as y increases time decreases. Without using the above equations we can ask if y goes to infinity can time go to zero? At this point they will find it is impossible for a particle to travel on a finite path with zero elapsed time and therefore there must be an optimal path. They can use the above kinematic equations or a numerical method to find the optimal path.

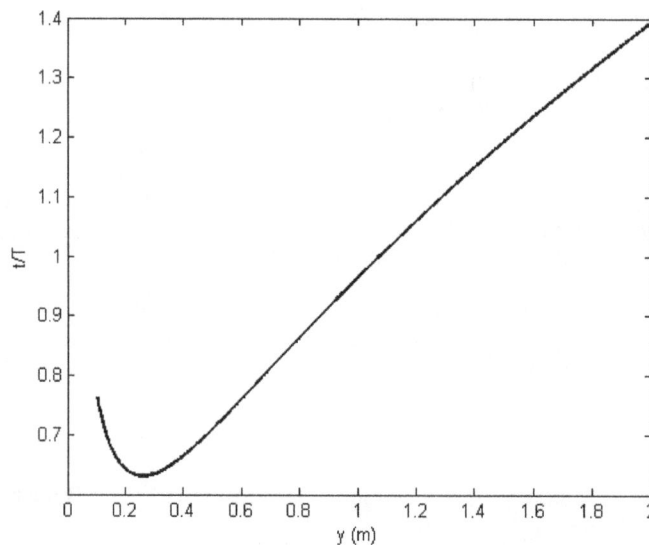

Figure 2. The ratio of elapsed time of path AMB to path AB, t/T, versus y: For $y_f = 0.1\ m$, $\theta = 10°, g = 9.8\ m/s^2$

Figure 2 shows the ratio of time for path AMB to the time for path AB, t/T, from (2) and (5) and it shows the minimum value for time. We used $y_f = 0.1\ m$, $\theta = 10°, g = 9.8$ m/s, for this case. If a different value for gravitational acceleration (*e.g.* the value for the moon) is used we will get the same results as in Figure 2. The value of time for path AB is $T = 0.82$ s and the minimum time for path AMB is $t = 0.52$ s that occurs at $y = 0.26$ m and with time ratio of $\frac{t}{T} = 0.63$. We can also find out from Figure 2 that if $y = 1.1$ m the elapsed time of both paths are equal $t = T = 0.83\ s$. Another useful question that can be asked of the students is the effect of the magnitude of gravity on $t(y)$. Figure 3 shows $t(y)$ for earth and moon gravities, $g_{Moon} = 1.6 \, m/s^2$ and $g_{Earth} = 9.8 \, m/s^2$, the minimum time path AMB on earth and moon are 0.52 s and 1.28 s respectively and occurs at $y = 0.26 \, m$. It is obvious that the ratio the minimum time for moon relative to of earth is $\frac{1.28}{0.52} = 2.46$ which is equal to $\sqrt{\frac{g_{Earth}}{g_{Moon}}} = \sqrt{6.1}$. Notice that if there is no gravity then an object with zero initial velocity stays at rest. If it starts with some initial velocity then because the net force on the object is zero its velocity remains constant and then the path of the shortest length is the same as the path of the shortest time.

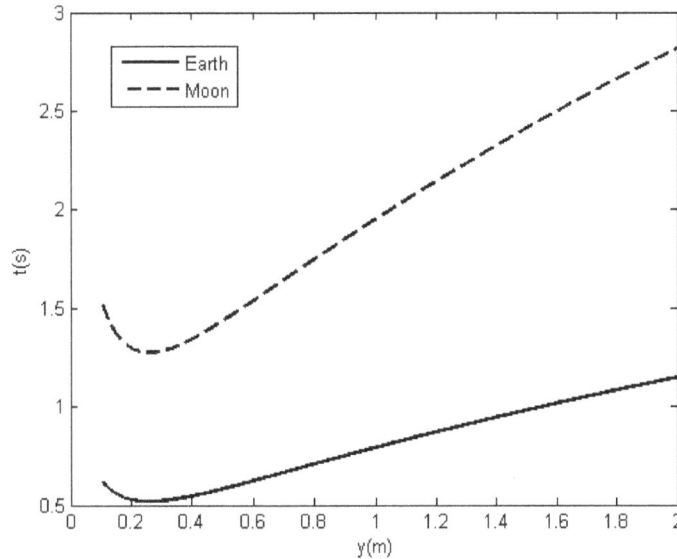

Figure 3. The results for time versus y for the earth and moon: $g_{Moon} = 1.6 \, m/s^2$ and $g_{Earth} = 9.8 \, m/s^2$, $y_f = 0.1 \, m$, $\theta = 10°$

Figure 4 shows another common example. The ABCD path is combination of two inclined surfaces AB and CD in addition to a horizontal path between them. We call this a trapezoidal-like path. We assume again that all surfaces are frictionless and that there is a point mass object that starts from rest at point A. Point D is at y_f position relative to point A and we again choose the vertical downward direction as the positive direction. We want to study the dependence of time on y and find the value of the minimum time and the corresponding y.

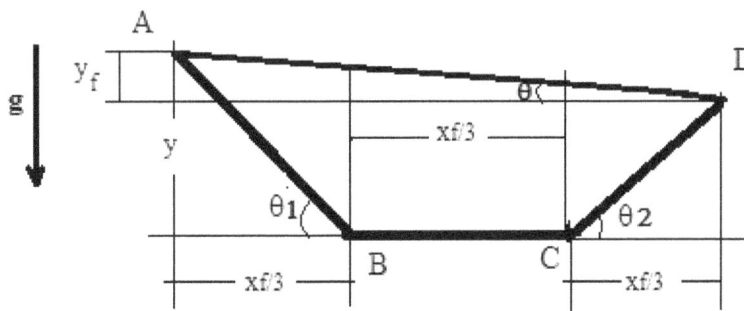

Figure 4. A path of two inclined and a horizontal surfaces ABCD track, a trapezoidal-like path

The corresponding time interval for Figure 4 can be found from simple kinematic equations. The velocity at points B and C are equal and are given by

$$v_B = v_C = \sqrt{2gy} \qquad (6)$$

The time from point A to B is given by

$$t_{AB} = \frac{v_B}{g \sin \theta_1} = \frac{1}{\sin \theta_1} \sqrt{\frac{2y}{g}} \tag{7}$$

where $\theta_1 = \tan^{-1}\left(\frac{3y}{x_f}\right)$. The time for the horizontal displacement from B to C is

$$t_{BC} = \frac{x_f}{3v_B} = \frac{x_f}{3\sqrt{2gy}} . \tag{8}$$

The velocity at point D is

$$v_D = \sqrt{2gy_f} \tag{9}$$

and the time for the final part is

$$t_{CD} = \frac{v_C - v_D}{g \sin \theta_2} = \frac{1}{\sin \theta_2} \sqrt{\frac{2}{g}} \left(\sqrt{y} - \sqrt{y_f}\right) \tag{10}$$

where $\theta_2 = \tan^{-1}\left(\frac{3(y - y_f)}{x_f}\right)$. Finally the total time is

$$t = t_{AB} + t_{BC} + t_{CD} = \frac{1}{\sin \theta_1} \sqrt{\frac{2y}{g}} + \frac{x_f}{3\sqrt{2gy}} + \frac{1}{\sin \theta_2} \sqrt{\frac{2}{g}} \left(\sqrt{y} - \sqrt{y_f}\right) \tag{11}$$

The derivative of t with respect to y for this case is also complicated and to find the value of the minimum time and corresponding y we need to use a numerical method to solve $\frac{dt}{dy} = 0$. Figure 5 shows the time as function of y for this case. The minimum time is $t = 0.5\ s$ and occurs at $y = 0.22$ m.

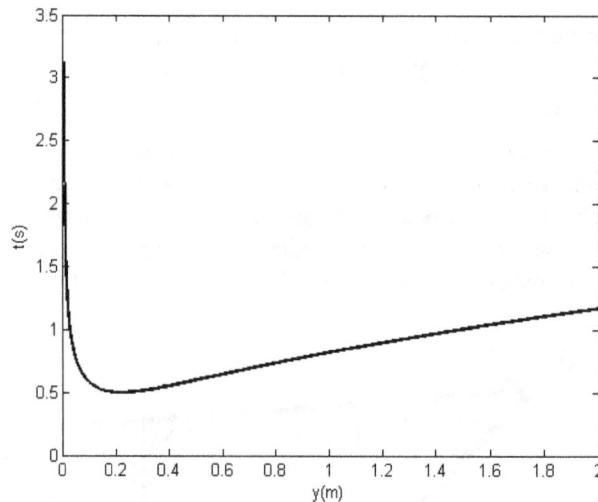

Figure 5. The results for the case that is shown in Fig 4, for $g = 9.8\ m/s^2$, $y_f = 0.1\ m$, and $\theta = 10°$

In these two examples we find the time analytically. To find the minimum time or optimal path we use a numerical method. In next section we will use a parabolic example and use numerical techniques to find time as a function of a parameter, the minimum time, and the corresponding optimal path.

3. Parabolic Paths

We will consider a family of parabolic paths that pass through two fixed points and we will compare them with a straight-line path between those points. As in the previous section the plan is to find the optimal path for the family of the parabolic paths. Figure 6 shows a linear and parabolic path between point A and B. Let point A be at origin $(0,0)$, then the coordinate of point B is $(x_f, y_f) = \left(\frac{y_f}{\tan \theta}, y_f,\right)$ and the equation of the parabola is given by

$$y = ax^2 + bx \tag{12}$$

$b = \tan\theta - a\left(\frac{y_f}{\tan\theta}\right)$, and a is a parameter that changes the depth of the parabolic equation and we will optimize the path by changing this parameter.

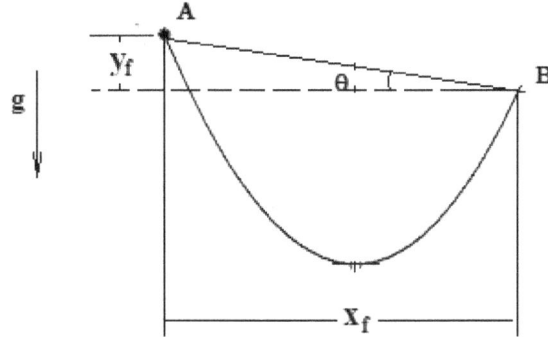

Figure 6. The parabolic and linear path between two points

The length of the probolic path is given by

$$\int_A^B ds = \int_A^B \sqrt{1 + \left(\frac{dy}{dx}\right)^2}\, dx = \int_0^{x_f} \sqrt{1 + (2ax + b)^2}\, dx \tag{13}$$

and the corresponding time is

$$t = \int_A^B \frac{ds}{v} = \int_A^B \sqrt{\frac{1 + \left(\frac{dy}{dx}\right)^2}{2gy}}\, dx = \int_0^{x_f} \sqrt{\frac{1 + (2ax + b)^2}{2g(ax^2 + b)}}\, dx \tag{14}$$

We evaluated this integral numerically. Figure 7 shows the results for time as function of a, which is the variable parameter. The minimum time occurs where the slope is zero, $\frac{dt}{da} = 0$. The minimum time is $t = 0.47$ s corresponding to $a = -2.2\ m^{-1}$.

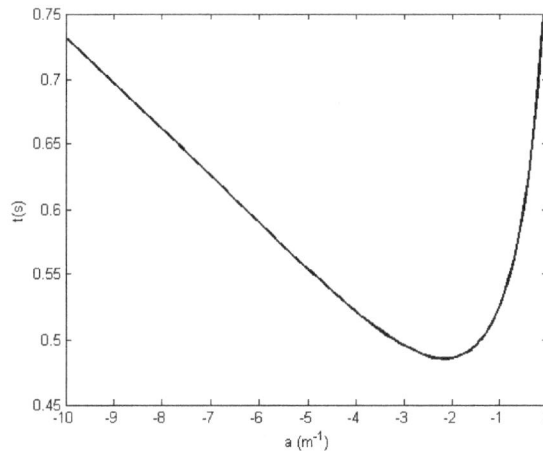

Figure 7. The results t versus a for: $g = 9.8\ m/s^2$, $y_f = 0.1\ m$, $\theta = 10°$, for the parabolic path

4. The Optimal Path, the Brachistochrone Problem

The examples in previous sections provide the optimal case for a given family of curves. In general there is an infinite family of the curves that can be defined between two points. We know that we need to use the calculus of variations to find the optimal path (path of shortest time) between two points. This path is known as the brachistochrone or cycloid. The equation of brachistochrone that passes through the origin *i.e.* point A $(0,0)$, and point B (x_f, y_f), is given by the following parametric equations

$$x = R(\theta - \sin\theta) \tag{15}$$

and

$$y = R(1 - \cos\theta) \tag{16}$$

where R is a parameter that can be determined by the coordinates of the point B and θ is a parameter that gives the relationship between y and x. The value of θ_f for point B (x_f, y_f), is given by

$$\frac{y_f}{x_f} = \frac{(1 - \cos\theta_f)}{(\theta_f - \sin\theta_f)} \tag{17}$$

The value of R can be found from (15) or (16)

$$R = \frac{x_f}{\theta_f - \sin\theta_f} = \frac{y_f}{(1 - \cos\theta_f)} \tag{18}$$

The time corresponding for the brachistochrone path is found numerically to be $t = 0.47$ s. Table 1 compares the results of the previous examples with the brachistochrone problem and Figure 8 shows all of these paths together.

Table 1. Comparison of the results of the previous examples with the brachistochrone problem

	Optimal parameter	Time/Minimum time	Path length
Inclined surface		$T = 0.82$ s	0.58 m
Triangular path	$y = 0.26$ m	$t = 0.52$ s	0.71 m
Trapezoid-like path	$y = 0.216$ m.	$t = 0.50$ s	0.698 m
Parabolic path	$a = -2.2\ m^{-1}$	$t = 0.49$ s	0.69 m
Brachistochrone path		$t = 0.47$ s	0.678 m

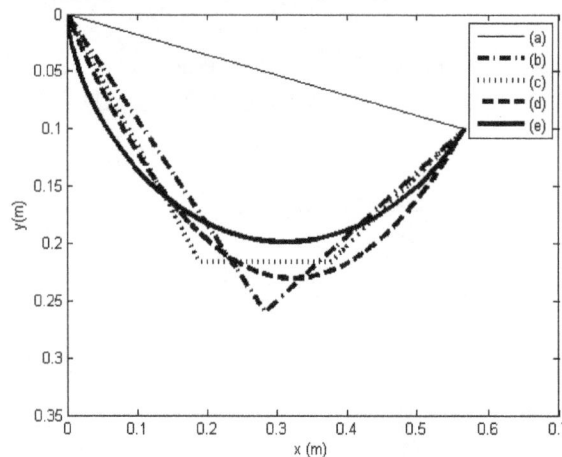

Figure 8. The paths for all five cases. (a) Linear, (b) Triangular, (c) Trapezoidal-like, (d) Parabolic and (e), the brachistochrone paths

The main purpose of this paper is to come up with a reasonable method to teach least action to students in introductory physics courses without directly using the calculus of variations. We can introduce it in the same manner as Snell's law. We believe that Snell's law also needs to be introduced in the light of both Huygens's and Fermat's principle. We should mention that even in the introductory physics courses Snell's law can be described as based on Fermat's principal. This important idea states that light follows the path of shortest time. In geometrical optics the laws of refraction and reflection are fundamental laws for ray tracing. Both refraction and reflection laws can be described by Fermat's principle. We know that the reason for the refraction of light is due to the change of the refractive index which causes the speed of light to change as it goes from one medium to another. Since the speed is changing, the direction of the light path corresponding to shortest time also changes. When light travels in a uniform medium then the index of refraction and speed of light are constant and the path of the shortest time and shortest length are the same. In this case light travels in straight line.

The same argument can be used to calculate the path of a particle that takes the least time where its speed depends on the particle position. In the presence of gravity the speed of a particle depends on its vertical coordinate, y. With the aid of numerical techniques we can apply this fundamental idea when speed varies with position. Numerical techniques allow students to study this topic, which is not easy to teach without a higher level of math than is expected for an introductory course. We believe that in the twenty first century students should know basic numerical techniques both in high school and in introductory college physics courses.

In analogy with Snell's Law the velocity of the particle depends on the spatial coordinate of the particle. The speed of the particle depends on the vertical coordinates, y, as $v = \sqrt{2gy}$.

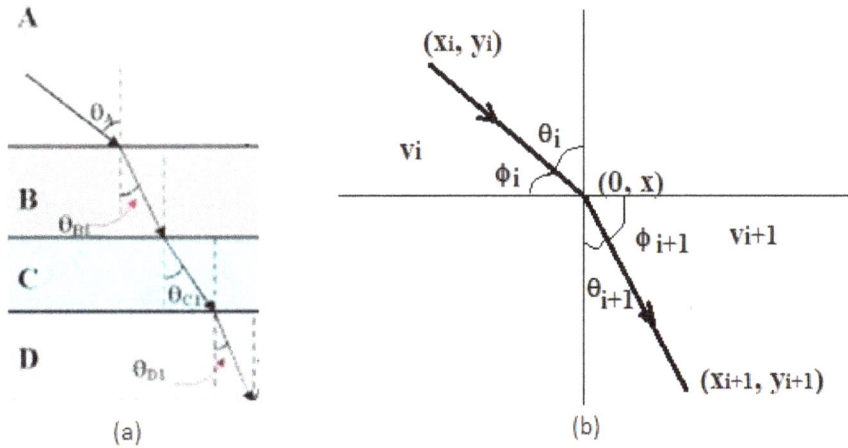

Figure 9. (a) Light traveling in discreet multimedia and the direction of light (or the motion of the particle) is changing as it travels from one layer to another. (b) Two media, with two different particle speeds

Figure 9-a shows layers of different media. As light (or in our example the particle) travels through this medium the particle experiences a change in the path as a result of change in the speed in order to follow the path of shortest time. Suppose an object has different speeds in two media as shown in Figure 9. In each medium the speed of the object is constant therefore the path corresponding for the shortest time in each medium is straight line. As the object is moving from the top medium to the second medium the speed changes and in order to minimize the time the direction of the motion of the object changes similar to the light ray going from one medium to another. This is similar to Snell's law in optics. We are looking for the path for the shortest time between points (x_i, y_i) and (x_{i+1}, y_{i+1}).

$$t = t_i + t_{i+1} = \frac{\sqrt{(x-x_i)^2+y_i^2}}{v_i} + \frac{\sqrt{(x_{i+1}-x)^2+y_{i+1}^2}}{v_{i+1}} \tag{19}$$

The only variable in the above equation is x. By taking derivative of time with respect to x and equating it to zero, $\frac{dt}{dx} = 0$, one will find

$$\frac{\sin\theta_i}{v_i} = \frac{\sin\theta_{i+1}}{v_{i+1}} \tag{20}$$

or

$$\frac{\cos\varphi_i}{v_i} = \frac{\cos\varphi_{i+1}}{v_{i+1}} \tag{21}$$

where as we see from Figure 8, $\sin\theta_i = \frac{x-x_i}{\sqrt{(x-x_i)^2+y_i^2}}$, $\sin\theta_{i+1} = \frac{x_{i+1}-x}{\sqrt{(x_{i+1}-x)^2+y_{i+1}^2}}$, $\theta_i + \varphi_i = \frac{\pi}{2}$ and

$\theta_{i+1} + \varphi_{i+1} = \frac{\pi}{2}$. But in the case of the motion of an object under the influence of gravity, speed changes continuously with y. With numerical methods either (20) or (21) can be used and a y interval that is very small is used so we can assume the change in the speed in that interval is negligible. The speed in ith interval is $v_i = \sqrt{2gy_i}$, where y_i is the distance of ith interval from the point at which we assume the object starts from rest. By substituting the speed in either (20) or (21) we will get:

$$\frac{\sin \theta_i}{\sqrt{y_i}} = \frac{\sin \theta_{i+1}}{\sqrt{y_{i+1}}} \tag{21}$$

$$\frac{\cos \varphi_i}{\sqrt{y_i}} = \frac{\cos \varphi_{i+1}}{\sqrt{y_{i+1}}} \tag{22}$$

As we seen in (21) or (22) the path does not depend on the value of g as we have seen in (15) and (16), but the time depends on g. As we have shown earlier the times ratio between two points on the surface of the moon with two points with the same vertical distance on the surface of earth is given by:

$$\frac{t_{Moon}}{t_{Earth}} = \sqrt{\frac{g_{Moon}}{g_{Earth}}} \approx \frac{1}{\sqrt{6}} \tag{23}$$

Figure 10 compares the results of using Snell's law with that of the brachistochrone path. We use a shooting method in the numerical Snell's law. The shooting method basically starts the ray with an initial angle at the initial point, A, and if the ray doesn't reach the target (final) point, B, then we aim it at a different angle until it reaches the neighborhood of the end point with an acceptable error.

Another advantage of using this method is that students can see a connection between a fundamental idea that works in different fields of physics, for example, optics and mechanics. Since calculus of variations is a mathematical foundation of Lagrangians and Hamiltonians (Boas, 2006; Taylor, 2005; Thornton & Marion, 2003), this concept is a reasonable bridge to that level of physics. This idea also provides students the basis of the path integral formulation in quantum mechanics that was introduced by Feynman (Feynman, Hibbs, & Styer, 2005; Sakurai, 2014). The technical advantage of this method is to expose students to basic numerical methods. This is an example of speed as function of position. We could provide many examples with the same application. One of the interesting examples is the relativistic brachistochrone path that can be discussed analytically and numerically (Goldstein & Bender, 1986). This is not only a useful physics example but also it can be used in modeling a lot of optimization problems. A simple conceptual problem is road traffic, where we usually follow the path of shortest time rather than shortest length. These can be different because of traffic conditions.

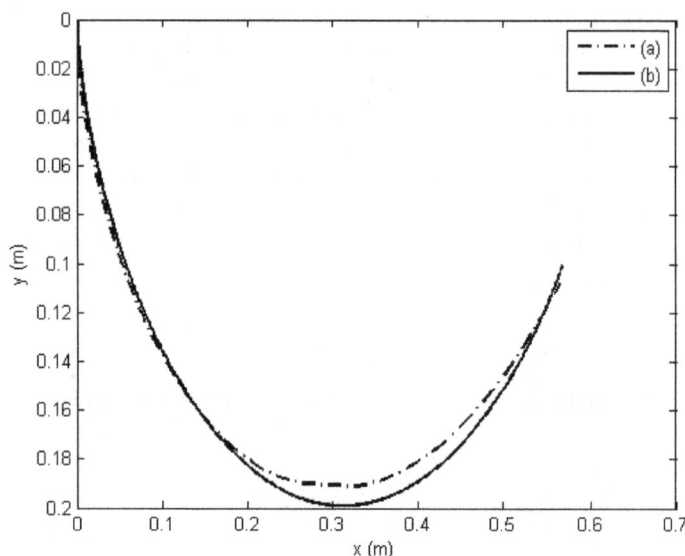

Figure 10. The brachistochrone path (solid line (b)) and the result of the numerical solution of Snell's law for a set of multilayers of media (dashed line (a))

5. Conclusion

In this paper we discuss several methods to teach the principal of the least action which is an important idea to be taught to students at different levels of math preparation. In these examples students also learn some basic computational techniques. We have discussed this topic with students at different levels and with high school teachers to understand how we can deliver this topic. We first approach this idea intuitively by showing some demonstrations without using any equations. Most of the audiences think the path of the shortest time and shortest length are the same. When they find out they are not the same, most of them think the path is a monotonic function of the corresponding changed path. If the time were a monotonic function then it would go to zero. For example, the audience thinks that as y goes to infinity for case 1 and 2, time would approach zero. In fact we know that it is impossible for time to go to zero and therefore the path is not a monotonic function and thus there is an optimal path. After the students understand the main idea then we analyze the problem with more mathematical detail by using the physics equations and numerical methods as discussed in the paper.

The main conclusion is that this is an important idea but it is a challenging problem for students. However we can teach this problem to a broad range of audiences intuitively, mathematically and numerically including physics formula and setting some simple experiments without using calculus of variations.

References

Aravind, P. K. (1981), Simplified approach to brachistochrone problems. *Am. J. Phys., 49*, 884. Retrieved from http://0-dx.doi.org.skyline.ucdenver.edu/10.1119/1.12389

Asadi-Zeydabadi, M. (2014). Bessel Function and Damped Simple Harmonic Motion. *JAMP, 2*, 26 http://dx.doi.org/10.4236/jamp.2014.24004

Asadi-Zeydabadi, M., & Sadun, A. C. (2013). One-dimensional relativistic dynamics with scaling formalism. *Eur. J. Phys., 34*, 75. http://dx.doi.org/10.1088/0143-0807/34/1/75

Asadi-Zeydabadi, M., & Sadun, A. C. (2014). How Can Non-Relativistic Projectile Motion Remain So in the Relativistic Limit? *Applied Physics Research, 6*, 42. http://dx.doi.org/10.5539/apr.v6n4p42

Babb, J., & Currie, J. (2008). The Brachistochrone Problem: Mathematics for a Broad Audience via a Large Context Problem. *TMME, 5*, 169-184.

Boas, M. L. (2006). Mathematical Methods in Physical Sciences (3rd ed., pp. 472-484). John Wiley & Sons.

Erlichson, H. (1999). Johann Bernoulli's brachistochrone solution using Fermat's principle of least time. *Eur. J. Phys., 20*, 299. http://dx.doi.org/10.1088/0143-0807/20/5/301

Feynman, R. P., Hibbs, A. R., & Styer, D. F. (2005). *Quantum mechanics and path integrals.* Courier Corporation.

Goldstein, H. F., & Bender, C. M. (1986). Relativistic Brachistochrone. *J. Math Phys, 27*, 507, http://0-dx.doi.org.skyline.ucdenver.edu/10.1063/1.527199

Gomez-Aiza, S., Gomez R. W., & Marquina, V. (2006). A simplified approach to the brachistochrone problem. *Eur. J. Phys., 27*, 1091. http://dx.doi.org/10.1088/0143-0807/27/5/008

McKinley, J. M. (1979). *Brachistochrones, tautochrones, evolutes, and tessalations. Am. J. Phys., 47*, 81, http://dx.doi.org/10.1119/1.11679

Sakurai, J. J. (2014). *Modern Quantum Mechanics* (2nd ed.). Pearson.

Taylor, J. R. (2005). Classical Mechanics (2nd ed., pp. 222-225). University Science Books.

Thornton, S. T., & Marion, J. B. (2003). Classical Dynamics of Particles and Systems (5th ed., pp. 211-213). Brooks/Cole.

Note

Note 1. The second author was a physics graduate student at Caltech and one of Feynman's teaching assistants. He was involved in preparing the three volume Feynman Lectures on Physics.

14

Speed of Light as an Emergent Property of the Fabric

Dirk J. Pons[1,], Arion D. Pons[2] & Aiden J. Pons[3]

[1] Department of Mechanical Engineering, University of Canterbury, New Zealand

[2] University of Canterbury, Christchurch, New Zealand

[3] Rangiora New Life School, Rangiora, New Zealand

Correspondence: Dirk Pons, Department of Mechanical Engineering, University of Canterbury, Private Bag 4800, Christchurch 8020, New Zealand. Email: dirk.pons@canterbury.ac.nz

Abstract

Problem- The theory of Relativity is premised on the constancy of the speed of light (c) in-vacuo. While no empirical evidence convincingly shows the speed to be variable, nonetheless from a theoretical perspective the invariance is an assumption. **Need**- It is possible that the evidence could be explained by a different theory. **Approach**- A non-local hidden-variable (NLHV) solution, the Cordus particule theory, is applied to identify the causes of variability in the fabric density, and then show how this affects the speed of light. **Findings**- Under these assumptions the speed of light is variable (VSL), being inversely proportional to fabric density. This is because the discrete fields of the photon interact dynamically with the fabric and therefore consume frequency cycles of the photon. The fabric arises from aggregation of fields from particles, which in turn depends on the proximity and spatial distribution of matter. Results disfavour the universal applicability of the cosmological principle of homogeneity and isotropy of the universe. **Originality**- The work proposes causal mechanisms for VSL, which have otherwise been challenging to ascertain. Uniquely, this theory identifies fabric density as the dependent variable. In contrast, other VSL models propose that c varies with time or some geometric-like scale, but struggle to provide plausible reasons for that dependency. This theory also offers a conceptually simply way to reconcile the refraction of light in both gravitational situations and optical materials.

Keywords: non-local hidden-variable solution, variable speed of light, cosmological principle

1. Introduction

The constancy of the speed of light in the vacuum was the central insight in Einstein's work on general and special relativity (Einstein, 1920), and forms the foundation of cosmological theory. There is no empirical evidence that convincingly shows the speed of light to be variable *in-vacuo*. Nonetheless, from a theoretical perspective the invariance is an assumption, and it is possible that the empirical evidence could be explained by a different theory. In this paper we develop such a theory.

We show that a variable speed of light (VSL) is a logical outcome for the non-local hidden-variable (NLHV) solution of the Cordus theory. Postulating a VSL theory is not new, but other VSL theories are focussed on cosmological physics, whereas this originates in fundamental physics.

2. Background

2.1 Invariance of the Speed of Light

The aberration of light, whereby the location of stars appeared to change depending on the velocity of Earth in its orbit, was an early indication that light had a finite speed (Bradley, 1729). In the late 1800s the propagation of light was believed to be through a medium called the luminiferous ether. One of these theories, the undulatory theory of de Broglie, Fresnel, Schrödinger and others (de Broglie, 1925; Schrödinger, 1926), proposed that the ether was generally at rest, except in transparent bodies where it caused refraction. The Michelson-Morley experiment (Michelson & Morley, 1887) directly tested the idea of Earth moving through a stationary ether, by looking for differences in the speed of light in orthogonal directions, and found no evidence to support such a theory.

Einstein subsequently built special relativity (SR) on the radical premise that the speed of light is the same for all inertial reference frames (Einstein, 1920). Specifically, that the speed of light is invariant under any velocity or orientation of the inertial frame of observation (Lorentz transformation). Yet the constancy of the speed of light is nonetheless an assumption. Einstein guessed rather than proved it, and designed it into special relativity. The invariance is therefore a premise of SR, rather than a proof thereof. However what is more convincing is that SR predicts time dilation and the relativity of simultaneity, and these do appear to be real. Furthermore, many subsequent measurements have supported the constancy of the speed of light. The most recent (Herrmann et al., 2009) set a lower limit on any local anisotropy Δc of the speed of light c as $\Delta c/c \sim 1 \times 10^{-17}$, this experiment being conducted over one year, using a block of fused silica.

2.2 Variable Speed of Light Theories

The constancy of the *in-vacuo* speed of light, at least on Earth, is well-established, as are the effects of SR. Nonetheless there has been ongoing interest in whether the speed of light really is invariant (Albrecht & Magueijo, 1999; Barrow & Magueijo, 1999; Bassett, Liberati, Molina-Paris, & Visser, 2000; Clayton & Moffat, 2003; J. Magueijo, 2000; Magueijo, 2001, 2003, 2008; Moffat, 2005; Singha & Debnath, 2007; Sisterna, 2011; Szydlowski & Krawiec, 2003; Unzicker, 2009). The primary purpose of these endeavours has been to explore for new physics at deeper levels, with a particular interest in quantum-gravity. One approach has been to look for violation of Lorentz invariance in other settings (João Magueijo, 2000). For example, it may be that the invariance of c breaks down at very small scales, or for photons of different energy. However such searches have been unsuccessful (Abdo et al., 2009).

Another approach is cosmological. If the speed of light was to be variable, it could solve certain problems. Specifically, it has been proposed that the horizon, inflation and flatness problems can be resolved if there were a faster c in the early universe, i.e. a time-varying speed of light (Sisterna, 2011). There are several other applications of variable speed of light (VSL) theories (Racker, Sisterna, & Vucetich, 2009; Sisterna, 2011) including branes (Youm, 2001) and particle creation (Harko & Mak, 1999). Another focus has been to use VSL to explain the Pioneer anomaly (Berman, 2011), e.g. using power law fitting (Shojaie, 2012).

In all these theories the difficulty is providing reasons for *why c* should vary with time or scale. They require the speed of light to be different at genesis, and then *somehow change* slowly or suddenly switch over at some time or event, for reasons unknown. None of these theories describe why this should be, nor do they propose underlying mechanics. This is ontologically problematic. Existing VSL theories cannot show a causal connection with existing physics, and therefore are not widely accepted.

3. Purpose and Approach

The purpose of this work was not initially to develop a VSL theory, though that is one of the outcomes achieved, but rather to identify whether there was a mechanism within the Cordus theory for c to have the value it does.

The Cordus theory is a specific type of NLHV solution (D. J. Pons, A. D. Pons, A. M. Pons, & A. J. Pons, 2012). Under this framework the vacuum is not entirely empty but comprises a fabric of discrete fields emitted by all particles (D. J. Pons & A. D. Pons, 2013). The approach taken here is to identify the causes of variability in the fabric density, and then show how this affects the speed of light.

4. Results

4.1 Fabric of the Vacuum

4.1.1 Structure of the Cordus Particule

The starting premise of the Cordus conjecture is that particles are *not* zero-dimensional (0D) points but rather have an internal structure. This structure is proposed to comprise two reactive ends, which are a small finite distance apart (span), and each behave like a particle in their interaction with the external environment (Pons et al., 2012), see Figure 1. A fibril joins the reactive ends and is a persistent and dynamic structure but does not interact with matter. It provides instantaneous connectivity and synchronicity between the two reactive ends. This structure was first used to explain wave-particle duality in the double-slit device (Pons et al., 2012), and has subsequently been applied to diverse other phenomena such as nuclide stability/instability (D. J. Pons, A. D. Pons, & A. J. Pons, 2013a, 2015c) and asymmetrical baryogenesis (D. J. Pons, A. D. Pons, & A. J. Pons, 2014a).

This is a non-local solution: the particule is affected by more than the fields at its nominal centre point. Each reactive end of the particle is energised in turn at the de Broglie frequency of that particle, which is dependent on its energy. The reactive ends are energised together for the photon, and in turn for matter particles (D. J. Pons, 2015). The span of massy particles shortens as the frequency increases, i.e. greater internal energy is associated with faster re-energisation sequence, hence also faster emission of discrete force and thus greater mass.

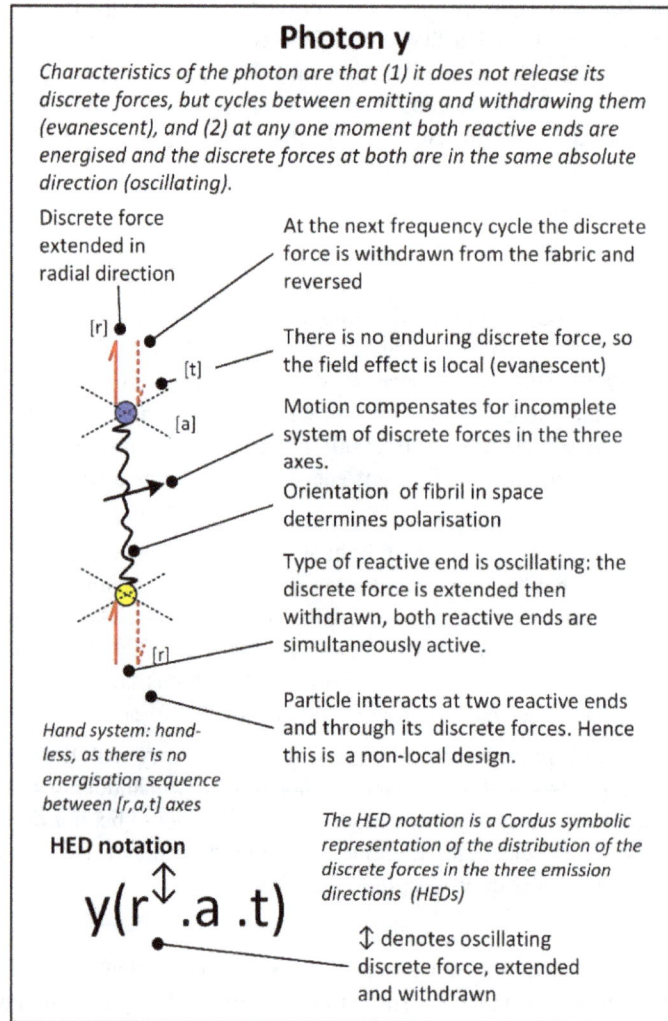

Figure 1. The structure of the photon as proposed by the Cordus theory. The particule has an internal structure comprising two reactive ends that energise together (whereas massy particules energise in sequence), and emit discrete forces into the fabric in the process. The emission is reversed at the next frequency cycle. Hence the energy is extracted from the fabric at one reactive end and instantaneously emitted at the other, and then this cycle is reversed. In contrast massy particules release their discrete forces, and this makes up a fabric of discrete forces in the vacuum. Image reproduced from (Pons, 2015) by permission of the author

The Cordus theory proposes the following lemma:

Each interaction of a photon with the fabric requires a frequency cycle. (Lemma 1)

Consequently the de Broglie frequency of a particle is inversely proportional to fabric density.

4.1.2 The Nature of Forces and Fields

The theory proposes that each matter particle emits discrete forces at each energisation, and these are propagated out into the external environment surrounding the particle. The particle continues to emit more discrete forces each time its reactive ends energise. This sequence of emissions makes up a flux tube of discrete forces that are propagated into the external environment. Furthermore the discrete forces are not consumed by interactions with other particles, but instead continue to travel away from the particle that emitted them. The aggregation of many such discrete forces from many neighbouring particules creates the electro-magnetic-gravitational (EMG) fields.

A physical explanation then arises for fields, namely that the direct lineal effect of the discrete force provides the electrostatic interaction, the bending of the flux line carries magnetism, and the torsion provides the gravitation interaction. These are all interactions of discoherent matter, whereas the strong force is predicted to be only applicable to coherent assemblies (D. J. Pons, A. D. Pons, & A. J. Pons, 2013b).

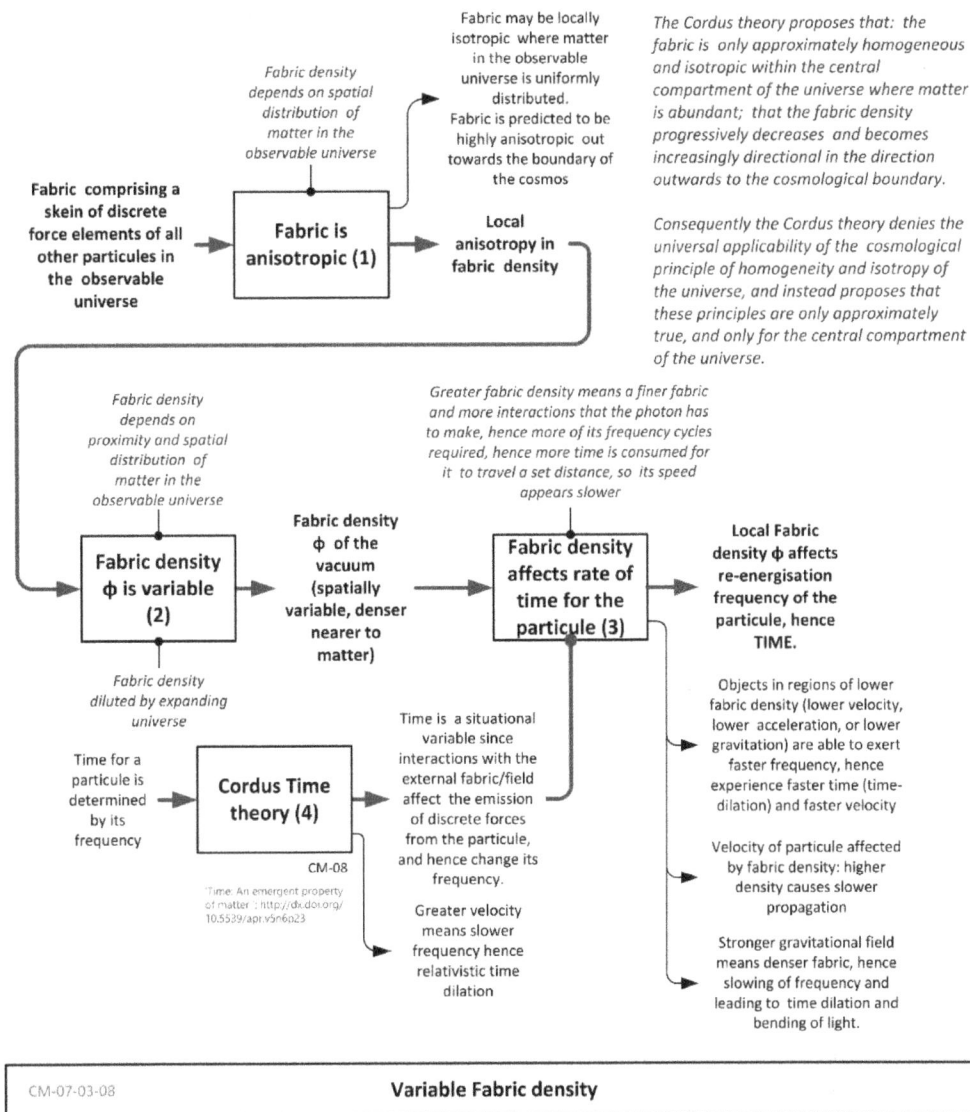

Figure 2. The fabric is proposed to have a variable density, due to the spatial distribution of matter within the universe. Consequently there are two effects to consider, the anisotropy and density of the fabric. It is the density effects that are of particularly interest here, because of the implications for the speed of light. This diagram represents the proposed relationships of causality and the logical structure of the argument, and uses the integration definition zero (IDEF0) (FIPS, 1993) systems engineering notation

4.1.3 Fabric

The overall result is that the space in and between matter particles is filled with a *fabric* of discrete forces (Pons & Pons, 2013). The fabric therefore corresponds in the first instance to the gross (rather than net) electrostatic field, more specifically to the local sum of the absolute values of the individual electric fields from negative and positive charges. Since matter in large quantities is most commonly neutral in charge, the net electrostatic field can be negligible small, but even in such situations the gross field may be large. It is therefore more helpful to associate the fabric with the gravitational field, which takes one sign only, and is more apparent at cosmological scales. In both cases the form of the dependency is the same, being $1/r^2$.

Thus in the Cordus theory the fabric comprises the moving discrete forces generated by matter particles within the observable universe (D. J. Pons & A. D. Pons, 2013). It is relevant to note that this fabric can be relativistic 'in that for an isotropic fabric the propagation speed of the photon is the same in any direction, regardless of the motion of the body emitting the photon' (D. J. Pons & A. D. Pons, 2013). Thus relativity can be recovered, at least in special inertial frames of reference where the fabric happens to be isotropic.

The fabric also distinguishes the vacuum within the universe from the void before genesis, and is the reason for the existence of electrical and magnetic constants of the vacuum. The Cordus theory further requires that the density of the fabric φ determines those constants and gives the speed of light a particular value.

4.2 Variable Fabric Density

The logical consequence of this hypothesis is that the fabric density of the vacuum is spatially variable, being determined by the proximity and local spatial distribution of matter, and in this regard is no different to the variability of EMG field strength. Hence also the fabric density is constrained by the observable universe. Thus matter that is too far away might be unable to send its discrete forces to a specific location of interest, and therefore does not contribute to the fabric density at that point. The complementary principle is that the fabric density is diluted by an expanding universe. These principles are summarised in the systems notation of Figure 2.

Furthermore, this means that the fabric will be anisotropic, since it depends on the spatial distribution of matter. It may be locally isotropic where the matter is uniformly distributed. However the theory predicts that the fabric will be highly anisotropic out towards the boundary of the cosmos (D. J. Pons & A. D. Pons, 2013).

Thus it is proposed (D. J. Pons & A. D. Pons, 2013):

(1) That the fabric is only approximately homogeneous and isotropic within the central compartment of the universe where matter is abundant;

(2) That the fabric density progressively decreases and becomes increasingly directional nearer the cosmological boundary.

Consequently the Cordus theory denies the universal applicability of the *cosmological principle of homogeneity and isotropy of the universe*, and instead proposes that these principles are only approximately true. Nonetheless, for privileged situations within the universe, such as inside galaxies or far from any galaxy, the night sky is approximately homogenous, and therefore the fabric density is also approximately isotropic and apparently constant in such situations.

Having proposed a means whereby the fabric may be inhomogeneous, we now consider the implications for the speed of light.

4.3 Dependency of Speed of Light on Fabric Density

4.3.1 Propagation of the Photon in the Fabric of the Vacuum

The Cordus theory proposes that the photon is different to other massy particules in that its discrete forces are emitted and withdrawn, as opposed to being released (Pons, 2015). The discrete forces of the photon can alternatively be considered to involve energy that is borrowed from the fabric. Consequently the photon can also be considered a type of electromagnetic transient within the fabric. This explains why many photons can share the same space: one photon imposes a distortion on the discrete forces of the fabric, and the distortions of multiple co-located photons are cumulative. This is consistent with electromagnetic (EM) wave theory, though the Cordus theory proposes a dipole arrangement rather than a point, and discrete forces rather than continuous EM fields.

Thus we propose that the photon interacts dynamically with the fabric. Each interaction requires a frequency cycle of the photon, since the photon's discrete forces are only generated when it re-energises. Hence frequency cycles of the photon are consumed with fabric interactions. This implies that denser fabric requires more frequency cycles of the photon per distance traversed (and less dense fabric requires fewer photon frequency cycles).

Complementary to this is the Cordus theory for time (D. J. Pons, A. D. Pons, & A. J. Pons, 2013c), which proposes that time at the fundamental level consists of the frequency oscillations of the particules concerned. Thus time at the fundamental level is locally generated and an emergent property of matter, rather than being universal or being a dimension of space-time. Nonetheless the space-time concept can be recovered for the macroscopic level. This is because a disturbance in one body (e.g. its movement in space) is communicated to other bodies via photons, massy particules, or discrete fields, Thus time at the macroscopic level is also universal and relative, which is consistent with the perspective of general relativity.

Consequently the Cordus theory predicts that the time taken for the photon to traverse space depends on the fabric density. This is because greater fabric density means more interactions that the photon has to make, hence more of its frequency cycles are required, hence time runs slower for the photon relative to a frame of reference with sparser fabric. Thus we propose a time dilation effect whereby the photon takes longer time to cross a unit distance when in a fabric that is denser than the Observer's frame of reference. This conclusion also relies on a lemma:

That fundamental spatial geometry is invariant, such that it is meaningful to consider a unit distance being the same whatever the fabric density.

This is consistent with the Lorentz length contraction being a kinematic effect associated with the time-dilation and the relativity of simultaneity, rather than being a static effect.

The outcome is that the Cordus theory predicts that the speed of light depends on the fabric density, which in turn depends on the proximity and spatial distribution of matter. Higher (lower) fabric density is expected to cause slower (faster) photon propagation. The system model for this is shown in Figure 3.

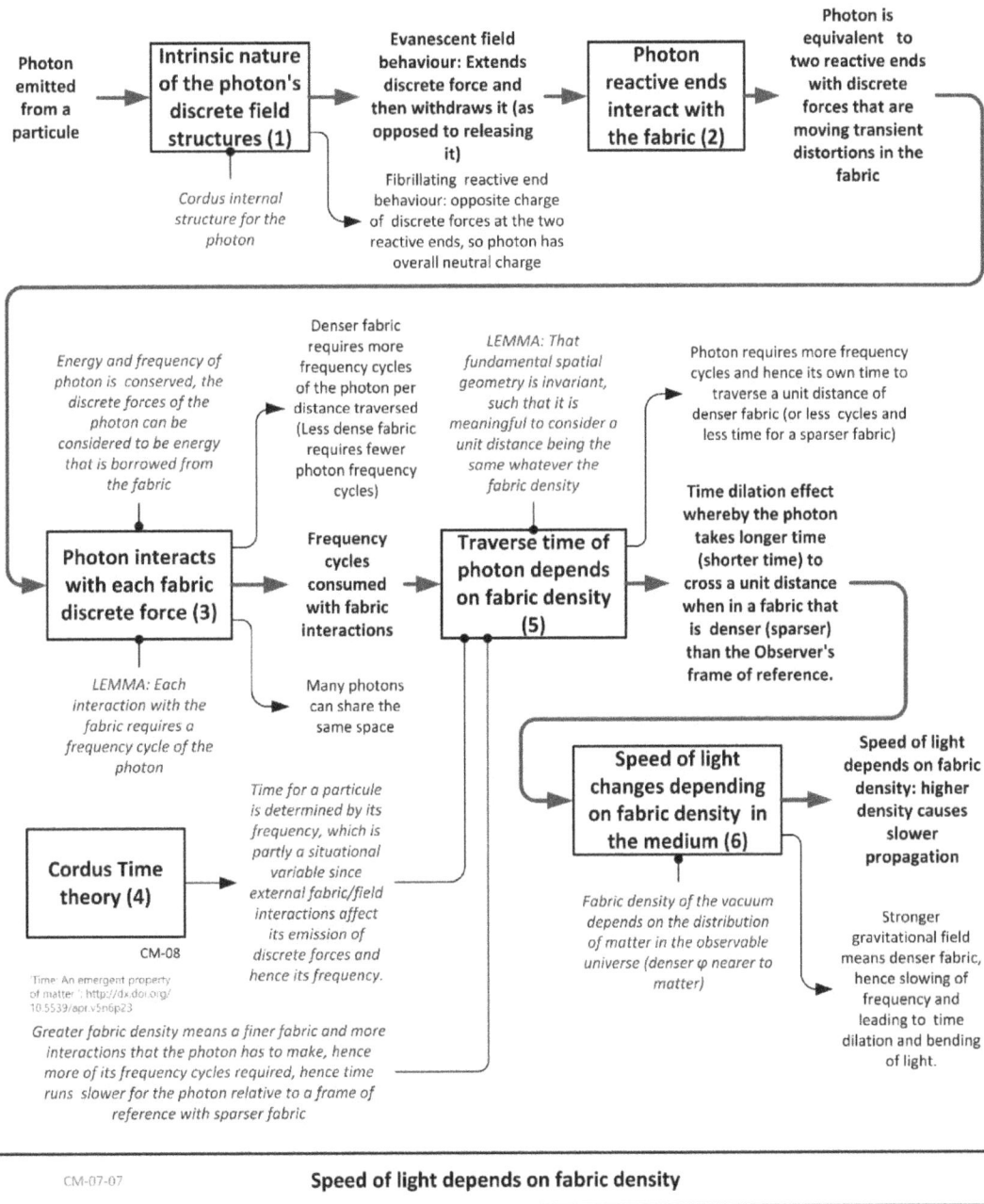

Figure 3. The Cordus theory predicts that the speed of light depends on the fabric density, which in turn depends on the proximity and spatial distribution of matter. Higher (lower) fabric density is expected to cause slower (faster) photon propagation

4.3.2 Fabric Density Transformations for Speed of Light In-Vacuo

We introduce the new term *situation* to refer to an inertial frame of reference with a specific fabric density φ. Thus in this theory the local fabric density is an important variable along with the motion of the observer. In contrast relativity is only concerned with the latter.

In the general case the particule B with velocity v_{B1} starts in situation 1 with fabric density φ_1 and frequency f_{B1}. It subsequently moves into situation 2 of different fabric density φ_2 where its velocity becomes v_{B2} and its frequency f_{B2} as measured in situation 2. The speed of light is the saturated speed of propagation of discrete forces. In turn that is affected inversely by the fabric density. Hence the speed of light in situation 2 becomes:

$$c_2 = c_1 \frac{\phi_1}{\phi_2} \text{ or } c_2 \, \phi_2 = c_1 \, \phi_1 \tag{1}$$

For example, if the Observer travelling with particule B moves into a situation φ_2 of *lower* fabric density, then the local speed of light c_2 increases, though it remains relativistic (is not affected by the velocity of the emitting particule) and is homogenous within that situation (providing there is no gradient to the fabric density). Consequently the Cordus theory is a variable speed of light (VSL) theory, in addition to being a non-local hidden-variable theory with discrete field emissions.

5. Discussion

5.1 Outcomes

This theory proposes that the speed of light is dependent on fabric density. In turn fabric density depends on the spatial distribution of matter in the local situation and the wider universe. This is a novel contribution to the VSL field, because (a) it proposes an underlying physical mechanism for the speed of light to be variable, which is otherwise exceedingly problematic, (b) it predicts a specific relationship between speed of light and fabric density, and this may be testable and falsifiable, (c) it recovers the Einsteinian assumption of the invariance of the speed of light, as a special case of isotropic fabric density, (d) it accommodates the relativity of light emissions, i.e. that the emitted speed of light is independent of the velocity of the emitting body. Thus the theory is not hostile to general and special relativity, but asserts that those theories are restricted to special situations where fabric density is constant.

5.2 Applications

5.2.1 Why does the In-Vacuo Speed of Light have the Value it Does?

The speed of light we observe is merely the value that occurs in our vicinity due to the background fabric density in this part of the universe. In turn, that is determined by the distribution of matter in the Milky Way plus contributions from the rest of the universe. Those other contributions are predicted to decrease as the universe expands, hence increasing the local speed of light as the universe evolves.

5.2.2 How does Gravitational Bending of Light Occur?

The gravitational field around a star creates a radial gradient in the fabric density. A photon with a Cordus particule structure has two reactive ends, and the discrete forces of these recruit a volume of space around each end. Where the density is higher, the recruitment process takes more time, i.e. there is a tendency for that reactive end to energise at a slower frequency. However the fibril mechanics of the particule does not permit different frequencies of the two reactive ends, as this would desynchronise the assembly and cause instability. Consequently the reactive end that experiences the greater fabric density, the one more medial to the star, recruits a smaller volume of space in the time available and thus moves forward a smaller increment. Hence the velocity of that reactive end is slower, and the locus of the particule is bent. This theory also predicts that the effect should depend on initial photon polarisation.

5.2.3 Refraction at Transparent Bodies

A body of matter, e.g. a glass prism, locally densifies the fabric density. The incident photon experiences a change in fabric density when it encounters the optical interface between the first medium (e.g. air) and the body of matter. That reactive end closest to the situation of higher fabric density will be retarded in its motion, as per the gravitational explanation above, and the locus with therefore bend. This is also consistent with the observation that transparent materials of higher density have higher refractive indices. Elsewhere the Cordus theory recovers the quantitative formalisms for reflection and refraction (Pons et al., 2012).

5.3 Implications

The outcome is that the Cordus theory rejects the concept of the invariance of the *in-vacuo* speed of light. Consequently this theory is contrary to special and general relativity. Nonetheless the theory accepts relativity as valid concept where the fabric density is sufficiently constant. Consequently the Cordus theory denies the universal applicability of the cosmological principle of homogeneity and isotropy of the universe, and instead proposes that these principles are only approximately true. The Cordus theory accepts the existing mathematical formulism of relativity and quantum mechanics as useful approximations in certain well-defined situations.

There are a number of implications of this theory. We have anticipated a mechanism which causes the speed of light to be situationally specific. This is conceptually consistent with gravitational bending of light and with relativity as a special case for situations where fabric density is constant. But what of situations where the fabric density is anisotropic? A variable speed of light has wide-ranging implications for theory development, since all of relativity is premised on its constancy. One example would be the black hole event horizon. The Cordus theory proposes that the speed of light is inversely proportional to fabric density, which in turn depends on the gravitational field strength. Consequently the speed of light slows down in a stronger gravitational field. This means that the light attempting to escape a black hole would be slowed, not stopped, thus there will be no event horizon. Instead the emitted light will be increasingly red-shifted and delayed in its escape. Furthermore the Cordus theory predicts that matter particles occupy a volume (the space between their reactive ends), and consequently there is no singularity at the centre of the black hole. These predictions have further implications for black hole cosmology. Obviously the Cordus theory is conjectural at this stage, so these implications are not definite. Nonetheless there are potentially profound implications if the theory were to be valid, and even if it were not there is still value in providing candidate new solutions to old unsolved-problems.

5.4 Limitations

While the implications may be profound, the concept still has many limitations. The work presented here is primarily of a theory-building type, and thus conceptual in nature. Consequently the mathematical formalism is not explicitly developed, and this is left for future work. It is also an exotic theory, on several counts: it proposes structure at the sub-particle level; it is non-local; it includes the concept of a fabric; it proposes that time is not a dimension but a property of matter; it is a variable speed of light theory; it denies the universal applicability of the cosmological principle of homogeneity and isotropy of the universe. It is thus appropriate to be cautious about the validity of the theory, and to consider it a candidate new physics rather than a proven fact.

5.5 Implications for Future Research

If this theory is true then the many applications in cosmology that assume invariance of *c* would need to be revisited. The Cordus theory started as a NLHV theory at the particle level, as evidenced by the multiple papers on particle related topics such as nuclear structure and decay processes, e.g. (D. J. Pons, A. D. Pons, & A. J. Pons, 2014b, 2015a, 2015b; Pons et al., 2015c; D. J. Pons, A. D. Pons, & A. J. Pons, 2015d). However the same principles have also been found to be applicable at the cosmological scale. Specifically, the theory provides candidate solutions for a genesis route from pair production (Pons et al., 2015d) through to asymmetrical genesis (Pons et al., 2014a). The asymmetrical genesis theory is a novel proposition in its own, in that it proposes a production route that has not previously been considered and which is falsifiable. It also solves both the baryogenesis and leptogenesis problems simultaneously, which is otherwise difficult to accomplish. Other aspects of the theory extend to the horizon problem (D. J. Pons & A. D. Pons, 2013) and to a novel formulation of the origins and arrow of time (Pons et al., 2013c). It is interesting that all these solutions do not conflict with either quantum mechanics or relativity. Instead those other theories become approximations of the deeper mechanics conceptualised by the Cordus theory. To the extent that the Cordus concept may be valid, it is therefore worth further developing and testing the theory for its potential to provide a deeper unification. Future research work streams could be directed at transforming the conceptual representation into mathematical formalism, which would be amenable to further investigation. There is also other conceptual work that might be undertaken, for example to explore gravitational-inertial mass equivalence, alternative explanations of dark matter, dark energy, and inflation.

6. Conclusion

The Cordus theory rejects the concept of the invariance of the *in-vacuo* speed of light, and instead proposes that the speed depends on the fabric density. The theory also provides a description of the composition of that fabric. This is in terms of an aggregation of discrete forces from many neighbouring particles to create an electro-magneto-gravitational (EMG) fabric.

There are several key concepts that differentiate this Cordus theory from other VSL theories. The first is the provision of a NLHV structure, the discrete force concept, and the idea that the fabric density at a location is variable being determined by the proximity and local spatial distribution of matter. The second is the idea that the discrete fields of the photon interact dynamically with each discrete force of the fabric and therefore consume frequency cycles of the photon. The third is that frequency cycles correspond to elapsed time, and therefore the traverse time of the photon depends on the fabric density. Thus we are proposing a time dilation effect whereby the photon takes longer time (shorter time) to cross a unit distance when in a fabric that is denser (sparser) than the Observer's frame of reference.

This work makes several original contributions. The first is to provide a variable speed of light theory with a systematic conceptual grounding. The argument starts with reasonable assumptions from a NLHV design, and builds an explanation for VSL behaviour. The originality is doing this from the fundamental level upwards. In contrast existing VSL theories are posterior constructs, which do not explain why c should vary.

A second contribution is the specific identification of fabric density as the dependent variable. In contrast, other VSL models propose that c varies with time or some geometric-like scale, and then struggle to provide naturally plausible reasons why c should depend on that variable. The Cordus theory is unique in proposing fabric density as the variable. This also means that compared to other VSL theories, the Cordus theory does not require an abrupt switch in c during cosmological epochs.

A third contribution is that this theory offers a conceptually simply way to reconcile the refraction of light in *ALL* settings, vacuum and optical materials. Optical theory already accepts that the speed of light is variable in any matter-based medium. Furthermore it is a general principle of optics that the refractive index increases with the mass density of the medium. The Cordus theory explains this as the fabric within a transparent object being denser, due to the proximity of the massy molecules, than the vacuum. Thus it is proposed that the fabric of the vacuum extends into and through all massy objects, and those objects densify it locally. Thus the Cordus theory provides a parsimonious way to reconcile the phenomenon of the speed of light varying with refractive index, and the *in-vacuo* behaviour of light.

A fourth contribution is that the Cordus theory reconciles many of the existing theories of light, despite their differences. Thus it accepts, albeit as approximations, the ideas from electromagnetic wave about light being an electromagnetic wave, that the vacuum is a medium, that the photon is a transient in the medium. It likewise accepts the quantum mechanics ideas of photons being individual items rather than waves. It also accepts, with caveats, the special relativity assumption that the speed of light is constant. The theory proposes that this is only approximately true because the mass density distribution of the Milky Way creates a uniform fabric density, hence a privileged location.

Falsifiable predictions

The Cordus theory is a conceptual work, built on a starting conjecture for the structure of matter. Falsifiable feature of the theory are:

1)   That speed of light depends on fabric density. This feature differentiates the Cordus theory from all other VSL models.

2)   Another falsifiable prediction is that the speed of light is *variant to this day.* By comparison, some of the other VSL models that are time-based propose that the speed of light is now invariant.

Author Contributions

All authors contributed to the creation of the underlying particule concept, development of the ideas, and editing of the paper. DP wrote the first draft of the paper.

Conflict of interest statement

The authors declare no funding sources external to their affiliations. The authors declare that there is no conflict of interests regarding the publication of this article. The research was conducted without personal financial benefit from any external funding body, nor did any such body influence the execution of the work or the decision to publish.

References

Abdo, A. A., Ackermann, M., Ajello, M., Asano, K., Atwood, W. B., Axelsson, M., ... & Bastieri, D. (2009). A limit on the variation of the speed of light arising from quantum gravity effects. *Nature, 462*(7271), 331-334. http://dx.doi.org/10.1038/nature08574

Albrecht, A., & Magueijo, J. (1999). Time varying speed of light as a solution to cosmological puzzles. *Physical Review D, 59*(4), 13. http://dx.doi.org/10.1103/PhysRevD.59.043516

Barrow, J. D., & Magueijo, J. (1999). Solving the flatness and quasi-flatness problems in Brans-Dicke cosmologies with a varying light speed. *Classical and Quantum Gravity, 16*(4), 1435-1454. http://dx.doi.org/10.1088/0264-9381/16/4/030

Bassett, B. A., Liberati, S., Molina-Paris, C., & Visser, M. (2000). Geometrodynamics of variable-speed-of-light cosmologies. *Physical Review D, 62*(10), 18. http://dx.doi.org/10.1103/PhysRevD.62.103518

Berman, M. S. (2011). General relativity with variable speed of light and Pioneers anomaly. *Astrophysics and Space Science, 336*(2), 327-329. http://dx.doi.org/10.1007/s10509-011-0839-y

Bradley, J. (1729). An account of the new discovered motion of the fixed stars. *Philosophical Transactions,* 637-661.

Clayton, M. A., & Moffat, J. W. (2003). Scale invariant spectrum from variable speed of light metric in a bimetric gravity theory. *Journal of Cosmology and Astroparticle Physics,* (7), 13.

de Broglie, L. (1925). Recherches sur la théorie des quanta (Researches on the quantum theory). *Annales de Physique, 3*(10). http://tel.archives-ouvertes.fr/docs/00/04/70/78/PDF/tel-00006807.pdf

Einstein, A. (1920). *Relativity: The special and general theory.* New York: Holt.

FIPS. (1993). Integration Definition for Function Modeling (IDEF0). Retrieved from http://www.itl.nist.gov/fipspubs/idef02.doc

Harko, T., & Mak, M. K. (1999). Particle creation in varying speed of light cosmological models. *Classical and Quantum Gravity, 16*(8), 2741-2752. http://dx.doi.org/10.1088/0264-9381/16/8/312

Herrmann, S., Senger A., Mohle K., Nagel M., Kovalchuk E. V., & A., P. (2009). Rotating optical cavity experiment testing Lorentz invariance at the 10^{-17} level. *Phys. Rev. D, 80*(10), 105011. http://dx.doi.org/10.1103/PhysRevD.80.105011

Magueijo, J. (2000). Covariant and locally Lorentz-invariant varying speed of light theories. *Physical Review D, 62*(10), 15. http://dx.doi.org/10.1103/PhysRevD.62.103521

Magueijo, J. (2000). Covariant and locally Lorentz-invariant varying speed of light theories. *Physical Review D, 62*(10), 103521. Retrieved from http://link.aps.org/doi/10.1103/PhysRevD.62.103521

Magueijo, J. (2001). Stars and black holes in varying speed of light theories. *Physical Review D, 63*(4), 14. http://dx.doi.org/10.1103/PhysRevD.63.043502

Magueijo, J. (2003). New varying speed of light theories. *Reports on Progress in Physics, 66*(11), 2025-2068. http://dx.doi.org/10.1088/0034-4885/66/11/r04

Magueijo, J. (2008). DSR as an explanation of cosmological structure. *Classical and Quantum Gravity, 25*(20), 8. http://dx.doi.org/10.1088/0264-9381/25/20/202002

Michelson, A. A., & Morley, E. W. (1887). On the Relative Motion of the Earth and the Luminiferous Ether. *American Journal of Science, 34*, 333-345.

Moffat, J. W. (2005). Variable speed of light cosmology and bimetric gravity: An alternative to standard inflation. *International Journal of Modern Physics A, 20*(6), 1155-1162. http://dx.doi.org/10.1142/s0217751x05024031

Pons, D. J. (2015). Inner process of Photon emission and absorption. *Applied Physics Research, 7*(4), 14-26. http://dx.doi.org/10.5539/apr.v7n4p24

Pons, D. J. (2015). Internal structure of the photon (Image licence Creative Commons Attribution 4.0). *Wikimedia Commons* (Creative Commons Attribution 4.0 International license). Retrieved from https://commons.wikimedia.org/wiki/File:Internal_structure_of_the_photon.jpg

Pons, D. J., & Pons, A. D. (2013). Outer boundary of the expanding cosmos: Discrete fields and implications for the holographic principle *The Open Astronomy Journal, 6*, 77-89. http://dx.doi.org/10.2174/1874381101306010077

Pons, D. J., Pons, A. D., & Pons, A. J. (2013a). Explanation of the Table of Nuclides: Qualitative nuclear mechanics from a NLHV design. *Applied Physics Research 5*(6), 145-174. http://dx.doi.org/10.5539/apr.v5n6p145

Pons, D. J., Pons, A. D., & Pons, A. J. (2013b). Synchronous interlocking of discrete forces: Strong force reconceptualised in a NLHV solution *Applied Physics Research, 5*(5), 107-126. http://dx.doi.org/10.5539/apr.v5n5107

Pons, D. J., Pons, A. D., & Pons, A. J. (2013c). Time: An emergent property of matter. *Applied Physics Research, 5*(6), 23-47. http://dx.doi.org/10.5539/apr.v5n6p23

Pons, D. J., Pons, A. D., & Pons, A. J. (2014a). Asymmetrical genesis by remanufacture of antielectrons. *Journal of Modern Physics, 5*, 1980-1994. http://dx.doi.org/10.4236/jmp.2014.517193

Pons, D. J., Pons, A. D., & Pons, A. J. (2014b). Differentiation of Matter and Antimatter by Hand: Internal and External Structures of the Electron and Antielectron. *Physics Essays, 27*, 26-35. http://dx.doi.org/10.4006/0836-1398-27.1.26

Pons, D. J., Pons, A. D., & Pons, A. J. (2015a). Asymmetrical neutrino induced decay of nucleons *Applied Physics Research, 7*(2), 1-13. http://dx.doi.org/10.5539/apr.v7n2p1 or http://vixra.org/abs/1412.0279

Pons, D. J., Pons, A. D., & Pons, A. J. (2015b). Hidden variable theory supports variability in decay rates of nuclides *Applied Physics Research, 7*(3), 18-29. http://dx.doi.org/10.5539/apr.v7n3p18

Pons, D. J., Pons, A. D., & Pons, A. J. (2015c). Nuclear polymer explains the stability, instability, and non-existence of nuclides. *Physics Research International, 2015*(Article ID 651361), 1-19. http://dx.doi.org/10.1155/2015/651361

Pons, D. J., Pons, A. D., & Pons, A. J. (2015d). Pair Production Explained in a Hidden Variable Theory *Journal of Nuclear and Particle Physics, 5*(3), 58-69. http://dx.doi.org/10.5923/j.jnpp.20150503.03

Pons, D. J., Pons, A. D., Pons, A. M., & Pons, A. J. (2012). Wave-particle duality: A conceptual solution from the cordus conjecture. *Physics Essays, 25*(1), 132-140. http://dx.doi.org/10.4006/0836-1398-25.1.132

Racker, J., Sisterna, P., & Vucetich, H. (2009). Thermodynamics in variable speed of light theories. *Physical Review D, 80*(8). http://dx.doi.org/10.1103/PhysRevD.80.083526

Schrödinger, E. (1926). An Undulatory Theory of the Mechanics of Atoms and Molecules. *Physical Review, 28*(6), 1049-1070. Retrieved from http://link.aps.org/doi/10.1103/PhysRev.28.1049

Shojaie, H. (2012). Variable speed of light cosmology as a solution to the Pioneer anomaly. *Canadian Journal of Physics, 90*(3), 229-234. http://dx.doi.org/10.1139/p2012-009

Singha, A. K., & Debnath, U. (2007). Varying speed of light, modified chaplygin gas and accelerating universe. *International Journal of Modern Physics D, 16*(1), 117-122. http://dx.doi.org/10.1142/s0218271807009358

Sisterna, P. D. (2011). Variable speed of light theories: A thermodynamic analysis of planetary and white dwarf phenomena. *International Journal of Modern Physics D, 20*(5), 805-820. http://dx.doi.org/10.1142/s0218271811019141

Szydlowski, M., & Krawiec, A. (2003). Dynamical system approach to cosmological models with a varying speed of light. *Physical Review D, 68*(6), 14. http://dx.doi.org/10.1103/PhysRevD.68.063511

Unzicker, A. (2009). A look at the abandoned contributions to cosmology of Dirac, Sciama, and Dicke. *Annalen Der Physik, 18*(1), 57-70. http://dx.doi.org/10.1002/andp.200810335

Youm, D. (2001). Brane world cosmologies with varying speed of light. *Physical Review D, 63*(12), 125011. Retrieved from http://link.aps.org/doi/10.1103/PhysRevD.63.125011

15

A New Kinetics Defect Diffusion Model and the Critical Current Density of Semiconductor Laser Degradation

Jack Jia-Sheng Huang[1,2] & Yu-Heng Jan[2,1]

[1]Source Photonics, 8521 Fallbrook Avenue, Suite 200, West Hills, CA 91304, USA

[2] Source Photonics, No.46, Park Avenue 2nd Rd., Science-Based Industrial Park, Hsinchu, Taiwan, R.O.C.

Correspondence: Jack Jia-Sheng Huang, Source Photonics, 8521 Fallbrook Avenue, Suite 200, West Hills, CA 91304, USA. E-mail: jack.huang@sourcephotonics.com

Abstract

Critical current density on the electromigration failure is a commonly observed phenomenon in the integrated circuit (IC) interconnect. The critical current density is important for the lognormal distribution and failure time extrapolation of IC metal conductors. In this paper, we report the critical current density (j_c) of semiconductor laser degradation for the first time. Despite of the different physical origin, the j_c of the laser degradation exhibits similar effect on the failure time distribution. We develop a new kinetic defect diffusion model that can account for the existence of j_c. We discuss the physical mechanism and its implication in the reliability extrapolation of diode lasers.

Keywords: semiconductor laser, reliability model, diffusion kinetics, critical current density, electromigration, IC interconnect

1. Introduction

Owing to the popularity of Facebook, Amazon, Netflix and Google, high-speed optical networks for datacenter, wireless and cloud computing are becoming increasingly popular. The hand-held mobile devices such as iPhone and iPad have further fueled the demand for ever-increasing internet speed and mega datacenter capacity. In order to enable social networking, web search and online purchasing, high-speed broadband transceivers have become the important building blocks in the modern communication network deployment (Basal, 2016; Bilal et al., 2013).

Reliability of diode laser is one of the most critical aspects to meet the environmental-friendly, energy-efficient and low-cost datacenters (Huang, 2015). One impact of reliability degradation is the higher power consumption related to increased bias current. The other is the waste of energy and cost to replace the failure parts. After a few decades of reliability studies on diode lasers, the optoelectronics industry has attained better understanding of degradation characteristics such as degradation behavior, failure criterion and temperature acceleration factor (Fukuda, 1988; Chin et al., 2003; Johnson, 2003; Sim, 1989; Huang, 2005; Huang et al., 2005; Lam et al., 2003; Chuang et al., 1998). On the other hand, there are still several outstanding reliability questions that remain challenging and debatable. The purpose of this study is to explore those unanswered fundamental questions that are important for product design and overall system robustness.

The first question concerns the burn-in behavior associated with rapid increase in the threshold current (I_{th}) during the first burn-in cycle, followed by a very small change in I_{th} during the second burn-in cycle. Engineers and scientisits often debate about whether the lasers are acceptable for production due to the concern of the large initial I_{th} changes. Some suggest to accept the diode lasers based on the nature of stable second burn-in, while others express concern about the large initial I_{th} increases. The physical mechanism has been largely unclear thus far. There was no model or theory developed in the past to explain why the lasers exhibited the large I_{th} increases, and suddenly became stable afterwards.

The second question is about the degradation behavior at low stress current regime. The laser diodes often exhibit little I_{th} changes when subjected to low stress conditions. Furthermore, the degradation at the low stress condition is usually slower than expected. One could observe a precipitously reduced degradation rate when the stress current is decreased to a certain level. There was no previous model established to account for such dramatic change in the degradation behavior for the low stress current.

The third question pertains the failure time distributions of the lasers subjected to the low stress current aging. Typically, the failure times for the low stress current are longer than those for the high stress current. At the meantime, the sigma of the failure times for the low stress current also tends to increase. The large sigma or dispersion in the failure distribution often results in a net increase in the failure rate, which makes the interpreation of reliability more complex. It has been unclear why the failure distributions of the lasers for the low stress current deviate from the normal stress current. For the latter, the failure distribution shows lognormal distribution. For the former, the failure distribution shows strong modification from lognormal.

2. Historical Background: Critical Current Density of IC Interconnect Electromigration

Electromigration is one of the key factors responsible for long-term interconnect wear-out. Under the high electric field, there is an electron wind force that imposes on the metal atoms to result in a net mass transport. When electromigration occurs, there is void formation at the cathode and extrusion at the anode (Ho et al., 1989; Tu et al., 1992). The electrical characteristics of electromigration degradation is typically associated with resistance increase (Huang et al., 2001; Hu et al., 1995). Failure time of interconnect is commonly defined as the time to reach 20% increase in resistance ($\Delta R = 20\%$).

Forty years ago in 1976, Blech and Herring reported a critical current density of electromigration (Blech, 1976; Blech et al., 1976). It was found that no electromigration occurred below a certain length due to the back stress generated in the metal stripe. For modern IC interconnect, the characterization of the critical current density for electromigration is important for reliability model development. The effect of critical current density on electromigration has been experimentally observed in Al, Al(Cu) and Cu dual-damascene vias (Thompson et al., 1993; Proost et al., 2000; Huang et al., 2000; Lee et al., 1995). Researchers have also shown the influence of j_c on the failure time distributions. Oates (Oates, 1996) and Huang (Huang et al., 2000) showed the deviations of failure distribution in the Al(Cu) interconnects as j approached j_c. Recently, Oates et al. (Oates et al., 2009) demonstrated similar modification of failure time distributions from lognormal in the Cu dual-damascene as the current densities approached j_c.

3. Experimental

Figures 1(a) and (b) shows the cross-sectional schematics of the ridge waveguide (RWG) and buried heterostructure (BH) laser structures, respectively. Both structures were grown by the metal organic chemical vapor deposition (MOCVD) technique. The single-mode distributed feedback (DFB) was defined by holographic grating. For the RWG structure in Figure 1a, one step regrowth after grating was performed to form the p-InP cladding and p^+-InGaAs contact. The active region consisted of InGaAlAs multi-quantum well (MQW) materials. For the BH structure in Figure 1b, three regrowth steps including grating overgrowth, blocking regrowth and final regrowth were performed to complete the epitaxial layers. The first regrowth was performed after the grating etch; the second regrowth was performed after the mesa etch to form the active region; the final regrowth consisted of the p-InP cladding and p^+-InGaAs contact.

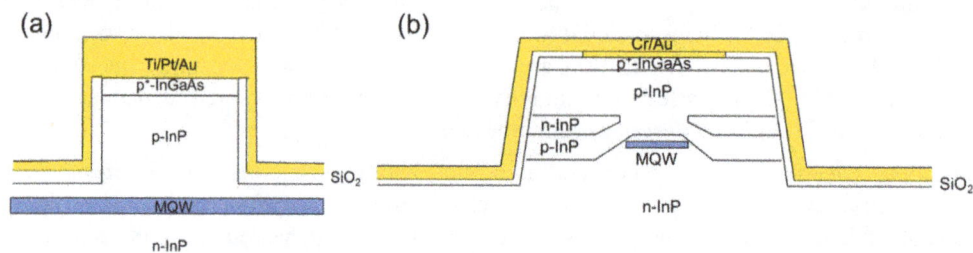

Figure 1. Schematics of the fabricated (a) ridge waveguide and (b) buried heterostructure lasers.

After the MOCVD regrowth, the wafers were processed with multiple lithography steps. The first lithography involved the ridge definition using the hard mask. The ridge structures (for both RWG and BH) were covered with the SiO$_2$ passivation for electrical isolation. The SiO$_2$ dielectric layer was then etched before the deposition of p-metal of Ti/Pt/Au to make ohmic contact. The wafer backside was thinned and deposited with n-metal of AuGe/TiAu.

To study the reliability, the laser diodes were stressed with different levels of currents to establish statistical data. The light versus current (LI) was tested before and after each reliability test interval. Due to the sample availability, RWG lasers of 1270nm lasing wavelength and BH lasers of 1550nm were used for the reliability study in this paper.

4. Results and Discussions

4.1 Kinetics model of defect diffusion

The laser degradation was related to defect formation (Endo et al., 1982; Fukuda, 1991; Hutchinson et al., 1975; Huang, 2012; Kallstenius, 1999). To study the degradation behavior, it is critical to understand the evolution of defect density during the aging process. Our previous model (Huang et al., 2016) assumed that the defect was formed at the surface damaged region and the defect diffusion involved propagation towards the active region. The defect concentration followed the Gauss error function where the surface concentration was the highest at the surface and decreased along the propagation direction.

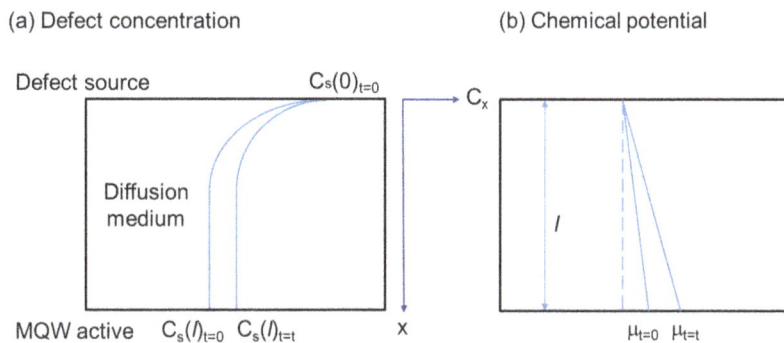

Figure 2. (a) Defect concentration and (b) chemcial potential at the active region before (time=0) and after aging (time=t). The defect souce could originate from the p-contact, BH interface or substrate dislocation network

In this paper, we expand the model to account for the kinetics of the defect diffusion process in general case, as illustrated in Figure 2a. First, the model assumes that the laser degradation is related to the diffusion from the defect source to the MQW active region. The defect source could be from the p-contact (Huang, 2012; Huang et al., 2016), the BH interface (Fukuda et al., 1987; Mawatari et al., 1997) or the substrate dislocation network (Petroff et al., 1976; O'Hara et al., 1977). Second, the model includes the kinetics of defect diffusion that is driven by the electric current and field. The flux of the defect from the source into the active region can be expressed as

$$J = C \frac{D}{kT}(Q^* E) \qquad \text{Equation (1)}$$

where C is the atomic defect concentration, D is the diffusivity of the defect, k is the Boltzmann constant, T is the temperature, Q^* is the effective defect charge and E is the electric field. From electromagnetics theory, the electric field can be related to the current density (Reitz et al., 1979). Thus, Equation (1) can be rewritten as

$$J = C \frac{D}{kT} Q^* \rho j \qquad \text{Equation (2)}$$

where ρ is the resistivity and j is the current density.

In the next step, we attempt to develop the mathematical model to account for the experimental observation that the degradation tends to become slower over time. Assuming that the distance from the defect source to the active region is l, the defect concentration at the source and the active region are denoted as $C_s(0)$ and $C_s(l)$, respectively. In the beginning (t=0), the defect concentration $C_s(l)_{t=0}$ follows the Gauss error curve. After the initial aging at time of t, the threshold current is increased, so the defect concentration at the active region can be expressed as

$$C_s(l)_{t=t} = C_s(l)_{t=0} \left(\frac{I_{th}(t)}{I_{th}(0)} \right) \qquad \text{Equation (3)}$$

When the defect concentration builds up at the active region, the chemical potential of the defect increases, as illustrated in Figure 2b. The chemical potential is defined by the thermodynamics terms as the increase in the Gibbs free energy of the system per mole (Gaskell, 1981). We speculate that there is a force of "backflow" due to the chemical potential, similar to the back stress effect in the case of electromigration. The backflow can be formulated by the second term in Equation (4)

$$J = C\frac{D}{kT}(Q*\rho j - \frac{\partial\mu}{\partial x})$$ Equation (4)

where is the chemical potential of the defect, x is the axis along the defect propagation direction. When the backflow and the electrical force reach the equilibrium, the two forces cancel each other. At the equilibrium state, there is no net flux of defect diffusion. Hence, the I_{th} value remains the same with no more increase.

$$Q*\rho j_c = \frac{\Delta\mu}{l}$$ Equation (5)

where j_c is the critical current density where no laser degradation occurs, and $\Delta\mu$ is the maximum chemical potential hat the defect can build up locally. The flux of defect can be rewritten in terms of critical current density

$$J = C\frac{D}{kT}Q*\rho(j - j_c)$$ Equation (6)

The flux of defect can be simplified as the drift velocity according to the relation of $v_d = J/C$.

$$v_d = \frac{D}{kT}Q*\rho(j - j_c)$$ Equation (7)

The failure time is proportional to the time to accumulate enough defects in the active region to cause the failure. The failure time can be expressed as

$$t_f = \frac{V_c}{Av_d}$$ Equation (8)

where V_c is the volume of defect needed to cause failure, and A is the cross-sectional area of the laser stripe.

$$t_f = \frac{(V_c/A)kT}{DQ*\rho(j - j_c)}$$ Equation (9)

Since the width of the active region is typically pre-determined by the process to match the performance design target, the current density for the laser diode is specified as the current density (j) times the active width (w) or the applied bias current (I) divided by the cavity length (L).

$$t_f = \frac{(V_c/L)kT}{DQ*\rho(j - j_c)w}$$ Equation (10)

4.2 Experimental Data

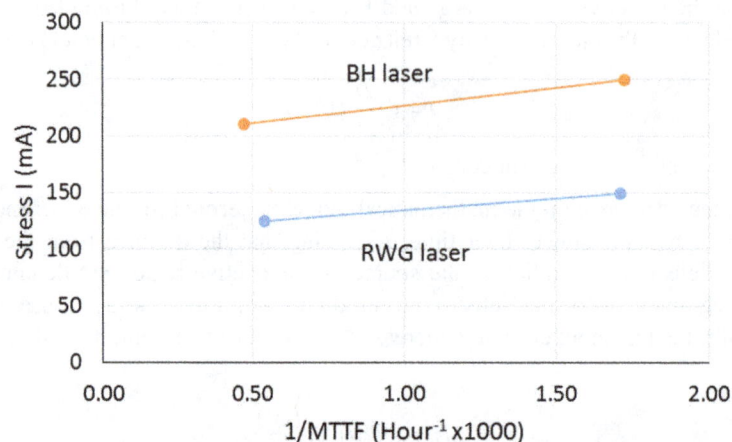

Figure 3. The plot of stress current versus reciprocal of failure time

For the semiconductor lasers, the stress current is typically specified based on the device operating condition. Hence, the stress bias is usually specified by the current as well. For example, the uncooled lasers typically operate at low bias around 30-60mA (Klotzkin et al., 2007; X. Ge et al., 2012; Lu et al., 2008), the analog cooled lasers tend to operate around 60-100mA (Huang et al., 2014), while the high power lasers typically operate at high bias around 350-600mA (Huang et al., 2008; Huang et al., 2012). Figure 3 shows the stress current plotted against the reciprocal of mean time-to-failure (MTTF). The MTTF is taken from the median value of the failure time distributions. For this study, the RWG lasers of 300μm were stressed with the current of 125-150mA, and the BH lasers of 550μm were stressed with 210-250mA. The stress current for the BH was higher than that of the RWG to accommodate the longer cavity length.

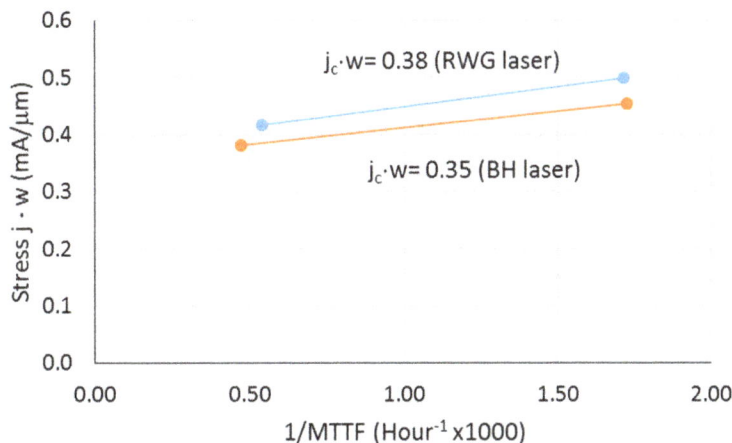

Figure 4. The plot of stress current density times active width versus reciprocal of failure time. The j_c is extracted from the intercept where 1/MTTF equals to 0

Next, we examine the relationship of stress current density times the active width (j·w) versus the reciprocal of MTTF to compare it with the model in Equation (10). By rearrangement of the variables, Equation (10) can be rewritten as Equations (11) and (12) where j·w is in the vertical axis and 1/MTTF is in the horizontal axis. Figure 4 shows the j·w plotted against 1/MTTF to determine the intercept. Several interesting features were observed. First, the MTTF between the RWG and BH was close to each other when normalized to the stress current density. Secondly, there existed a "critical current density" from the experimental data. Finally, the critical current density $j_c·w$, extracted from the intercept where 1/MTTF equaled to zero, was similar between the RWG and BH lasers. The values of $j_c·w$ for RWG and BH lasers were determined to be 0.38 and 0.35mA/μm, respectively. Using the active width of 1.8μm, the corresponding j_c values for RWG and BH were 0.21 and 0.19mA/μm², respectively.

$$(j - j_c)w = \frac{(V_c / L)kT}{DQ^* \rho}(\frac{1}{t_f})$$

Equation (11)

$$jw = j_c w + \frac{(V_c / L)kT}{DQ^* \rho}(\frac{1}{t_f})$$

Equation (12)

The existence of j_c implies that there is a stress current density threshold below which no degradation would occur. In the following, we attempt to explain the fundamental unanswered questions about the observed degradation phenomena mentioned in the Introduction section.

The first myth concerned the burn-in behavior associated with rapid increase in the threshold current (I_{th}) during the first burn-in cycle, followed by a very small change in I_{th} during the subsequent burn-in cycles (see Type-A in Figure 5). The subsequent smaller changes can be attributed to the backflow effect. After the large initial I_{th} increase occurred, the defect concentration at the active region increased. As a result, the chemical potential gradient ($\partial\mu/\partial x$) that could oppose the diffusion from the defect source became higher. According to Equation (4), the net flux or drift velocity of the defect diffusion became lower. In constrast, some lasers exhibited smaller initial I_{th} increase followed by gradual I_{th} stabilization during the subsequent burn-in, as illustrated in Type-B in Figure 5. The smaller I_{th} increase led to lower chemical potential that would allow continued defect diffusion due to the net flux. The I_{th} increase would follow gradual pattern until the chemical potential (the second term in Equation 4) eventually became large enough to offset the electric current force (the first term in Equation 4).

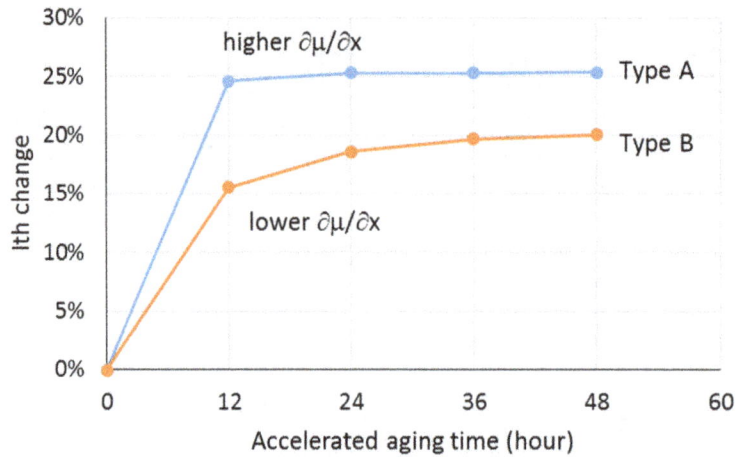

Figure 5. The relative I_{th} change as a function of accelerated aging time for Type-A and Type-B. Type-A is characterized with larger initial I_{th} increase followed by full stabilization. Type-B is featured with smaller initial I_{th} increase followed by gradual stabilization

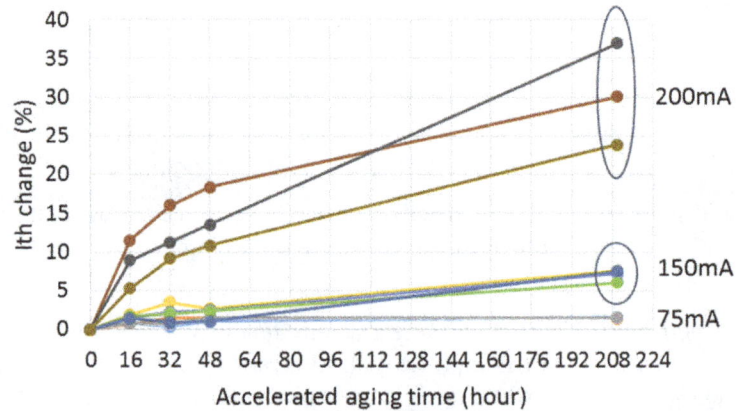

Figure 6. The relative I_{th} change of laser devices as a function of stress current. For the low stress current (75mA), the degradation was very little due to small j-j_c. For the medium stress current (150mA), the degradation was observable as j moved away from j_c. For the high stress current (200mA), the degradation became pronounced due to large j-j_c

The second myth was about the degradation behavior at low stress current regime. The laser diodes often exhibited little I_{th} changes when subjected to low stress conditions. Furthermore, the degradation at the low stress condition was usually slower than expected. According to Equation 11, 1/MTTF became infinitely small as j approached j_c. In other word, the MTTF became infinitely high as j approached j_c. The effect of j_c could account for the experimentally observed slow degradation for the low stress regime. Figure 6 illustrates the effect of $(j$-$j_c)$ from the same type of devices that were subjected to different levels of stress currents. For the low stress current (75mA in this case), the degradation curve was very small due to the small value of j-j_c. When extrapolated to the end-of-lifetime, the failure time would be very large that would lead to a very small value of 1/MTTF. For the medium stress current (150mA), the degradation curves were more observable as j moved away from j_c. The reliability data based on the medium stress current regime was typically suitable for the determination of j_c. For the high stress current (200mA), the degradation was pronounced due to the large j with reference to j_c. In this regime, the degradation rate was likely larger than expected due to Joule heating. Care needed to be taken when using the degradation data of the high stress current regime to estimate the device lifetime.

The third myth pertained the failure time distributions of the lasers subjected to the low stress current aging. As shown in Figure 7, the failure times for the low stress current were typically longer than those for the high stress current. At the meantime, the sigma of the failure times for the low stress current also tended to increase. The deviation of the low failure distribution of the low stress regime from lognormal could be attributed to the effect of j_c. According to Equation 10, t_f was inversely proportional to j-j_c. The Monte-Carlo simulation would result in the strong modification of failure distributions as j approaches j_c (Huang & Oates, 2000). This was likely due to the smaller value of $(j-j_c)$ in the denominator.

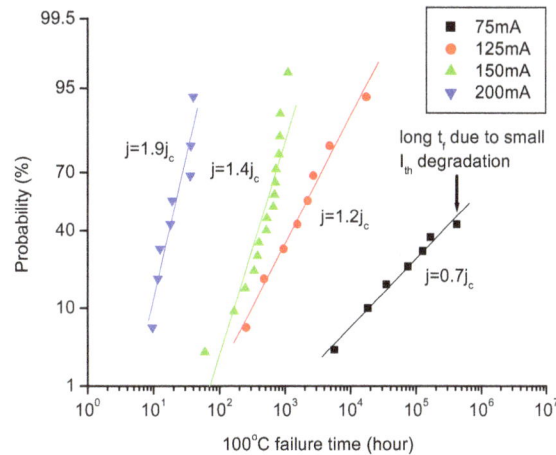

Figure 7. Failure time distributions of laser devices that were subjected to different stress current density with respect to j_c. The symbols are taken from experimental data, and the lines are from regression fitting

5. Conclusion

We developed the new kinetics model of defect diffusion processes for the RWG and BH semiconductor lasers. The model was based on the kinetics process of the defect diffusion from the source to the active region. The defect source could be from the p-contact, BH interface or substrate dislocation. The model introduced the chemical potential of the defect at the active as the "backflow" force to oppose the defect diffusion driven by the electric current. The model suggested that there existed a critical current density (j_c) below which no degradation would occur. The j_c represented the back flow effect due to the chemical potential of the defect at the active region. Experimental data showed that $j_c \cdot w$ was close to $0.36 mA/\mu m$, corresponding to the j_c estimation value of $0.20 mA/\mu m^2$.

The effect of j_c seemed to provide good explanation for several unanswered questions: (1) The larger the initial I_{th} increase was, the smaller the subsequent I_{th} change would occur due to the higher chemical potential of the defect at the active region. (2) The degradation was usually slower than expected for the low stress current due to the "backflow" effect of j_c. (3) The deviation of failure distribution from lognormal for the low stress current due to the effect of j_c. Despite of the different physical origin, the j_c of the laser degradation exhibited similar effect on the failure time distribution compared to electromigration in the IC interconnect.

Acknowledgment

The authors would like to thank Jesse Chang for the assistance in test data collection, Emin Chou for support of this work and wafer fab team in Source Photonics, Taiwan for chip processing and testing.

References:

Basal, J. (2016). Source Photonics announces shipment of 10,000 single mode 100G QSFP28 modules. *Optical Fiber Communications Conference* (OFC, Anaheim, CA).

Bilal, K., Khan, S. U., & Zomaya, A. Y. (2013). Green Data Center Networks: Challenges and Opportunities. In 11th *IEEE International Conference on Frontiers of Information Technology (FIT)*, Islamabad, Pakistan, 229-234. http://dx.doi.org/10.1109/fit.2013.49

Blech, I. A. (1976). Electromigration in thin aluminum films on titanium nitride, *J. Appl. Phys., 47*(4), 1203-1208. http://dx.doi.org/10.1063/1.322842

Blech, I. A., & Herring, C. (1976). Stress generation by electromigration, *Appl. Phys. Lett., 29*(3), 131-133. http://dx.doi.org/10.1063/1.89024

Chin, A. K., Zipfel, C. L., Geva, M., Gamlibel, I., Skeath, D., & Chin, B. K. (1984). Direct evidence for the role of gold migration in the formation of dark-spot defects in 1.3μm InP/InGaAsP light emitting diodes. *Appl. Phys. Lett., 45*(1), 37–39. http://dx.doi.org/10.1063/1.94995

Chuang, S. L., Nakayama, N., Ishibashi, A., Taniguchi, S., & Nakano, K. (1998). Degradation of II-VI blue-green semiconductor lasers. *IEEE J. Quantum Electron., 34*(5), 851–857. http://dx.doi.org/10.1109/3.668773

Endo, K., Matsumoto, S., Kawano, H., Sakuma, I., & Kamejima, T. (1982). Rapid degradation of InGaAsP/InP double heterostructure lasers due to <110> dark line defect formation. *Appl. Phys. Lett., 40*(11), 921–923. http://dx.doi.org/10.1063/1.92979

Fukuda, M. (1988). Laser and LED reliability update. *J. Lightw. Technol., 6*(10), 1988, 1488–1495. http://dx.doi.org/10.1109/50.7906

Fukuda, M. (1991). *Reliability and Degradation of Semiconductor Lasers and LEDs* (Chapter 4). Norwood, MA, USA, Artech House.

Fukuda, M., Noguchi, Y., Motosugi, G., Nakano, Y., Tsuzuki, N., & Fujita, O. (1987). Suppression of interface degradation in InGaAsP/InP buried heterostructure lasers, *J. Lightw. Technol., LT-5*(12), 1778-1781. http://dx.doi.org/10.1109/JLT.1987.1075475

Gaskell, D. R. (1981). *Introduction to metallurgical thermodynamics* (Chapter 5). New York, NY, USA, McGraw-Hill.

Ge, X., Moat, P., Xie, J., Hu, J., Huang, J. S., Su, X., … Klotzkin, D. (2012). Thermal conductivity of 1.3μm InAs/GaAs quantum dot laser active material from chirp and 3ω measurements. *Appl. Phys Lett., 100*(8), 082108. http://dx.doi.org/10.1063/1.3687160

Ho, P. S., & Kwok, T. (1989). Electromigration in metals. *Rep. Prog. Phys., 52*, 301–348. http://dx.doi.org/10.1088/0034-4885/52/3/002

Hu, C. K., Luther, B., Kaufman, F. B., Hummel, J., Uzoh, C., & Person, D. J. (1995). Copper interconnection integration and reliability, *Thin Solid Films, 262*(1-2), 84-92. http://dx.doi.org/10.1016/0040-6090(94)05807-5

Huang, J. S. (2005). Temperature and current dependences of reliability degradation of buried heterostructure semiconductor lasers. *IEEE Trans. Device Mater. Reliab., 5*(1), 150-154. http://dx.doi.org/10.1109/TDMR.2005.843834

Huang, J. S. (2012). Design-in reliability for modern wavelength-division multiplex (WDM) distributed feedback (DFB) InP lasers. *Appl. Phys. Res., 4*(2), 15-28. http://dx.doi.org/10.5539/apr.v4n2p15

Huang, J. S. (2015). *Reliability of optoelectronics* (Chapter 6, J. Swingler (Ed.)). Cambridge, UK, Woodhead Publishing. http://dx.doi.org/10.1016/b978-1-78242-221-1.00006-x

Huang, J. S., Oates, A. S., Obeng, Y. S., & Brown, W. L. (2000). Asymmetrical critical current density and its influence on electromigration of two-level W-plug interconnection, *J. Electrochem. Soc., 147*(10), 3840-3844. http://dx.doi.org/10.1149/ 1.1393982

Huang, J. S., & Oates, A. S. (2000). Monte-Carlo simulation of electromigration failure distribution of submicron contacts and vias: a new extrapolation methodology for reliability estimate. *Proc. International Interconnect Technology Conference (IITC)*, 208-210. http://dx.doi.org/10.1109/iitc.2000.854327

Huang, J. S., He, X., Blauvelt, H., Chin, H., Lomeli, M., & Zendejaz, R. (2014). Cost-effective O-band high-power, low distortion CWDM analog lasers, *IEEE Photonics Conference* (San Diego, CA, USA), 461-462.

Huang, J. S., Isip, E., & Carson, R. F. (2012). High-resolution wavelength stability aging study of C-band 100mW high-power DWDM laser modules, *IEEE Photonics Conference* (Burlingame, CA, USA), 538-539.

Huang, J. S., Jan, Y. H., Ren, D., Hsu, Y., Sung, P., & Chou, E. (2016). Defect diffusion model of InGaAs/InP semiconductor laser degradation, *Appl. Phys. Research, 8*(1), 149-157. http://dx.doi.org/10.5539/apr.v8n1p149

Huang, J. S., Nguyen, T., Hsin, W., Aeby, I., Ceballo, R., & Krogen, J. (2005). Reliability of Etched-Mesa Buried-Heterostructure Semiconductor Lasers. *IEEE Trans. Device Mater. Reliab., 5*(4), 665-674. http://dx.doi.org/10.1109/TDMR.2005.860562

Huang, J. S., Shofner, T. L., & Zhao, J. (2001). Direct observation of void morphology in step-like electromigration resistance behavior and its correlation with critical current density. *J. Appl. Phys., 89*(4), 2130-2133. http://dx.doi.org/10.1063/1.1340004

Huang, J.S., Lu, H., & Su, H. (2008). Ultra-high power, low RIN and narrow linewidth lasers for 1550nm DWDM 100km long-haul fiber optic link, *LEOS Conference* (Newport Beach, CA, USA), 894-895. http://dx.doi.org/10.1109/leos.2008.4688910

Hutchinson, P. W., & Dobson, P. S. (1975). Defect structure of degraded GaAlAs-GaAs double heterojunction lasers. *Phil. Mag., 32*, 745–754. http://dx.doi.org/10.1080/14786437508221617

Johnson, L. A. (2003). Reliability counts for laser diodes. *Photonics, 37*(7), 55-56.

Kallstenius, T., Bäckström, J., Smith, U., & Stoltz, B. (1999). On the degradation of InGaAsP/InP-based bulk lasers. *J. Lightw. Technol., 17*(12), 2584. http://dx.doi.org/10.1109/50.809681

Klotzkin, D., Huang, J. S., Lu, H., Nguyen, T., Pinnington, T., Rajasekarem, R., & Tsai, C. (2007). An optimized structure for InGaAlAs spot size converted lasers for direct fiber coupling fabricated without overgrowth, *IEEE Photonics Tech. Lett., 19*(13), 975-977. http://dx.doi.org/10.1109/LPT.2007.898824

Lam, S. K. K., Mallard, R. E., & Cassidy, D. T. (2003). Analytical model for saturable aging in semiconductor lasers. *J. Appl. Phys., 94*(3), 1803–1809. http://dx.doi.org/10.1063/1.1589594

Lee, K. L., Hu C. K., & Tu, K. N. (1995). *In-situ* scanning electron microscope comparison studies on electromigration of Cu and Cu(Sn) alloys for advanced chip interconnects, *J. Appl. Phys., 78*, 4428-4437. http://dx.doi.org/10.1063/1.359851

Lu, H., Huang, J. S., & Su, H. (2008). Ultra-high speed 1.3μm complex-coupled DFB lasers for future uncooled 10Gb/s GPON, *IEEE Int'l Semiconductor Laser Conference (ISLC)* (Sorrento, Italy), 15-16.

Mawatari, H., Fukuda, M., Matsumoto, S., Kishi, K., & Itaya, Y. (1997). Reliability and degradation behaviors of semi-insulating Fe-doped InP buried heterostructure lasers fabricated by RIE and MOVPE. *J. Lightw. Technol., 15*(3), 534-537. http://dx.doi.org/10.1109/50.557570

O'Hara, S., Hutchinson, P. W., & Dobson, P. S. (1977). The origin of dislocation climb during laser operations. *Appl. Phys. Lett., 30*(8), 368-371. http://dx.doi.org/10.1063/1.89432

Oates, A. S. & Lin, M. H. (2009). Electromigration failure distributions of Cu/low-k dual-damascene vias: impact of the critical current density and a new reliability extrapolation methodology. *IEEE Trans. Dev. Mater. Reliab., 9*(2), 244-254. http://dx.doi.org/10.1109/TDMR.2009.2015767

Oates, A. S. (1996). Electromigration failure of contacts and vias in sub-micron integrated circuit metallizations, *Microelectron. Reliab., 36*(7/8), 925-953. http://dx.doi.org/10.1016/0026-2714(96)00102-3

Petroff, P. M., & Kimerling, L. C. (1976). Dislocation climb model in compound semiconductors with zinc blende structure. *Appl. Phys. Lett., 29*(8), 461-463. http://dx.doi.org/10.1063/1.89145

Proost, J., Maex, K., & Delaey, L. (2000). Electromigration-induced drift in damascene and plasma-etched Al(Cu). II. Mass transport mechanisms in bamboo interconnects. *J. Appl. Phys., 87*, 99–109. http://dx.doi.org/10.1063/1.372389

Reitz, J. R., Milford, F. J., & Christy, R. W. (1979). *Foundations of electromagnetic theory*. Reading, MA, USA, Addison-Wesley Publishing.

Sim, S. P. (1989). A review of the reliability of III-V opto-electronic components, in *Proc. Semiconductor Device Reliability: Advanced Workshop II, NATO International Scientific Exchange Program*, Crete, Greece, 301.

Thompson, C. V., & Lloyd, J. R. (1993). Electromigration and IC interconnects, *MRS Bulletin*, 19-25.

Tu, K. N., Mayer, J. W., & Feldman, L. C. (1992). *Electronic thin film science for electrical engineers and materials scientists* (Chapter 14). New York, NY, McMillan Publishing.

16

Predictive Reliability Model of 10G/25G Mesa-Type Avalanche Photodiode Degradation

Jack Jia-Sheng Huang[1,2], Yu-Heng Jan[2,1], H. S. Chen[2], H. S. Chang[2], C. J. Ni[2] & Emin Chou[2]

[1] Source Photonics, 8521 Fallbrook Avenue, Suite 200, West Hills, CA 91304, USA

[2] Source Photonics, No.46, Park Avenue 2nd Rd., Science-Based Industrial Park, Hsinchu, Taiwan, R.O.C.

Correspondence: Jack Jia-Sheng Huang, Source Photonics, 8521 Fallbrook Avenue, Suite 200, West Hills, CA 91304, USA. E-mail: jack.huang@sourcephotonics.com

Abstract

Avalanche photodiodes (APDs) are important building blocks for high-sensivity, low-noise receivers deployed in the datacenter, wireless and cloud computing networks. Maintaining stable dark current is a crucial task for overall robust sysem reliability. To achieve design-in low dark current stability, good knowledge of reliability physics is indispensable. In this work, we study the physical mechanisms of 10G/25G mesa-type APD degradation. We institute a predictive reliability model to account for the degradation processes. A comprehensive comparison of APD and IC transistor is also illustrated in terms of dielectric breakdown, mobile ion migration and hot carrier injection. The model suggests that surface leakage current is the dominant factor for the mesa-type APD degradation. Based on the model, it is predicted that highly reliable 10G/25G APD can be achieved with the suppression of weak links at the surface/interface states.

Keywords: semiconductor, reliability model, avalanche photodiode, reliability physics, dielectric breakdown, mobile ion, hot carrier injection

1. Introduction

Semiconductor photodiodes are important building blocks for the high-sensitivity receivers deployed in the 2.5 and 10G optical communication systems such as passive optical network (PON) and local area networks (LAN) (Ishimura et al., 2007). The recent development of datacenter networking, wireless and cloud computing have further fueled the demand for the high-performance receivers.

Among the photodiode portfolio, APDs are attractive devices due to the signficiant improvement in photoreceiver sensivity compared with traditional p-i-n (PIN) photodiodes (Achouche et al., 2010). By adding the multiplication layer, the avalanche photodiodes combine the detection and amplification properties simultaneously. However, the high electric field in the avalanche region often imposes a reliability concern.

Recently, 10G/25G APD has drawn increasing interest in the market place due to the high-bandwidth and low-noise performance advantages. In order to achieve the high bandwidth, a mesa structure with both P and N electrodes on the top surface is typically employed (Kim et al., 2001; Takeshita et al., 2006; Watanabe et al., 1996; Smith et al., 2009). One of the key challenges though is to maintain low dark current over time. Compared to semiconductor lasers (Huang et al., 2016; Twu et al., 1993; DeChiaro et al., 1992), there was relatively scarce amount of reliability work conducted on the APD (Ishimura et al., 2007; Kim et al., 2001; Watanabe et al., 1996). In this paper, we study the physical degradation mechanisms of 10G/25G mesa-type APD. We present the predictive reliability model for the degradation processes and suggest that the surface leakage current is dominant for the mesa-type APD. The model predicts that highly reliable 10G/25G APD devices can be achieved with the suppression of the weak links.

2. Experimental

The reliability model was based on the schematic of the 10G/25G mesa-type APD structure shown in Figure 1. The APD structures considered in the model included the N-mesa at the bottom, the active region in the middle and the P-mesa at the top. The N-mesa consisted of the N-InP buffer and contact layers. The active region consisted of the InAlAs multiplication, the graded quaternary (Q) layer and the absorption layer. The P-mesa consisted of the p-InAlAs window and p-InGaAs contact layers. To achieve the 10G high-speed, the mesa-type

APD was formed where both P and N electrodes were on the top surface. The p-contact was made by forming the metal ring immediately outside the anti-reflective (AR) window. The p-ring was connected to the outside p-pad by the metal bridge. The n-metal contact was connected to the N-mesa. For the passivation, the model considered the low-k dielectric material such as polyimide.

Since the mesa-type APD typically exhibited higher degradation rate than the planar type (Kim et al., 2001; Watanabe et al, 1996), the model focused on the surface leakage at the semiconductor/dielectric interface. The reliability model discussed the effects of dielectric breakdown, mobile ion migration and hot carrier injection.

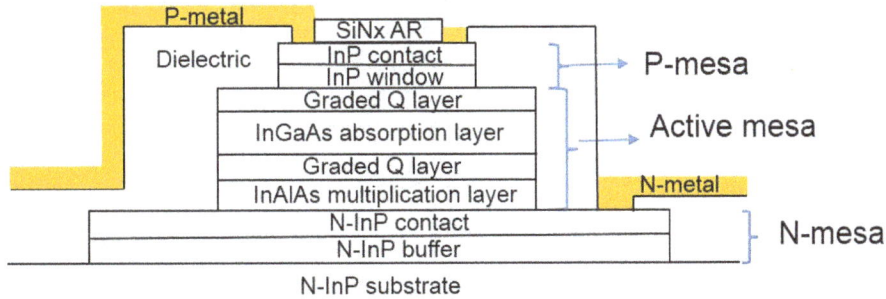

Figure 1. Schematic of 10G/25G mesa-type APD structure used in the reliability model

3. Results and Discussions

The typical APD degradation for long-term wear-out is characterized with the increase in dark current over time. The dark current variation, $[I_d(t)-I_d(0)]/I_d(0)$, can be expressed in the following empirical terms shown in Equation (1). The first term accounts for the stress current dependence where the K_I is the normalization constant, I_r is the reverse current applied to the APD during aging and N is the current exponent for the acceleration factor. The second term relates to the stress temperature dependence where E_a is the activation energy, k is the Boltzmann's constant and T is the temperature. The third term describes the aging curve where a is the fitting constant that determines the failure time, t is the aging time and m is the fitting parameter. The temperature and current dependences in Equation (1) are similar to the Black's equation where the mean-time-to-failure (MTTF) follows the reciprocal of the first two terms (Black, 1969; Huang, 2005).

$$\frac{I_d(t)-I_d(0)}{I_d(0)} = K_I I_r^{N} \exp(-\frac{E_a}{kT})at^m \tag{1}$$

For the current exponent, two stress currents can theoretically determine the N value. In practice, estimates based on three currents are more reliable since the confidence level is higher. However, one has to ensure that no new failure mechanism is introduced at the increased current density to avoid electrical overstress (Huang et al., 2007; Huang, 2015). The other alternative method of bias aging involves reverse voltage bias. The empirical equation can be modified as shown in Equation (2) where K_V is the normalization constant and V_r is the reverse voltage bias.

$$\frac{I_d(t)-I_d(0)}{I_d(0)} = K_V V_r^{N} \exp(-\frac{E_a}{kT})at^m \tag{2}$$

For the activation energy, two stress temperatures in principle can determine the E_a value. In practice, estimates based on three temperatures are more reliable due to the higher confidence level. Again, the elevated temperature for the aging test needs to be carefully designed in order to avoid unwanted new failure mechanism. The typical stress temperature is in the range of 60 to 200°C for APD aging.

For the fitting of degradation curves, the polynomial function has been employed to describe the aging behaviors of both lasers (Sim, 1989; Huang et al., 2005) and photodiodes (Kuhara et al., 1986). Typically, the sublinear model provides more accurate fit. In the sublinear model, the experimental curve is fitted by the equation where the exponent *m* is varied until the best correlation is found. The constant *a* is then deduced to determine the failure time.

Owing to the larger amount of reliability studies for the laser diodes, there have been several models established to describe the degradation behavior over aging time. For example, the Sim model (Sim, 1989) was based on a polynomial function. The polynomial expression provided good description for the late stage of degradation, and it has also been widely accepted by Telcordia and telecommunication communities. The Chuang (Chuang et al., 1998) and Lam (Lam et al., 2003) models were based on exponential functions and offered good description for the early stage of degradation. The exponential term was derived from the rate equation assuming that the degradation rate was proportional to the defect density.

$$I_d = I_{dd} + I_{dt} + I_{dg} + I_{ds} \tag{3}$$

For the APD degradation, the most common feature is associated with the increase in dark current. The total dark current can be expressed in Equation (3) where I_{dd} is the diffusion current, I_{dt} is the tunneling current, I_{dg} is the generation current and I_{ds} is the surface leakage current (Ohnaka et al., 1987; Fukuda, 1999). The diffusion current ideally corresponds to the saturation current and is typically determined by the epitaxial layers and design. The tunneling and generation currents are also largely pre-determined by the epitaxial layers. For the mesa-type APD structures, the surface leakage current (I_{ds}) is the dominant factor due to the presence of the interface states or traps between the dielectric film and the semiconductor. As shown in Equation (4), the surface leakage current becomes more pronounced after aging due to the weak-link nature near the surface/interface of the passivation/semiconductor. The surface and interface states could easily trap defects or carriers to cause the leakage current to increase. Any surface leakage current flows via the p-n junction at the defective surface or interface site would contribute to the dark current.

$$I_d(t) \sim I_{ds}(t) \tag{4}$$

In the following, we discuss the degradation mechanisms that could be responsible for the surface leakage current. Due to the larger number of research papers in the ICs (Fang et al., 2014; Lee et al., 1995; Kufluoglu et al., 2006; Heremans et al., 1988; Hu, 2009), we will also compare the APD degradation components with those of the IC transistor in order to shed some light on the APD reliability. We highlight the similarities in terms of dielectric breakdown, mobile ion migration and hot carrier injection in Sections (a)-(c).

(a) Dielectric breakdown (DB)

For the high-speed APD, low-k dielectric material is often employed in the passivation. However, the low-k passivation imposes a reliability concern due to two main reasons. First, the low-k dielectrics such as polyimide and benzocyclobutene (BCB) exhibit lower dielectric strength than that for the SiO_2 or SiN_x (Franssila, 2010; El-Kareh, 1995). For example, the dielectric strengths for the low-k polyimide and BCB are around 1.2-2.5 and 3.0MV/cm, respectively. On the other hand, the dielectric strengths for SiO_2 and SiN_x are around 10-12MV/cm. Secondly, the dielectric strength for the passivation thin film on the APD device is expected to be lower than that rated for the bulk. For the passivation film, the breakdown failure may be enhanced by the high electric field near the depletion region as well as the localized inhomogeneity on the mesa surface, as shown in Figure 2a. During the APD aging, the traps are initially formed in scattered manner in the dielectric film. As more and more traps are created, they start to connect and form conduction path. Once the conduction path is developed, localized heat would be generated, leading to thermal damage (Ma, 2009). The heat would further accelerate additional formation of traps and development of conduction path in a positive feedback loop, eventually leading to thermal runaway.

The surface leakage current associated with the dielectric breakdown could be expressed by the thermochemical "E" model or the anode hole injection "1/E" model (McPherson, 2012). For the "E" model, the dark current associated with the dielectric breakdown can be described in Equation (5) where A is a constant, γ is the field acceleration parameter, E is the dielectric field and ΔH is the enthalpy or activation energy required to break or coordinate the bonding of dielectric molecules. The model assumes that the defect generation in the dielectric film results from phonon-driven bond breakage/coordination process.

$$I_{ds}(DB_E) = A \exp(\gamma E - \frac{\Delta H}{kT}) \tag{5}$$

For the "1/E" model, the dielectric damage was assumed to result from the electrical conduction through the dielectric due to Fowler-Nordheim (F-N) mechanism (Fowler et al., 1928). The ionization process involving electron and hole injections was responsible for the dielectric damage or breakdown. The surface leakage current based on the "1/E" model followed the exponential dependence in Equation (6) where B is a constant, G is the tunneling factor and E is the dielectric field.

$$I_{ds}(DB_{1/E}) = BE^2 \exp(-\frac{G}{E})$$ (6)

The high electric field near the depletion region of the avalanche layer or the localized mesa surface inhomogeneity may further aggravate the dielectric damage. The field-assisted conduction mechanism is called Poole-Frenkel effect (Frenkel, 1938). The P-F conduction in the dielectric can be generally expressed as Equation (7) where C is a constant, ϕ_B is the electron energy barrier, ε is the dynamic permittivity. The exponential term describes that the electron does not need as much thermal energy ($q\phi_B$) to get into the conduction band under the influence of electric field. In the high electric field, the energy barrier is significantly reduced to $q(\phi_B - \sqrt{qE/(\pi\varepsilon)})$.

$$I_{ds}(DB_{P-F}) = CE \exp(-\frac{q(\phi_B - \sqrt{qE/(\pi\varepsilon)})}{kT})$$ (7)

For the complementary metal oxide semiconductor (CMOS) transistor, the gate dielectric leakage also occurs as a result of gate voltage stress. Time-dependent dielectric breakdown (TDDB) is commonly used terminology to describe the increase in the gate dielectric leakage due to the irreversible formation of electrically active defects. Those active defects form conductive percolation in the dielectric film that shorts the cathode and the anode, as shown in Figure 2b. The early phase of the dielectric breakdown is typically associated with the leakage current increase, characterized as "soft" breakdown. The later stage of the dielectric breakdown involves short failure, characterized as "hard" breakdown (Oates, 2015).

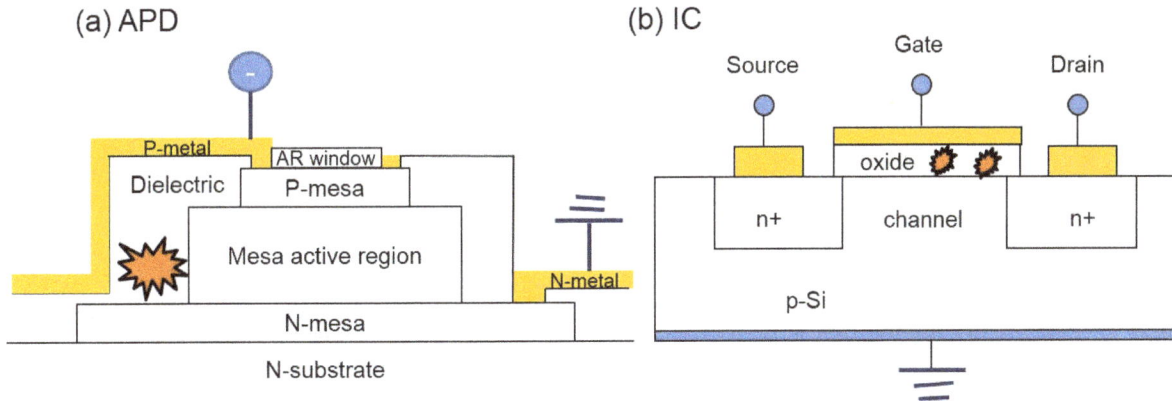

Figure 2. Schematics of degradation mechanisms of (a) APD dielectric breakdown and (b) IC gate oxide breakdown

As shown in Table I, the mechanism of the APD dielectric breakdown share some commonality with the ILD or gate dielectric breakdown in the integrated circuit (IC) transistor devices. For the ILD breakdown in the IC, the breakdown occurs along the interface between the low-k dielectric and its capping layer (Oates, 2015).

Table I. Dielectric breakdown of APD and its comparison with IC failure mechanism.

	APD	IC
Degradation pattern	Increase in dark current or short-circuit failure	Increase in the gate dielectric leakage or the interlevel dielectric (ILD) leakage
Physical mechanism	Low-k dielectric passivation breakdown	Low-k ILD breakdown Gate dielectric breakdown
Aging method	Reverse bias, typically involving negative stress current of 100µA at elevated temperature in the range of 60-200°C.	TDDB typically involving constant stress voltage applied to capacitor or transistor test structure at elevated temperature such as 150°C.

(b) Bias temperature instability (BTI)

The second potential root cause for the increase in dark current was related to the surface leakage current associated with mobile ion migration. For the polyimide passivation, there was some trace amount of impurities such as Na and Cu contained in the film (Watanabe et al., 1996; Pereira et al., 1993). Those impurities were likely to diffuse along the interface or into the passivation film driven by the high electric field at the depletion layer, as illustrated in Figure 3a. The mobile ion migration along the interface of the passivation and the semiconductor would result in the surface leakage current, leading to the increase in dark current. Brown (Brown, 1982) has reported that only 0.8ppm of sodium ions was enough to make polyimide a leaky dielectric. The density of the surface and interface states strongly depended upon device process and passivation method. The surface and interface states could easily trap carriers as generation-recombination centers.

The surface leakage current associated with the mobile ion migration can be expressed in Equation (8) where q is the charge of the mobile ion, A_s is the surface (interface) area of the p-n junction depletion region, n_m is the mobile ion density, σ_m is the capture cross section of mobile ions, υ_m is the thermal velocity of the mobile ion and N_{sm} is the trap density of mobile ions at the surface (interface).

$$I_{ds}(BTI) = (\frac{qA_s}{2})n_m\sigma_m\upsilon_m N_{sm} \tag{8}$$

In the metal oxide semiconductor field effect transistor (MOSFET), negative bias temperature instability (NBTI) manifests as an increase in the threshold voltage and consequent decrease in the drain current. The NBTI is most pronounced in the p-channel MOS (PMOS) due to the operation of negative gate-to-source voltage. The mechanism of NBTI is not completely clear. One plausible explanation is the hydrogen model that involves the hydrogen release from the Si dangling bonds at the SiO_2/Si interface (see Figure 3b), following the capture of a hole from the Si channel (Chakravarthi et al., 2004; Drapatz et al., 2009). The accumulation of the mobile species (H^0 and H^+ shown in Equations 9 and 10) can form the interface states or traps that are responsible for the NBTI degradation.

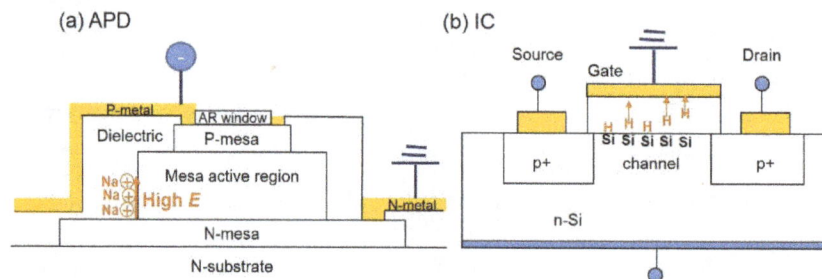

$$SiH + h^+ \Leftrightarrow Si^+ + H^0 \tag{9}$$

$$SiH + 2h^+ \Leftrightarrow Si^+ + H^+ \tag{10}$$

Figure 3. Schematics of degradation mechanisms of (a) APD mobile ion migration and (b) IC bias temperature instability

Table II. Mobile ion migration of APD and its comparison with IC failure mechanism.

	APD	IC
Degradation pattern	Dark current increase or short-circuit failure	Threshold voltage shift of PMOS due to NBTI or NMOS due to PBTI
Physical mechanism	Mobile ion accumulation (Na, Cu, etc.) from the polyimide and migration by the high electric field in the depletion layer	Positive charge generation due to hydrogen release from passivated Si dangling bond at the SiO_2/Si interface in the transistor
Aging method	Reverse bias, typically negative stress current of 100μA at 60-200°C.	DC stress involving the gate voltage and drain voltage at elevated temperature such as 150°C.

As shown in Table II, the mechanism of mobile ion migration such as Na in the polyimide passivation of the APD is similar to the mobile species migration such as H^0 and H^+ in the gate dielectric of the PMOS. The other similarity is the polarity of bias applied to the devices. For the APD, the p-contact is subjected to the negative bias. For the PMOS, the gate voltage is set in negative polarity relative to the source and drain voltages.

(c) Hot carrier injection (HCI)

The third mechanism for the increase in dark current of APD was related to hot carrier injection, as shown in Figure 4a. Under the relatively high electric field in the depletion layer, part of the thermally excited holes (hot holes) acquired sufficient energy to overcome the energy barrier and enter into the dielectric film (Sudo et al., 1988). When the holes entered the dielectric, they formed interface states and create positive surface charges. Upon the accumulation of the positive charge, the depletion region shrank gradually that made the field become more intense. The enhanced electric field would accelerate more hole injection, leading to positive thermal runaway.

The surface leakage current associated with the hot carrier injection can be expressed as Equation (11) where q is the carrier charge, n_i is the intrinsic carrier density, σ_i is the carrier capture cross section, υ_i is the carrier thermal velocity and N_{si} is the carrier trap density at the surface (interface).

$$I_{ds}(HCI) = (\frac{qA_s}{2})n_i\sigma_i\upsilon_iN_{si} \tag{11}$$

For the IC transistor, hot carrier degradation occurred as a result of high lateral field in the source/drain channel (see Figure 4b). As a result of vertical field in the gate, the holes and electrons were separated and either injected into the gate dielectric or ejected into the transistor substrate (Duncan et al., 1998).

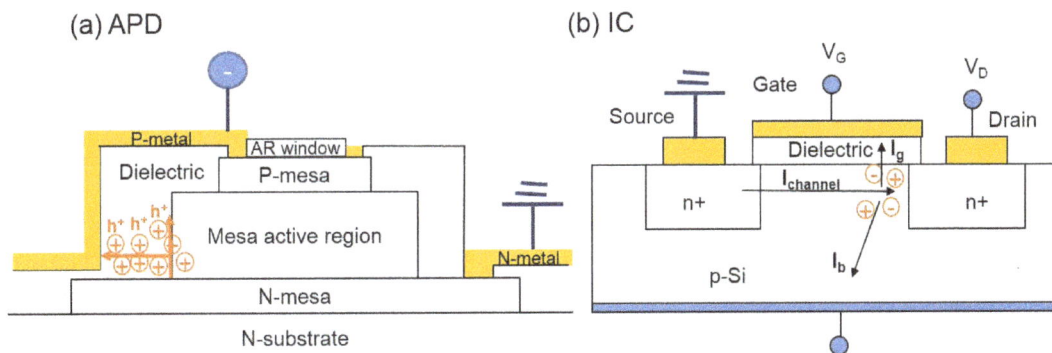

Figure 4. Schematics of degradation mechanisms of (a) APD hot hole injection and (b) IC hot carrier injection

As shown in Table III, the hot hole injection into the passivation dielectric of the APD is similar to the hot carrier injection into the gate oxide of the IC transistor. The carrier injection would result in the formation of the interface states, leading to the drift in device performance. For the APD, the drift is manifested as the increase in dark current or short failure. For the IC, the drift shows up as the shift in the threshold voltage.

Table III. Hot carrier injection of APD and its comparison with IC failure mechanism.

	APD	IC
Degradation pattern	Increase in dark current or short-circuit failure	Shift in the threshold voltage of CMOS transistor
Physical mechanism	Hot hole generation by the high electric field in the depletion layer and injection into the passivation dielectric	Hot carrier generation by the high lateral electric field in the source/drain channel and injection into the gate dielectric or into the transistor substrate
Aging method	Reverse bias, typically negative stress current of 100μA at 60-200°C.	DC stress involving the gate voltage and drain voltage at elevated temperature such as 150°C.

Based on the APD reliability model, it is predicted that highly reliable 10G/25G mesa-type APD devices may be feasible with the suppression of the weak links such as dielectric breakdown, mobile ion surface migration and hot carrier injection. The elimination of the weak links would be helpful to maintain stable surface/interface states and reduce the surface leakage current, resulting in low dark current for the long-term reliability performance.

4. Conclusion

We developed a predictive reliability model to account for the degradation behavior of 10G/25G mesa-type APD. For the mesa-type structure, surface dark current became dominant. We discussed the components of surface dark current in terms of dielectric breakdown, mobile ion migration and hot carrier injection. We also compared the degradation mechanisms of the APD degradation with those observed in the IC devices. The model predicted that highly reliable mesa-type APD were achievable with the suppression of weak links at the surface/interface states.

References

Achouche, M., Glastre, G., Caillaud, C., Lahrichi, M., Chitoui, M., & Carpentier, D. (2010). InGaAs communication photodiodes: from low- to high-power-level designs. *IEEE Photonics Journal, 2*(3), 460-468. http://dx.doi.org/10.1109/JPHOT.2010.2050056

Black, J. R. (1969). Electromigration failure modes in aluminum metallization for semiconductor devices. *Proc. IEEE, 57*(9), 1587–1594. http://dx.doi.org/10.1109/PROC.1969.7340

Brown, G. A. (1982). Implications of electronic and ionic conductivities of polyimide films in integrated circuit fabrication. Polymer Mater. Electronic Applications. American Chemical Society, Washington, D.C. http://dx.doi.org/10.1021/bk-1982-0184.ch012

Chakravarthi, S., Krishnan, A.T., Reddy, V., Machala, C. F., & Krishnan, S. (2004). A comprehensive framework for predictive modeling of negative temperature instability. International Reliability Physics Symposium Proceedings (Phoenix, AZ, pp.273-282).

Chuang, S. L., Nakayama, N., Ishibashi, A., Taniguchi, S., & Nakano, K. (1998). Degradation of II-VI blue-green semiconductor lasers. *IEEE J. Quantum Electron., 34*(5), 851–857. http://dx.doi.org/10.1109/3.668773

DeChiaro, L. F., & Sandroff, C. J. (1992). Improvements in electrostatic discharge performance of InGaAsP semiconductor lasers by facet passivation. *IEEE Trans. Electron Devices, 39*(3), 561–565. http://dx.doi.org/10.1109/16.123478

Drapatz, S., Georgakos, G., & Schmitt-Landsiedel, D. (2009). Impact of negative and positive bias temperature stress on 6T-SRAM cells. *Adv. Radio Sci., 7*, 191-196. http://dx.doi.org/10.5194/ars-7-191-2009

Duncan, A., Ravaioli, U., & Jakumeit, J. (1998). Full-band Monte-Carlo investigation of hot- carrier trends in the scaling of metal-oxide-semiconductor field-effect transistors. *IEEE Trans. Electron Devices, 45*(4), 867-876. http://dx.doi.org/10.1109/16.662792

El-Kareh, B. (1995). *Fundamentals of semiconductor processing technologies*. Springer, New York, NY, USA. http://dx.doi.org/10.1007/978-1-4615-2209-6

Fang, Y.-P., & Oates, A. S. (2014). Thermal neutron-induced soft errors in advanced memory and logic devices. *IEEE Trans. Device Mater. Reliab., 14*(1), 583-586. http://dx.doi.org/10.1109/TDMR.2013.2287699

Fowler, R. H., & Nordheim, L. (1928). Electron Emission in Intense Electric Fields. *Proceedings of the Royal Society A, 119*(781), 173–181. http://dx.doi.org/10.1098/rspa.1928.0091

Franssila, S. (2010). Introduction to microfabrication (2nd Ed.). John Wiley & Sons, West Sussex, United Kingdom. http://dx.doi.org/10.1002/9781119990413

Frenkel, J. (1938). On pre-breakdown phenomena in insulators and electronic semiconductors. *Phys. Rev., 54*, 647-648. http://dx.doi.org/10.1103/PhysRev.54.647

Fukuda, M. (1999). *Optical semiconductor devices* (Chapter 4 "Photodiodes"). Wiley, New York, NY.

Heremans, P., Bellens, R., Groeseneken, G., & Maes, H. E. (1988). Consistent model for the hot-carrier degradation in n-channel and p-channel MOSFETs". *IEEE Trans. Electron Devices, 35*(12), 2194-2209. http://dx.doi.org/10.1109/16.8794

Hu, C. M. (2009). *Modern semiconductor devices for integrated circuits*. Pearson Education, New York, NY, USA.

Huang, J. S. (2005). Temperature and current dependences of reliability degradation of buried heterostructure semiconductor lasers. *IEEE Trans. Device Mater. Reliab., 5*(1), 150-154. http://dx.doi.org/10.1109/TDMR.2005.843834

Huang, J. S. (2015). *Reliability of optoelectronics* (Chapter 6, J. Swingler (Ed.)). Cambridge, UK, Woodhead Publishing.

Huang, J. S., Jan, Y. H., Ren, D., Hsu, Y., Sung, P., & Chou, E. (2016). Defect diffusion model of InGaAs/InP semiconductor laser degradation. *Appl. Phys. Research, 8*(1), 149-157. http://dx.doi.org/10.5539/apr.v8n1p149

Huang, J. S., Nguyen, T., Hsin, W., Aeby, I., Ceballo, R., & Krogen, J. (2005). Reliability of Etched-Mesa Buried-Heterostructure Semiconductor Lasers. *IEEE Trans. Device Mater. Reliab., 5*(4), 665-674. http://dx.doi.org/10.1109/TDMR.2005.860562

Huang, J. S., Olson, T., & Isip, E. (2007). Human-Body-Model Electrostatic-Discharge and Electrical-Overstress Studies of Buried-Heterostructure Semiconductor Lasers. *IEEE Trans. Device Mater. Reliab., 7*(3), 453-461. http://dx.doi.org/10.1109/TDMR.2007.907425

Ishimura, E., Yagyu, E., Nakaji, M. N., Ihara, S., Yoshiara, K., Aoyaji, T., Tokuda, Y., & Ishikawa, T. (2007). Degradation mode analysis on highly reliable guardring-free planar InAlAs avalanche photodiode. *J. Lightwave Tech., 25*(12), 3686-3693. http://dx.doi.org/10.1109/JLT.2007.909357

Kim, H. S., Choi, J. H., Bang, H. M., Jee, Y., Yun, S. W., Burm, J., Kim, M. D., & Choo, A. G. (2001). Dark current reduction in APD with BCB passivation. *Electron. Lett., 37*(7), 455-457. http://dx.doi.org/10.1049/el:20010318

Kufluoglu, H., & Adam, M. A. (2006). Theory of interface-trap-induced NBTI degradation for reduced cross section MOSFETs. *IEEE Trans. Electron Devices, 53*, 1120-1130. http://dx.doi.org/10.1109/TED.2006.872098

Kuhara, Y., Terauchi, H., & Nishizawa, H. (1986). Reliability of InGaAs/InP long-wavelength p-i-n photodiodes passivated with polyimide thin film. *J. Lighwave Tech., LT-4*(7), 933-937. http://dx.doi.org/10.1109/JLT.1986.1074785

Lam, S. K. K., Mallard, R. E., & Cassidy, D. T. (2003). Analytical model for saturable aging in semiconductor lasers. *J. Appl. Phys., 94*(3), 1803–1809. http://dx.doi.org/10.1063/1.1589594

Lee, K. L., Hu, C. K., & Tu, K. N. (1995). In-situ scanning electron microscope comparison studies on electromigration of Cu and Cu(Sn) alloys for advanced chip interconnects. *J. Appl. Phys., 78*, 4428-4437. http://dx.doi.org/10.1063/1.359851

Ma, J. (2009). Study of gate oxide breakdown and hot electron effect of CMOS circuit performances (Ph.D. dissertation, U. Central Florida).

McPherson, J. W. (2012). Time dependent dielectric breakdown physics- models revisited. *Microelectron. Reliab., 52*, 1753-1760. http://dx.doi.org/10.1016/j.microrel.2012.06.007

Oates, A. S. (2015). Reliability of silicon integrated circuits (Chapter 7, J. Swingler (Ed.)). Cambridge, UK, Woodhead Publishing.

Ohnaka, K., Kubo, M., & Shibata, J. (1987). A low dark current InGaAs/InP p-i-n photodiode with covered mesa structure. *IEEE Trans. Electron Devices, ED-34*(2), 199-204. http://dx.doi.org/10.1109/T-ED.1987.22907

Pereira, J. S., Shieh, P. J., Gobbi, A. L., Sato, E., Malberti, P., Santos, T., Borin, F., & Patel, N. (1993). Reliability of InGaAs/InP photodiodes passivated with polyimide. International Reliability Physics Symposium Proceedings, (Atlanta, GA), 372-374. http://dx.doi.org/10.1109/relphy.1993.283273

Sim, S. P. (1989). A review of the reliability of III-V opto-electronic components. In *Proc. Semiconductor Device Reliability: Advanced Workshop II, NATO International Scientific Exchange Program, Crete, Greece* (p. 301).

Smith, G. M., McIntosh, K. A., Donnely, J. P., Funk, J. E., Mahoney, L. J., & Verghese, S. (2009). Reliable InP-based Geiger-mode avalanche photodiode array. *Proceedings of SPIE, 7320*, 1-10. http://dx.doi.org/10.1117/12.819126

Sudo, H., & Suzuki, M. (1988). Surface degradation mechanism of InP/InGaAs APD's. *J. Lightwave Tech., 6*(10), 1496-1501. http://dx.doi.org/10.1109/50.7907

Takeshita, T., Hirota, Y., Ishibashi, T., Muramoto, Y., Ito, T., Tohmori, Y., & Ito, H. (2006). Degradation behavior of avalanche photodiodes with a mesa structure observed using a digital OBIC monitor. *IEEE Trans. Electron Devices, 53*(7), 1567-1574. http://dx.doi.org/10.1109/TED.2006.875820

Twu, Y., Cheng, L. S., Chu, S. N. G., Nash, F. R., Wang, K. W., & Parayanthal, P. (1993). Semiconductor laser damage due to human-body-model electrostatic discharge. *J. Appl. Phys., 74*(3), 1510–1520. http://dx.doi.org/10.1063/1.354850

Watanabe, I., Tsuji, M., Hayashi, M., Makita, K., & Taguchi, K. (1996). Reliability of mesa-structure InAlGaAs-InAlAs superlattice avalanche photodiodes. *IEEE Photonics Tech. Lett., 8*(6), 824-826. http://dx.doi.org/10.1109/68.484263

17

Thought Experiment Revealing a Contradiction in the Special Theory of Relativity

Koshun Suto[1]

[1] Chudaiji Buddhist Temple, Isesaki, Japan

Correspondence: Koshun Suto, Chudaiji Buddhist Temple, 5-24, Oote-Town, Isesaki, 372-0048, Japan. E-mail: koshun_suto129@mbr.nifty.com

Abstract

In the thought experiment in this paper, we considered inertial frames M and A moving at a constant velocity relative to each other. A light signal emitted from inertial frame A, when time of a clock in inertial frame A was 1(s), arrived at inertial frame M when time of a clock in inertial frame M was 2(s). In this paper, the time in inertial frame A when the time in inertial frame M was 2(s) was predicted by observers in inertial frames M and A by applying the special theory of relativity (STR). Predictions of the two observers did not match. Einstein regarded all inertial frames as equivalent, but there are cases where a velocity vector is attached to some inertial frame. Einstein overlooked this fact, and thus a discrepancy appeared in the values predicted by the two observers. It is not the case that all inertial frames are equivalent. This paper concludes that the STR is a theory incorporating a contradiction which must be corrected.

Keywords: Special Theory of Relativity, Minkowski Diagram, Velocity Vector

1. Introduction

STR is not just a single theoretical system. It is composed of two theories of different types. The first is a theory derived from Lorentz transformations which has full symmetry, and the second is Einstein's energy-momentum relationship which holds in free space.

Consider a rod A (inertial frame A) and rod B (inertial frame B) moving at constant velocity relative to each other. First, let us regard inertial frame A as a stationary system, and treat inertial frame B as a moving system in motion at constant velocity v in the x-axis direction of inertial frame A.

According to the STR, when length in the direction of motion of rod B, moving at constant velocity, is measured from inertial frame A, the rod contracts in the direction of motion. Also, the time which elapses on clock B in inertial frame B is delayed compared to the time which elapses on clock A in inertial frame A.

If, conversely, inertial frame A is measured from inertial frame B, rod A contracts in the direction of motion, and the time which elapses on clock A is delayed.

According to Einstein's "principle of relativity," the two inertial frames are equivalent, and thus the same results are obtained no matter which inertial frame measurement is carried out from. The essence of STR is the symmetry of the theory.

Theoretically, there is no problem with the STR, as indicated below:

1) It is mathematically complete.

2) It can be explained using Minkowski diagrams.

It is also thought that the correctness of the STR has been demonstrated based on the following two types of experiments:

1) Extended life of elementary particles.

2) When the velocity of a moving object increases, the mass (or energy) of the object increases.

Experiment (1) is recognized even by physicists who have doubts about the STR. However, to demonstrate the correctness of the STR, one must observe lengthening of the life of stationary elementary particles from a moving system. Experiments carried out thus far have not demonstrated the symmetry of time delay.

Next is Einstein's energy-momentum relationship, which holds in free space.

$$\left(mc^2\right)^2 = p^2c^2 + \left(m_0c^2\right)^2 . \tag{1}$$

Here, m_0c^2 is the rest mass energy of a particle or object, mc^2 is the relativistic energy, and p is the momentum. According to Equation (1), when the velocity of a moving object increases, the mass (or energy) of the object also increases. However, even if physical quantities of a stationary system are measured from a moving system, the STR does not assert that the same results are obtained. That is, there is no symmetry in Equation (1). Even if we assume that an increase in the mass of a moving object has been detected, that does nothing more than demonstrate the correctness of Equation (1).

Incidentally, Equation (1) is not applicable in the atom where potential energy exists (Suto, 2011: Suto, 2014: Suto, 2015). However, the equation definitely holds in free space. The STR which this paper views as a problem is the former theory which has perfect Lorentz symmetry.

2. Thought Experiment Indicating a Contradiction in the Special Theory of Relativity

Thought experiment: Rocket A is moving at a constant velocity of $3c/5$ in the x-axis direction of "Stationary system." (In the following, "Stationary system may be indicated as S, and the coordinate system of rocket A as S'_A.)

There is an observer M at the origin O of the x-axis of S, and M has a stopwatch W. In addition, there is an observer A at the origin O'_A of the x'_A-axis of S'_A, and A has a stopwatch W_A. (In the following "stopwatch W" may be abbreviated as W, and "stopwatch W_A" as W_A.)

Now, when rocket A passes in front of observer M in S, observer M starts W, and observer A starts W_A.

According to the STR, an observer in S, finds the following relationship between the time t which elapses on W and the time t'_A which elapses on W_A.

$$t'_A = \frac{t}{\gamma} = t\left(1 - \frac{v^2}{c^2}\right)^{1/2} . \tag{2}$$

Here, when 1(s) is substituted for t,

$$t'_A = \frac{4}{5} \text{ (s).} \tag{3}$$

Here, this thought experiment is explained using Minkowski diagram 1 (see Figure 1).

Point O indicates both origins: $x=0$, $t=0$ and $x'_A=0$, $t'_A=0$. The point event M_0 of the point light source O and the point event A_0 of the point light source O'_A are at the origin O. (Here, the subscripts "$_0$" of the point events M_0 and A_0 mean, respectively, $t=0$ and $t'_A=0$.)

The x-axis indicates the x-axis of the inertial frame S when $t=0$. In addition, the x'_A-axis indicates the x'_A-axis of the inertial frame S'_A when $t'_A=0$.

The ct-axis is the path for $x=0$. Put another way, it is the world line of the origin of S. The ct'_A-axis is the world line of the origin of S'_A.

In addition, the straight line extending at a 45° angle from the origin O indicates the light signal emitted from the two light sources at the instant that O and O'_A pass by each other.

OE is the distance over which the light signal emitted from O propagates in the x-axis direction while 1(s) elapses on the stopwatch W in S.

OE′ is the distance over which the light signal emitted from O'_A propagates in the x'_A-axis direction while 1(s) elapses on the stopwatch W_A in S'_A.

Oe is the value when an observer in S measures the distance OE′, and Oe′ is the value when the distance OE is measured by an observer in S'_A. However, Ee′ is parallel to the ct-axis, and eE′ is parallel to the ct'_A-axis. Therefore, the relationship between OE, OE′, Oe and Oe′ is as follows.

$$\frac{\text{Oe}}{\text{OE}} = \frac{\text{Oe}'}{\text{OE}'} = \frac{1}{\gamma}, \qquad \gamma = \left(1 - \frac{v^2}{c^2}\right)^{-1/2} . \tag{4}$$

Here, when the position of the point E is determined, it is possible to determine the positions of the points e', e and E' based on the relationship in Equation (4).

Furthermore, if a point is plotted on the ct-axis at a distance equal to OE from O, that is the point event M_1 for O at $t = 1(s)$.

Also, if a point is plotted on the ct'_A-axis at a distance equal to OE' from O, that is the point event A_1 for O'_A at $t'_A = 1(s)$.

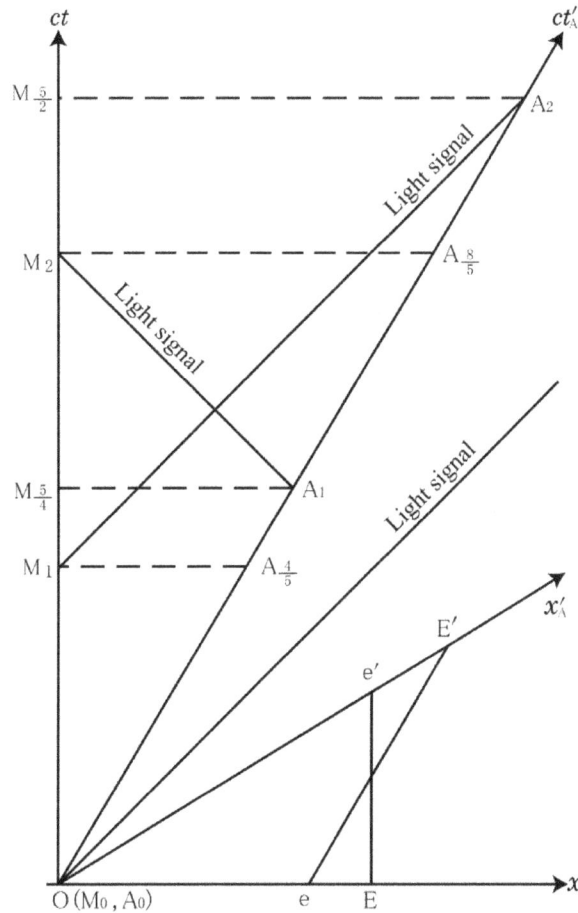

Figure 1. Minkowski diagram 1: This diagram corresponds to thought experiment

Now, how should we find the relationship between the times which elapse in the stationary system and in the coordinate system of rocket A?

To find that, it is enough to compare the times when the straight line parallel to the x-axis intersects with the ct-axis and ct'_A-axis.

For example, among the lines which pass through M_1, the straight line parallel with the x-axis intersects the ct'_A-axis at point $A_{4/5}$, and this is the point event of W_A when $t = 1(s)$. Therefore t'_A matches with Equation (3).

Now when W in S is at 1(s), a light signal is emitted from O to O'_A in S'_A. That light propagates isotropically with respect to O. Then it arrives at O'_A when W_A on rocket A is 2(s). (This light signal corresponds to the world line M_1A_2.)

In the inverse case, when W_A on rocket A is 1(s), a light signal is emitted from O'_A to O. That light arrives at O when W of the stationary system is 2(s). (This light signal corresponds to the world line A_1M_2.)

These results also seem to show there is symmetry between the two inertial systems. In this paper, the propagation situation of the two light signals (M_1A_2 and A_1M_2) is expressed as follows.

$$t = 1 \text{ (s)} \rightarrow t'_A = 2 \text{ (s)}, \tag{5a}$$

$$t' = 2 \text{ (s)} \leftarrow t_A = 1 \text{ (s)}. \tag{5b}$$

Now, are the two inertial systems truly equivalent, as claimed by the STR? Next let's try having observer M and A predict the time of W_A on rocket A when W is 2(s).

3. Discussion

A. Prediction of observer M (prediction based on the STR) (see Figure 2 (a))

In this paper, the moving object is taken to be rocket A, which has passed through an acceleration stage. Therefore, the time t'_A of W_A can be found from Equation (2).

To find t'_A when $t = 2(s)$, it is enough to substitute 2(s) for t and $3c/5$ for v in Equation (2). This yields:

$$t'_A = \frac{t}{\gamma} = 2\left[1 - \frac{(3c/5)^2}{c^2}\right]^{1/2} = 1.6 \text{ (s)}. \tag{6}$$

The observer in S concludes that the time t'_A of W_A when the time of W is 2(s) is 1.6(s).

B. Prediction of observer A (prediction based on the STR) (see Figure 2 (b))

The observer in rocket A regards his own coordinate system as the stationary system. With the STR, the observer in S_A predicts the value of the time t_A of W_A as follows when the time of W in S' is 2(s).

$$t_A = \gamma t' = 2\left[1 - \frac{(3c/5)^2}{c^2}\right]^{-1/2} = 2.5 \text{ (s)}. \tag{7}$$

The observer in S_A concludes that the time t_A of W_A when the time of W is 2(s) is 2.5(s).

In the end, if observers M and A predict the time of W_A at a certain instant by applying the STR, different values are obtained.

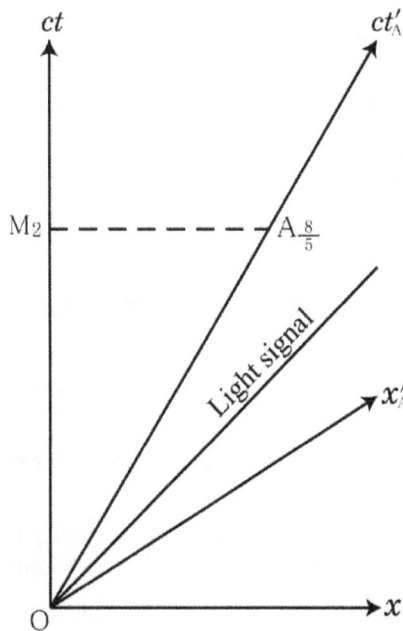

Figure 2(a). Minkowski diagram 2: Prediction of observer M applying the STR, $t = 2$ (s) $\leftrightarrow$ $t'_A = 1.6$ (s)

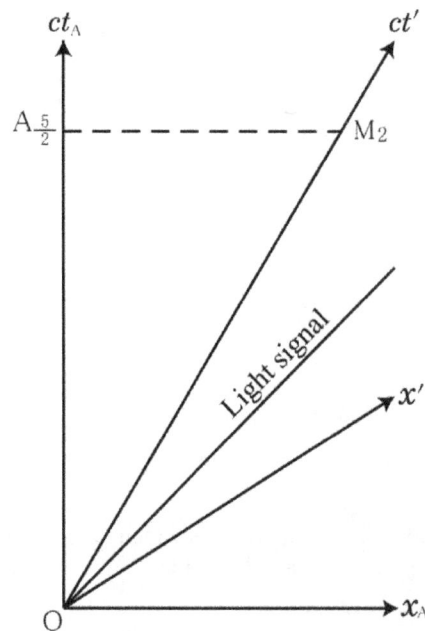

Figure 2(b). Minkowski diagram 3: Prediction of observer A applying the STR, $t' = 2$ (s) $\leftrightarrow$ $t_A = 2.5$ (s)

4. Conclusion

In the thought experiments in this paper, the light signal emitted from O'_A of rocket A when the time of W_A was 1(s) arrived at O of S when the time of W was 2(s). In this paper, the observer M in S and observer A on rocket A predicted the time of W_A in rocket A when the light signal arrived at O.

(1) Prediction of observer M applying the STR

Observer M predicts 1.6(s) as the time t'_A of W_A when $t = 2$ (s) . That is,

$$t = 2 \text{ (s)} \leftrightarrow t'_A = \underline{1.6 \text{ (s)}}. \tag{8}$$

(2) Prediction of observer A applying the STR

Observer A predicts 2.5(s) as the time t_A of W_A when $t' = 2$ (s). That is,

$$t' = 2 \text{ (s)} \leftrightarrow t_A = \underline{2.5 \text{ (s)}}. \tag{9}$$

If STR is applied, there is no match between the times of W_A predicted by observers M and A. This means that at least one of these predictions is wrong.

However, in the thought experiment in this paper, it is not possible to determine the correctness of the predictions of the two observers. If a conclusion based on experiment is required, a more complex thought experiment will be necessary (Suto, 2010, 2015).

Einstein regarded all inertial frames as equivalent, but there are cases where a velocity vector is attached to some inertial frame (Suto, 2010: Suto, 2015). Einstein overlooked this fact, and thus a discrepancy appeared in the values predicted by the two observers. It is not the case that all inertial frames are equivalent. This paper concludes that the STR is a theory incorporating a contradiction which must be corrected.

Acknowledgments

I would like to express my thanks to the staff at ACN Translation Services for their translation assistance. Also, I wish to express my gratitude to Mr. H. Shimada for drawing figures.

References

Suto, K. (2010). Violation of the special theory of relativity as proven by synchronization of clocks. *Physics Essays, 23*(3), 511-519. http://dx.doi.org/10.4006/1.3474836

Suto K. (2011). An energy-momentum relationship for a bound electron inside a hydrogen atom. *Physics Essays, 24*(2), 301-307. http://dx.doi.org/10.4006/1.3583810

Suto K. (2014). Previously unknown ultra-low energy level of the hydrogen atom whose existence can be predicted. *Applied Physics Research, 6*(5), 109-115. http://dx.doi.org/10.5539/apr.v6n5p64

Suto K. (2015). Demonstration of the existence of a velocity vector missing from the special theory of relativity. *Physics Essays, 28*(3), 345-351. http://dx.doi.org/10.4006/0836-1398-28.3.345

Suto K. (2015). Presentation of strong candidates for dark matter. *Global Journal of science frontier research: A, 15*(7), 1.0, 1-6.

18

Superconducting String Theory (Gravity Explanation)

Sergio P. F.[1]

[1] Independent Researcher, graduated from Computer Science, Madrid, Spain

Correspondence: Sergio P. F., Independent Researcher, graduated from Computer Science, Madrid, Spain. E-mail: sergiopf@gmail.com

Abstract

Gravity explained by a new theory, 'Superconducting String Theory (SST)', completely opposite from current field emission based and inspired on originals string theories. Strengths are decomposed to make strings behave as one-dimensional structure with universe acting as a superconductor where resistance is near 0 and the matter moves inside. Strong nuclear force, with an attraction of 10.000 Newtons is which makes space to curve, generating acceleration, more matter more acceleration. Electromagnetic moves in 8 decimals, gravity is moved to more than 30 decimals to work as a superconductor.

Keywords: superconducting string theory, string theory, strong nuclear force, gluon, unified field theory, gravity, relativity

1. Introduction

The 'Theory of Everything' is a hypothetical theory of physics that explains and connects all known physical phenomena into one. There is a possible solution to the origin of gravity force, postulating it as angular piece of this theory, this solution erases gravity as one of the fundamental forces of nature and unifies it with strong nuclear force.

Let's analyze the forces that occur in the universe transforming string theory. It allows to explain many physical behaviors that without its existence would be practically impossible to understand, even so, these strings have not been able to be discovered and are only that, a theory that serves as an important support to the world of physics. One of the best known theoretical applications about them is how their vibration can provoke the creation of matter, but this is not about theories already written, we are going to place these strings in a simpler way to answer some doubts in subatomic world.

This theory uses 4 dimensions in space and a behavior as one dimension in strings with superconducting capacities. Like an elastic band between V-shaped sticks where the elastic band slides down, the strong nuclear force, forces these strings to bend to fall dawn.

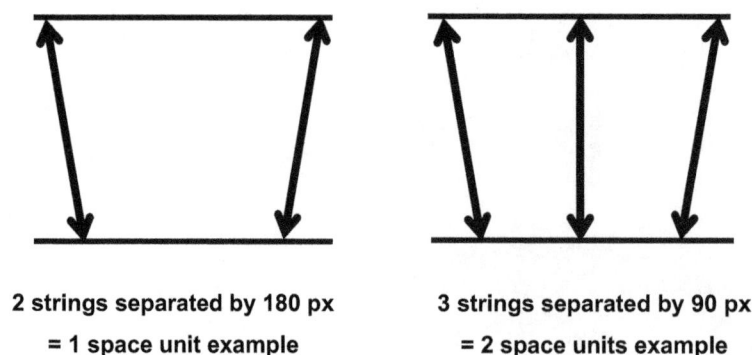

2 strings separated by 180 px

= 1 space unit example

3 strings separated by 90 px

= 2 space units example

Figure 1. String example

It's not directly related to electromagnetism.

2. Actors

2.1 Sring Theory

String theory is a theoretical framework in which the point-like particles of particle physics are replaced by one-dimensional objects called strings. Each string that we cross would be the minimum distance that can be traversed during a displacement.

We can note two important qualities of strings:

·   Distance to the most distant object detected by the human being is more than 30 billion light years, that means there are beams of light which are able to travel that distance without decreasing its speed (they modify only its wavelength). Like light, an object can move into space for a practically unlimited period, as long as it doesn't find a force to stop it. If strings exist, they act as a superconductor of matter with a resistance near 0.

·   In order to generate waves it's easier into a strongly linked structure. Gravitational waves behave like ocean waves which are similar to an uptight net, these tensions can be decomposed as one-dimensional structure for its study. Strings, at same time, could be one or zero-dimensional, like points under extreme bound forces, think about them as something tenser than any cable that holds the heaviest bridge in the world.

The new framework we have drawn would be a set of extremely tense strings, with a practically infinite matter conduction capacity.

Figure 2. Example of quantum entanglement. Gravity waves behave as an uptight net

2.2 Strong Nuclear Force

Strong nuclear force is another variable. This force allows the atomic nucleus to remain together, being the strongest of the so-called fundamental interactions (gravitational, electromagnetic, strong nuclear, and weak nuclear). Gluon is in charge of this interaction, it has a scope not greater than 10 to the power of -15 meters, preventing matter to separate by a constant attraction force between quarks of maximum 10.000 N (F).

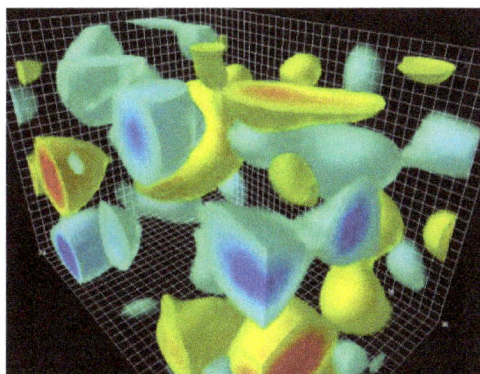

Figure 3. Gluon into vacuum (Special Research Centre for the Subatomic Structure of Matter)

This real picture illustrates the three-dimensional structure of gluon-field configurations, describing the vacuum properties. The volume of the box is 2,4 by 2,4 by 3,6 fm. Contrary to the concept of an empty vacuum, this induces chromo-electric and chromo-magnetic fields in its lowest energy state. The frame rate into this example is billions of billions frames per second (FPS).

3. Superconducting String Theory (SST)

3.1 Fundamentals

We have created a scenario with a superconductor of matter interacting with a force that makes that matter hold together, but, how can they interact? The most simple is to think about two V-shaped sticks (simulating the strings), and an elastic band that tight them at the most opened side (it would simulate the gluon, with size 10 to the power of -15 meters). If sticks are sufficiently lubricated and tense, what does the elastic band do? It will slide to the thinnest side. More elastic bands, more force will be exerted on the sticks to join them, so next bands will slide even faster (equally, more mass causes more attraction).

We are talking about unknown limits in known world, such as infinite conduction or tensions never seen in materials.

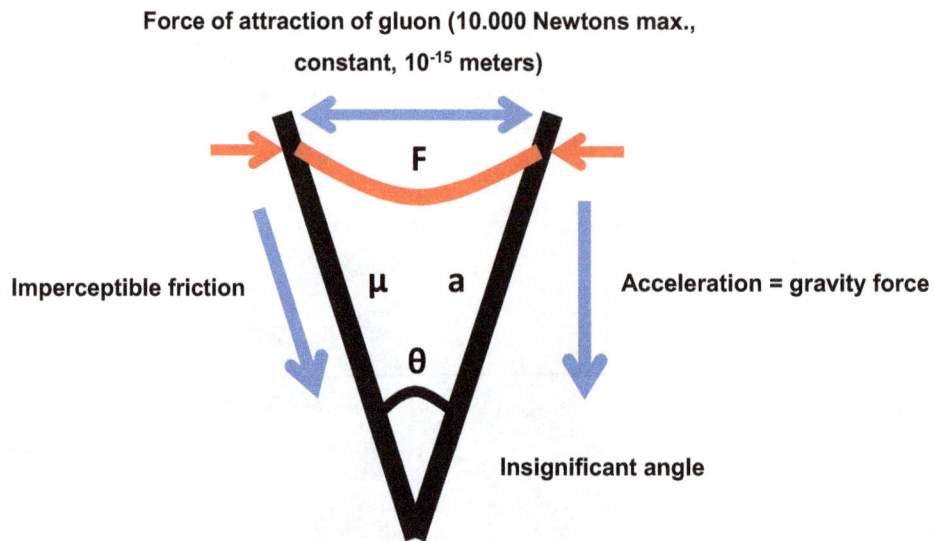

Force of attraction of gluon (10.000 Newtons max.,

constant, 10^{-15} meters)

Imperceptible friction   μ   a   **Acceleration = gravity force**

θ

Insignificant angle

Figure 4. Implied forces

Suddenly, we have erased one of the fundamental forces of nature, gravity force doesn't really exist, exists the strong nuclear force interacting with strings. I have called this theory 'Superconducting String Theory (SST)'.

3.2 Calculations

Apply formulas from inclined planes (Newton's second law). Simulation is in horizontal direction.

Friction is imperceptible and acceleration down the plane is matched with gravity acceleration in our planet. Vertical force is not gravity force, it is gluon force. It can be considered vertical angle, but is depreciable and gluon force is estimated, so we keep 10.000 N.

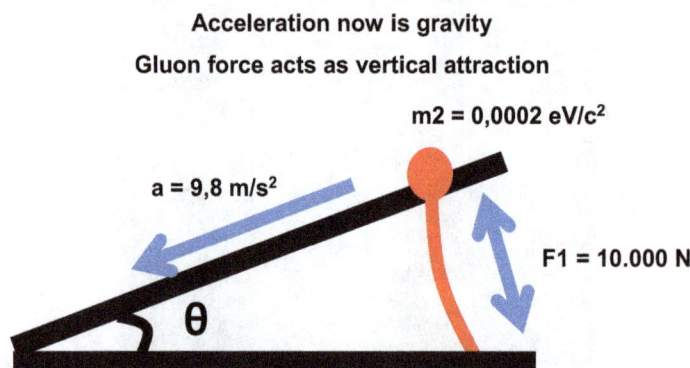

Acceleration now is gravity

Gluon force acts as vertical attraction

$m2 = 0,0002$ eV/c^2

$a = 9,8$ m/s^2

$F1 = 10.000$ N

θ

Figure 5. Inclined plane forces (gluon catching strings)

$$m_1 \times g \times sen(x) = m_2 \times a$$

Match vertical force (F_1) with the attraction force from one gluon (estimated max. constant 10.000 N).

$$F_1 = m_1 \times g = 10.000\ N \qquad (1.1)$$

$$F_1 \times sen(\,x\,) = m_2 \times a$$

Convert variables to metric system considering the mass of the gluon (0,0002 eV/c²).

$$1eV/c^2 = 1,782662 \times 10^{-36}\ kg \qquad (1.2)$$

$$m_2 = 0,0002 \times 1,782662 \times 10^{-36} = 3,565324 \times 10^{-40}\ kg$$

$$a = 9,8\ m/s^2$$

$$F_2 = m_2 \times a = 3,565324 \times 10^{-40} \times 9,8 = 3,49401752 \times 10^{-39}\ N$$

Angle calculation of all strings under gluon influence.

$$x = arcsen(\,F_2\,/\,F_1\,) \qquad (1.3)$$

$$F_2\,/\,F_1 = 3,49401752 \times 10^{-43}\ N$$

$$x = arcsen(\,3,49401752 \times 10^{-43}\,) = 3,49401752 \times 10^{-43}\ ^\circ$$

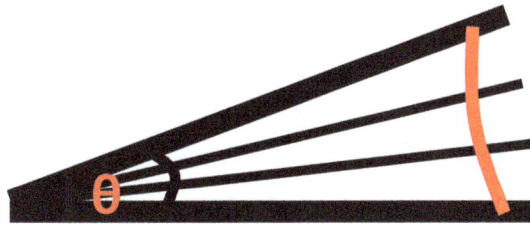

Figure 6. θ angle of all strings gluon can catch

4. Conclusions

It can explain:

·   What is gravity force.

·   Unified field theory between gravity and strong nuclear force.

·   New behavior in dark matter because of differences in density of the superconductor. Lower density generates bigger angle, this implies bigger attraction force. If threads are separated, matter (m_2) becomes energy (F_1). Some places at universe could have bigger accelerations because of this effect; this means much less dark matter (which is stimated at approximately 27% of the mass and energy in the observable universe).

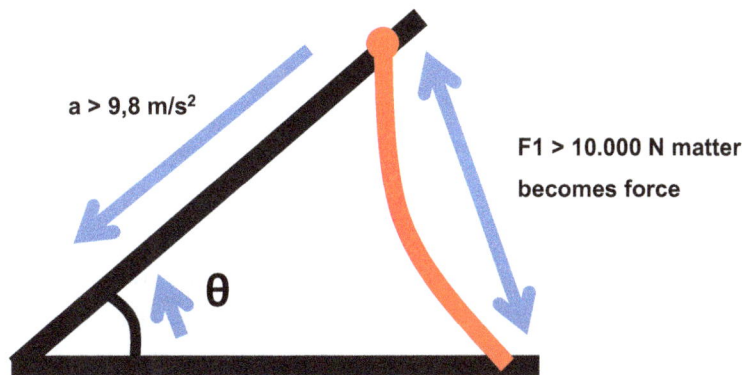

Figure 7. Increase the separation, increase the force exerted. Matter is transformed into energy

$$F_1 \times sen(\,x\,) = m_2 \times a \qquad (2.1)$$

$$1eV/c^2 = 1,782662 \times 10^{-36}\ kg$$

$$m_2 = 0,0002 \times 1,782662 \times 10^{-36} = 3,565324 \times 10^{-40} \ kg$$

Calculate acceleration in relation to angle.

$$F_2 = m_2 \times a = 3,565324 \times 10^{-40} \times a \ kg/(m/s^2) \qquad (2.2)$$

$$x = arcsen(\ F_2 / F_1 \ )$$

$$F_2 / F_1 = (\ 3,565324 \times 10^{-40} \times a \ ) / 10.000 = (\ 3,565324 \times 10^{-44} \times a \ )$$

$$x = arcsen(\ 3,565324 \times 10^{-44} \times a \ ) = (\ 3,565324 \times 10^{-44} \times a \ ) \ ^{\circ}$$

Bigger angle generates more acceleration.

$$a = (\ x \ / \ 3,565324 \times 10^{-44} \ ) \ m/sg^2 \qquad (2.3)$$

- Einstein field equations. Apply Hooke's law to gluon force as a vertical spring to calculate tensions.

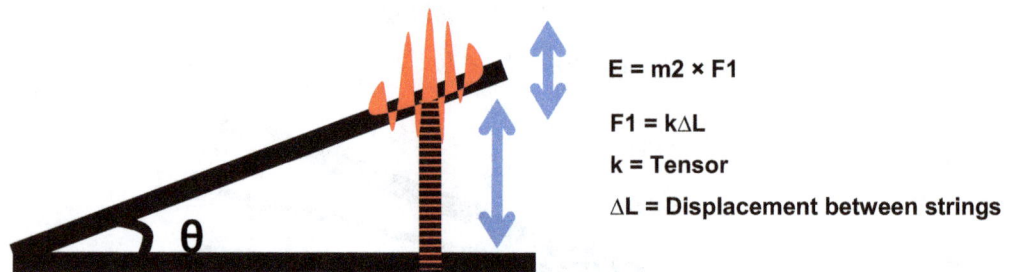

E = m2 × F1

F1 = kΔL

k = Tensor

ΔL = Displacement between strings

Figure 8. Energy don't disappear becomes matter

Gluon force has negative correlation when transforming into matter, decrease F1 or increase m_2 implies less acceleration.

$$a = F_1 \times sen(\ x \ ) / m_2 \qquad (3.1)$$

$$a = 5.000 \ x \ sen \ (\ 3,49401752 \times 10^{-43} \ ) / 3,565324 \times 10^{-40}$$

$$a = 4,9 \ m/s^2$$

- Spherical and circular movements at planets and galaxies. Space deforms not proportionally to create more acceleration near accumulation of matter, behaving like an elastic material; this behavior can be quantified by Young's modulus which represents the factor of proportionality in Hooke's law at non-linear systems.

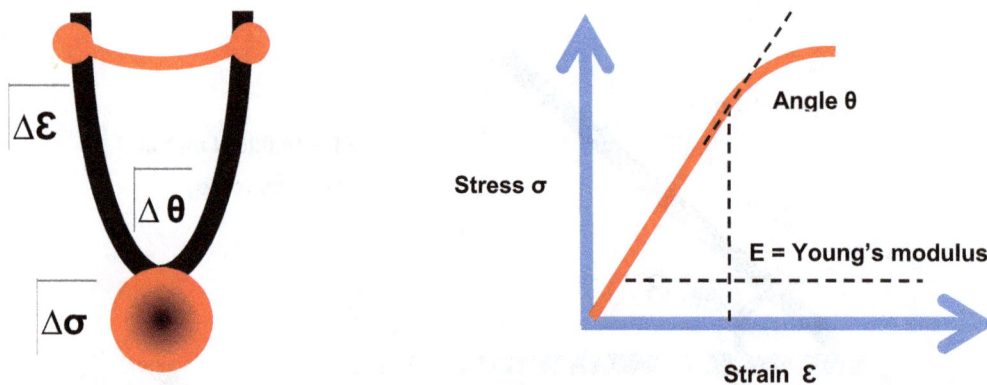

ΔƐ

Δθ

Δσ

Stress σ

Angle θ

E = Young's modulus

Strain Ɛ

Figure 9. Angle originated by big amounts of matter has an interaction over very large distances

The elastic modulus or Young's modulus (E) depends on the force exerted by matter (σ) and the deformation at each point of the resulting vector (Ɛ).

$$E = \Delta\sigma / \Delta\mathcal{E}$$

Force exerted by the angle ($F\theta$), increases (Δ) faster than force exerted by gluon (F_1) and its relation with matter (m_2).

$$\Delta F\theta > \Delta F_1 / \Delta m_2 \qquad\qquad (4.1)$$

This gluon-matter relation also modifies the density of the superconductor in space, since it induces their approach, therefore we can speak of the existence of a bulk modulus (K).

$$K = -V \left(\Delta p / \Delta V\right)$$

The bulk modulus (K) depends on pressure changes (p) and volume (V).

Other properties as volume viscosity also called bulk viscosity can be applied.

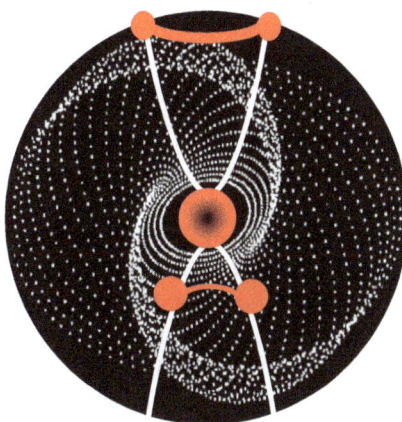

Figure 10. All these variables help to create shapes in galaxies like the golden spiral ($\varphi = 1,6180$)

·   Gravitational constant ($G = 6,67408 \times 10^{-11}$ m^{3}kg^{-1}s^{-2}) and its problem to measure with high accuracy since it can be related to the density exposed.

·   Schrödinger equation, to describe how the quantum state of a quantum system changes with time, similar to Newton's second law.

·   Planck's length ($1,616229 \times 10^{-35}$ m) which can indicate the distance between strings.

·   Gluon size and its larger size far from earth.

·   Black holes.

Rerefences

Newton, I. (1687). *Philosophiae naturalis principia mathematica.*

Hooke, R. (1678). *The extension is proportional to the force.*

Euler, L. (1768). *Institutionum calculi integralis* (Vol. 1). imp. Acad. imp. Saènt.

19

Rotating Space Elevators: A New Venue in Space Elevator Physics

Leonardo Golubovic[1] & Steven Knudsen[1]

[1] Physics and Astronomy Department, West Virginia University, Morgantown, WV 26506-6515, USA

Correspondence: Leonardo Golubovic, Physics and Astronomy Department, West Virginia University, POB 6315, Morgantown, WV 26506, USA. E-mail: lgolubov@wvu.edu

Abstract

The physics of Space Elevators connecting the Earth with outer space has recently attracted increased attention, in part due to the discovery of ultra-strong materials such as carbon nanotubes and diamond nano-thread structures. In this article we review a new venue in space elevator physics: Rotating Space Elevators (RSE) [Golubović, L. & Knudsen, S. (2009). Classical and statistical mechanics of celestial scale spinning strings: Rotating space elevators. *Europhysics Letters 86(3)*, 34001.]. The RSE is a double rotating system of strings reaching outer space. Objects sliding along the RSE string (sliding climbers) do *not* require internal engines or propulsion to be transported far away from the Earth's surface. The RSE thus solves a major problem in the space elevator technology which is how to supply the energy to the climbers moving along the string. RSE strings exhibit interesting nonlinear dynamics and statistical physics phenomena. Satellites and spacecraft carried by sliding climbers can be released (launched) along RSEs. RSE strings can host space stations and research posts. Sliding climbers can be then used to transport useful loads and humans from the Earth to these outer space locations.

Keywords: Space Elevator, Inertial Forces, Classical Mechanics, Statistical Physics, Space Travel, Nonlinear Dynamics, Instabilities and Transitions, Chaos

1. Introduction

The physics of Space Elevators connecting the Earth with outer space has attracted increased attention in this millennium (Edwards & Westling, 2003). This interest emerged in part due to the discovery of ultra-strong materials such as carbon nanotubes (Yu et al., 2000a, Yu et al., 2000b) and diamond nano-thread structures (Fitzgibbons et al., 2014). Space elevators are celestial scale examples of physical systems with reduced dimensionality such as the strings, polymers, and membranes (Kardar, 2007; Nelson, 2002; Nelson, Piran & Weinberg, 1988). The classical and statistical mechanics of space elevators represents a bold extension of previous studies of satellite dynamics such as those of Beletskii (1965) and Hughes (2012).

Recently, a new concept has emerged in this applied physics area: Rotating Space Elevator (RSE) (Golubovic & Knudsen, 2009; Knudsen & Golubovic, 2014; Knudsen & Golubovic, 2015). In this review article we discuss this new venue in space elevator physics. The RSE is a double rotating floppy string reaching extraterrestrial locations. Interestingly, objects sliding along the RSE string (climbers) do not require internal engines or propulsion to be transported far away from the Earth's surface. The RSE thus solves a major problem in space elevator physics which is how to supply energy to the climbers moving along space elevator strings.

2. Historic Background: The Conceptual Development of the Traditional Space Elevator

Dreams of traveling to the heavens have entranced men since the early times of civilization. The story of the "Tower of Babel" in Genesis 11 of the Bible connects the notion of human cooperation for space travel to "heaven" to the multiplying of human languages, which frustrates the effort. In modern history, the fable "Jack and the Beanstalk," from 1807 (and a burlesque version named *The Story of Jack Spriggins and the Enchanted Bean* from 1734) presents a young boy whose mother plants foolishly obtained seeds which then grow into a great tower that can even hold a giant! Neither of these stories addresses the physics questions of how the towers can remain upright under compressive and buckling (bending) forces.

It was therefore up to the famous Russian scientist Konstantin Tsiolkovsky in 1895 to integrate the vision of the space elevator with the realities of physics (Edwards & Westling, 2003). Tsiolkovsky was considered to be a rocket scientist and the father of spaceflight and he had spent considerable time thinking about the limitations and alternatives of rocket flight. He was inspired by the Eiffel Tower in Paris to conceptualize a tower that reached from ground zero all the way into deep space, above the geosynchronous satellite orbit. This "celestial castle" would orbit the Earth in a geosynchronous fashion meaning that it would be directly overhead one spot on Earth's surface at all times. An object released at the tower's top would also have the orbital velocity necessary to remain in geosynchronous orbit. Thus, the Tsiolkovsky's tower can be used to deploy satellites into orbits around the Earth.

The centrifugal force due to Earth rotation acting on celestial towers has an interesting effect: A tall enough tower is under *tension* rather than compression, and therefore is not subject to the sorts of buckling that limits the height of skyscrapers. In the case of skyscrapers, the centrifugal force is negligible, but for the celestial size objects envisioned by Tsiolkovsky and his followers, both scientists and science fiction writers such as Arthur C. Clarke (1978, 1982), the gravitational force and centrifugal force play equally significant roles. Because the internal force is a tension rather than compression, the space elevator can be a floppy non-rigid object ("string").

It wasn't until 1960 that someone suggested a feasible method for building the space elevator. Another Russian scientist, Yuri N. Artsutanov, conceived a scheme for building a space tower (Artsutanov, 1960, July 31; Lvov, 1967). Artsutanov proposed using a geosynchronous satellite as the starting point from which to construct the tower. By using a counterweight, a cable would be lowered from the satellite down to the Earth surface while the counterweight was extended from the satellite away from Earth, keeping the center of gravity of the cable motionless relative to the rotating Earth. This construction scheme is still the standard (Edwards & Westling, 2003).

Jerome Pearson (1975) brought the idea of the space elevator to the scientific community in the U.S. In his careful and detailed design of a workable space elevator while at the U.S. Air Force Flight Dynamics Laboratory he outlined, mathematically and physically, the implications of a space elevator string designed to have the constraint of constant stress (tension/cross sectional area) throughout, while maintaining an external force balance. Artsutanov independently proposed the same idea (Artsutanov, 1960, July 31; Lvov, 1967). The two balancing external forces in the earth frame are the centrifugal and gravitational forces. The Pearson-Artsutanov constant stress elevator provides a simple way to handle the high tensions present in space elevators: The elevator can be designed for *any* given value of the constant stress. This value is can be chosen to be smaller than the critical breaking stress of the material used. Hence, from the materials science point of view, real space elevators *can* be made. This spurred a lot of recent interest in building space elevators out of novel materials such as carbon nanotubes (Yu et al., 2000a, Yu et al., 2000b, Edwards & Westling, 2003).

It is very easy to understand the advantages of the space elevator concept over conventional rocket propulsion. With chemical propulsion, a rocket carries its own fuel that it needs to overcome gravitational forces, leading to intrinsic energy inefficiency. Because of earth's deep gravitational well, the load-to-fuel ratios are typically very small (e.g., ~10^{-2} for the Apollo/Saturn V missions to the Moon), so that essentially all fuel energy is used to accelerate the fuel itself. On the other side, within the space elevator concept, a spaceship climbs along the elevator via an internal electrical engine which uses externally supplied electric energy. Since there is no fuel carried by the climber, the supplied energy is 100% used to lift the climber. So, the space elevator concept is immensely more energy efficient than the rocket propulsion.

The problem however remains on how to externally supply the energy to the climber. Naively, one may think of running an electrical transmission line along the space elevator, until realizing just how long this structure is compared to transmission lines on earth, so that power losses will be close to 100%. To remedy this, Edwards proposes that laser power be beamed up the elevator from the ground to the climber (Edwards & Westling (2003)). The beam energy would be absorbed by climbers and converted into electrical energy driving their engines. For any of these schemes climbing is typically slow and it may take several months for the climber to travel along the space elevator from the Earth to the geosynchronous level. During such a long climb, the useful load (including possibly humans) would be exposed for a very long time to dangerous cosmic radiation, which is especially strong in this range of altitudes above the Earth. Even without this problem with cosmic rays, typically long travel time itself is certainly not a satisfactory feature of space elevators, especially if a rapid deployment of objects into outer space is desired.

3. Rotating Space Elevator (RSE): Solution of the Climbers Energy Supply Problem

In this review article we discuss a novel class of nonlinear dynamical systems, Rotating Space Elevators (RSE). The RSE concept has been introduced for the first time by Golubovic and Knudsen (2009) and elaborated in detail in their subsequent studies (Knudsen & Golubovic, 2014; Knudsen & Golubovic, 2015).

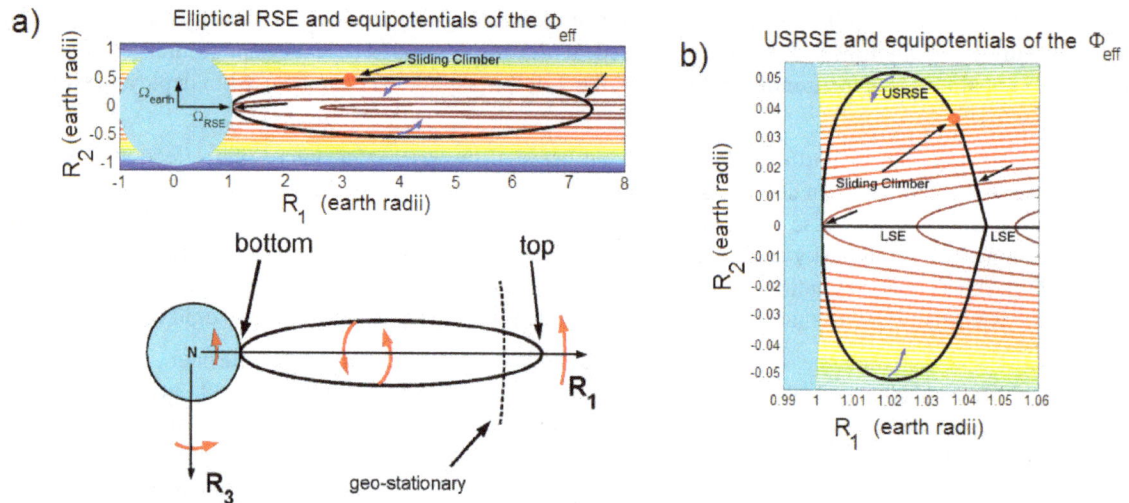

Figure 1. In (a) upper panel, the elliptical RSE with minor semi-axis $b = 0.5$ Earth radii and major semi-axis $a = 3.2107$ Earth radii (so its top is about 0.8 Earth radii above the geo-stationary level). In (a) lower panel, a different view on the RSE. The coordinate system (R_1, R_2, R_3) rotates together with the Earth around the R_2-axis pointing through the north pole N. Indicated are the internal (nearly around the R_1-axis) and geo-synchronous (together with the Earth) rotations of the RSE. The RSE is tied to the Earth at its bottom point. In (b), we show the USRSE (attached to a LSE) with $T_{RSE} = 4.22$ min (discussed in sect. 4). In these figures we also include the equipotentials of the effective potential in eq. (7). Sliding climbers oscillate between two turning points (indicated by straight arrows) that are on the same equipotential. Adapted from Golubovic and Knudsen (2009)

The RSEs are multiply rotating systems of strings. Remarkably, useful loads and humans *sliding* along the RSE strings do *not* require internal engines or propulsion to be rapidly transported (sled) into space far away from the Earth's surface; see section 4 discussions. Thus, the RSE concept solves the major problem of energy supply to climbers that troubles the ordinary LSE concept.

The RSE is a *double* rotating floppy string typically having the shape of a loop as in Figure 1. Due to its special kind of motion (see below), the RSE becomes pre-tensioned due to gravitational and inertial forces. Due to the tension, the floppy RSE maintains its loopy shape.

The special RSE motion, ensuring the persistence of its shape, is a nearly a geometrical superposition of: (a) geosynchronous (one day period) rotation around the Earth, i.e. the R_2-axis in Figure 1, and (b) yet another rotational motion of the string which is typically much faster (with period ~ tens of minutes) and goes on around a line perpendicular to the Earth at its equator (the R_1 axis in Figure 1).

This second, *internal rotation* plays a very special role: It provides the dynamical stability of the RSE shape and, importantly, it also provides a mechanism for the climbing of objects free to slide along the RSE string.

RSE can be used to elevate climbers from the surface of the Earth to remote outer space locations in a simple way; see Sec. 4. Remarkably, the climbers in Figure 1 do not need any internal engine to execute their motion. Rather, they *spontaneously* slide along the RSE string from the Earth to outer space locations. This unusual climber sliding motion is facilitated by the inertial force (centrifugal force) acting on climbers due to the RSE's internal rotation. In section 6 we describe possible use of RSE to launch satellites and interplanetary spaceships.

The RSE can be made in various shapes. The simple double rotating geometrical motion can be made to represent an approximate yet exceedingly accurate solution to the exact equations of the RSE string dynamics. This RSE feature is accomplished by a special ('magical') choice of mass distribution of the RSE cable; see Sec. 4. This RSE feature is corroborated by numerical simulations showing that (under the conditions discussed in Sec. 5) the RSE double rotation motion as well as nearly constant RSE shape can both persist indefinitely in time (Golubovic & Knudsen, 2009; Knudsen & Golubovic, 2014; Knudsen & Golubovic, 2015).

The elliptical RSE in Figure 1(a) exhibits very high tensile stresses at its points near mid-height. Therefore, other shapes were described whose mass distribution yields a Uniform Stress RSE (USRSE) (Golubovic & Knudsen, 2009, Knudsen & Golubovic, 2014). The USRSE, displayed in Figure 1(b), can be made by using technologically available materials such as carbon nanotubes; see section 4.

4. The physics of RSE

In this section we review the basic physics of RSE (Golubovic & Knudsen, 2009). For simplicity, let us consider inextensible limit in which the RSE floppy string obeys the Newtonian equation of motion

$$\mu(s)\frac{\partial^2 \vec{R}}{\partial t^2} = \frac{\partial}{\partial s}T(s,t)\frac{\partial}{\partial s}\vec{R} + \vec{f}_{ext} \tag{1}$$

In Eq. (1), $\vec{R}(s,t)$ are 3-D space positions of string points parametrized by their arc-length distances s [thus, $|\partial\vec{R}(s,t)/\partial s| = 1$]. $\mu(s)$ in Equation (1) is the string's local mass line density, while $T(s,t)$ is the local value of the string's tension field, and $\vec{f}_{ext}$ are external forces (per unit length) acting on the string. Prominent among them is the force of the Earth gravity, for which,

$$\vec{f}_{ext}{}^{grav} = \mu(s)\ \vec{a}_{grav} \,,$$

with

$$\vec{a}_{grav} = -\frac{\partial\Phi_{grav}}{\partial\vec{R}}.$$

Here,

$$\Phi_{grav}(\vec{R}) = -\frac{GM_{earth}}{|\vec{R}|}$$

is the Earth's gravitational potential with M_{earth} = the Earth's mass. $T(s,t)$ is obtained by solving Eq. (1) combined with the local constraint $|\hat{t}(s,t)| = 1$, with $\hat{t}(s,t) = \partial\vec{R}/\partial s$. In the presence of a sliding climber of mass m_{cl} at the arc-length distance $s(t)$, i.e., at the 3-D position $\vec{R}^{cl}(t) = \vec{R}(s(t),t)$, the external force density acting on the RSE, $\vec{f}_{ext}$ in Eq. (1) includes also the term

$$\vec{f}_{ext}{}^{cl} = -\delta(s - s(t))\ \vec{N} \,,$$

where $\vec{N}$ is the normal force exerted by the climber on the string; $\vec{N}\cdot\hat{t} = 0$. The climber dynamics is governed by the second Newton's law,

$$m_{cl}\frac{d^2\vec{R}^{cl}}{dt^2} = \vec{N} + \vec{F}_{ext} \,,$$

with $\vec{F}_{ext}$ signifying (other than $\vec{N}$) external forces acting on the climber, e.g., the Earth's gravity,

$$\vec{F}_{ext}{}^{grav} = m_{cl}\ \vec{a}_{grav} = -m_{cl}\frac{\partial\Phi_{grav}}{\partial\vec{R}}.$$

Using the constraint $|\hat{t}(s,t)| = 1$, from the above equations one finds

$$\ddot{s}(t) = \hat{t}\cdot\left(\frac{\vec{F}_{ext}}{m_{cl}} - \frac{\partial^2\vec{R}}{\partial t^2}\right), \tag{1'}$$

and

$$\vec{N} = -(\vec{F}_{ext})_{\perp} + m_{cl}\left(\frac{\partial^2\vec{R}}{\partial t^2}\right)_{\perp} + m_{cl}\frac{\partial\hat{t}}{\partial s}[\dot{s}(t)]^2 + 2m_{cl}\frac{\partial\hat{t}}{\partial t}\dot{s}(t) \,,$$

with $(\vec{V})_{\perp} = \vec{V} - \hat{t}(\hat{t} \cdot \vec{V})$ for any vector $\vec{V}$.

Space elevators such as RSE are most naturally discussed by using *non-inertial reference frames*. In a non-inertial frame rotating with the angular velocity $\vec{\Omega}$, inertial forces have to be included into $\vec{f}_{ext}$ and $\vec{F}_{ext}$ in Eqs. (1) and (1'), yielding

$$\vec{f}_{ext} = \mu(s) (\vec{a}_{grav} + \vec{a}_{inert}), \quad \vec{F}_{ext} = m_{cl} (\vec{a}_{grav} + \vec{a}_{inert}), \tag{2}$$

with,

$$\vec{a}_{inert} = \vec{\Omega} \times (\vec{R} \times \vec{\Omega}) + \vec{R} \times \frac{d\vec{\Omega}}{dt} + \frac{d\vec{R}}{dt} \times 2\vec{\Omega}, \tag{3}$$

(Landau & Lifshitz, 1976). In the geosynchronous frame (used in the simulations and the figures displayed here) rotating with the period $= 2\pi / \Omega_{earth}$ = one day, Eq. (3) is employed with

$$\vec{\Omega} = \vec{\Omega}_{earth} = \hat{e}_2 \Omega_{earth},$$

with unit vector $\hat{e}_2$ along the Earth polar axis and the equator in the (R_1, R_3) plane; see Figs. 1(a) and (b). In the simulations, at $t = 0$ the RSE is initially in the (R_1, R_2) plane. To initiate the double rotation motion, the RSE is given initial spin around the R_1-axis, with the angular velocity Ω_{RSE}. The RSE bottom point is tied to the Earth to provide access for the sliding climbers starting there their trip into outer space. Other than this, *the RSE moves purely under the influence of inertia and gravity.*

A remarkable effect of the RSE double rotation motion is that it facilitates a physical mechanism which efficiently moves sliding *engine free* climbers from the surface of the Earth to remote extraterrestrial locations. As evidenced by the simulations in Figure 2, a sliding climber starting at rest close to the Earth *spontaneously* oscillates between its initial position and a turning point in outer space. The nearly periodic character of the climber's motion is explained later on in this section.

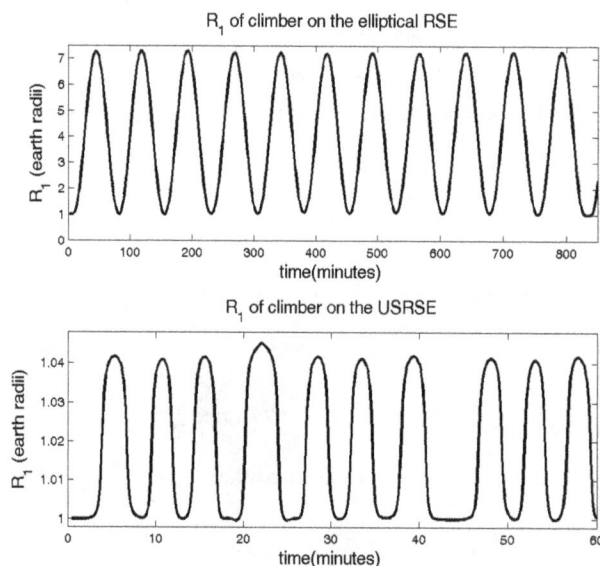

Figure 2. From the simulations of Knudsen and Golubovic (2014), the upper panel: The $R_1(t)$ coordinate of the climber which slides with no friction along the floppy RSE with the (initial) shape in Figure 1(a) and T_{RSE}=10.83 min. The lower panel: The $R_1(t)$ coordinate of the climber on the floppy RSE with initial shape in Figure 1(b) with T_{RSE}=4.22 min. See Sec. 4 for the analytic explanation of the nearly periodic character of climber motion. Note: With a weak sliding friction, climbers would eventually stop near the RSE point minimizing the $U(s) = \Phi_{eff}(\vec{R}(s))$. From the equipotentials of the effective potential labeled in Figure 1, one can see that this point occurs close to the RSE point maximizing its R_2 coordinate in Figure 1

Another remarkable effect is an enduring stability of the RSE sizes and orientation and the persistence the RSE's double rotation motion which is provided by a specially chosen form of the mass line density $\mu(s)$; see Eq. (8) below and Figure 3. This effect is documented by the simulations results displayed in Figure 4 (Golubovic & Knudsen, 2009; Knudsen & Golubovic, 2014).

These two outstanding RSE effects are revealed by considering the system in the (natural for the RSE) *double rotating frame* (DRF) obtained from the geosynchronous (single rotating) frame by adding to it the rotation around the R_1-axis in Figs. 1(a) and (b). The net angular velocity of the DRF is thus $\vec{\Omega}(t) = \vec{\Omega}_{RSE} + \vec{\Omega}_{earth}(t)$. Here, $\vec{\Omega}_{RSE} = \Omega_{RSE}\,\hat{e}_1$ corresponds to the rotation around the R_1-axis while $\vec{\Omega}_{earth}(t)$ is the Earth's angular velocity vector which in the DRF rotates with the angular velocity $-\Omega_{RSE}$ (and thus acquires a time-dependence). With this $\vec{\Omega}(t)$, in DRF the Eq. (3) yields

$$\vec{a}_{inert} = -\frac{\partial \Phi_{inert}}{\partial \vec{R}} + \vec{a}_{res}, \tag{4}$$

with

$$\Phi_{inert}(\vec{R}) = -\frac{1}{2}\Omega_{earth}^2 R_1^2 - \frac{1}{2}(\Omega_{RSE}^2 + \frac{1}{2}\Omega_{earth}^2)(R_2^2 + R_3^2), \tag{5}$$

being a *time-independent* effective potential generating inertial forces sensed in the DRF. The residual, $\vec{a}_{res}$ term in Eq. (4) includes velocity dependent terms that vanish for an object *at rest in the DRF*, as well as *fast time-dependent* oscillatory terms of $\vec{a}_{inert}$ (with frequencies Ω_{RSE} and $2\Omega_{RSE} \gg \Omega_{earth}$) that have *zero* time average over one RSE period ($T_{RSE} = 2\pi / \Omega_{RSE}$); see Knudsen and Golubovic (2014). For the here interesting situations with $\Omega_{RSE} \gg \Omega_{earth}$ [$T_{RSE} \sim 10\,min \ll T_{earth} = 2\pi/\Omega_{earth} = 1\,day$], the potential term in Eq. (4) dominates over the $\vec{a}_{res}$ term. Thus, Eq. (2) reduces to

$$\vec{f}_{ext} \cong -\mu(s)\frac{\partial \Phi_{eff}}{\partial \vec{R}}, \qquad \vec{F}_{ext} \cong -m_{cl}\frac{\partial \Phi_{eff}}{\partial \vec{R}}, \tag{6}$$

with the net effective potential,

$$\Phi_{eff}(\vec{R}) = \Phi_{grav}(\vec{R}) + \Phi_{inert}(\vec{R}) = -G\frac{M_{earth}}{|\vec{R}|} + \Phi_{inert}(\vec{R}). \tag{7}$$

The time *independence* of the $\Phi_{eff}(\vec{R})$ allowed Golubovic and Knudsen (2009) to find a special ('magical') mass distribution $\mu(s)$ with which the RSE string *indefinitely* maintains its initial shape by remaining at rest in the DRF. For a given flat RSE string shape specified by a 2-d curve $\vec{R}(s) = [R_1(s), R_2(s), R_3(s) = 0]$, this special mass line density $\mu(s)$ is obtained from Eq. (1) with $\partial^2 \vec{R} / \partial t^2 = 0$ and $\vec{f}_{ext}$ as in Eq. (6). The resulting differential equations for $T(s)$ and $\mu(s)$ can be integrated exactly, yielding the magical mass line density $\mu(s)$,

$$\ln\left[\frac{\mu(s)}{\mu(0)}\right] = -\int_0^s ds'\,\frac{(\vec{a}_{net}(s')\cdot\hat{t}(s'))}{K(s')} + \ln\left[\frac{K(s=0)}{K(s)}\right], \tag{8}$$

(Golubovic & Knudsen, 2009). Here,

$$\vec{a}_{net} = -\frac{\partial \Phi_{eff}}{\partial \vec{R}}$$

is evaluated at the RSE point $[R_1(s), R_2(s)]$, whereas

$$K(s) = -\frac{(\vec{a}_{net}(s)\cdot\hat{n}(s))}{C(s)} = \frac{T(s)}{\mu(s)}. \tag{8'}$$

In Eq. (8'), $C(s)$ is the local RSE string curvature; $C(s) = d\theta / ds$ with $\theta(s)$, the angle between the tangent unit vector $\hat{t}(s) = \partial\vec{R}/\partial s$ and the R_1 – axis in Figs. 1(a) and (b). The unit vector $\hat{n}(s)$ makes the angle $\theta(s) + \pi / 2$ with the R_1 – axis.

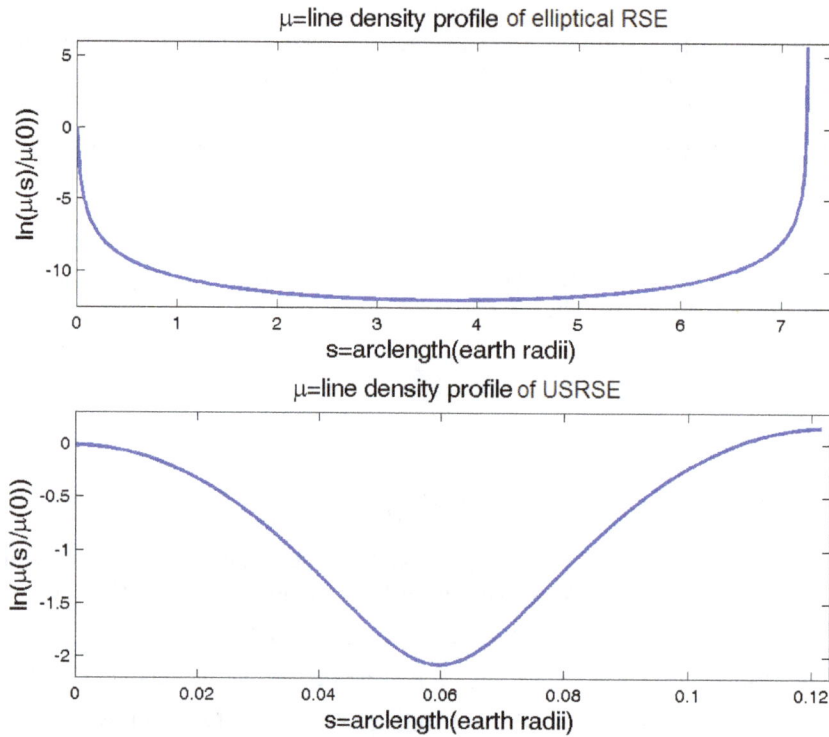

Figure 3. The upper panel: the magical mass distribution [i.e., line density obtained by Eq. (8)] of the RSE with the shape in Figure 1(a) and $T_{RSE} = 10.83\,\text{min}$. The lower panel: the magical mass distribution (line density) of the RSE with the shape in Figure 1(b) and $T_{RSE} = 4.22$ min. Adapted from Golubovic and Knudsen (2009)

The magical mass distributions obtained by applying Eq. (8) to the (initial) RSE shapes in Figure 1, which are used in the simulations discussed in this paper, are shown in Figure 3. The simulations of Knudsen and Golubovic (2014), which are free of the approximation Eq. (6) employed in Eq. (8), indeed show (under the conditions discussed in Sec. 5 in the following) a remarkable stability of the RSE sizes and orientation provided by the magical mass distribution in Eq. (8). We note that $\mu(s) = \rho\ A(s)$, with ρ the density of the RSE material and $A(s)$=the string cross-sectional area (that can be made to vary along the RSE by tapered cable design). Thus, by Eq. (8'), the tensile stress obeys the relation,

$$p(s) = \frac{T(s)}{A(s)} = \rho\ K(s) \tag{8''}$$

The RSE displayed in Figure 1(b) is a *uniform stress* RSE (USRSE) for which the tensile stress $p(s)$ is s-independent. For an USRSE, by Eq. (8''), the $K(s) = K = const$. With this condition, Eq. (8') yields the second order differential equation

$$\frac{d^2 R_2}{dR_1^{\ 2}} = -\frac{1}{K}\left[1 + \left(\frac{dR_2}{dR_1}\right)^2\right]\left[a_2 - a_1\frac{dR_2}{dR_1}\right], \tag{8'''}$$

with $a_i = -\partial\Phi_{eff}/\partial R_i$; $i = 1,\ 2$. Differential equation Eq. (8''') can be used to obtain a USRSE shape for *any*

given K and T_{RSE}. Figure 1(b) shows thus obtained USRSE shape for $T_{RSE} = 4.22$ min and $K = \tilde{K}\ v_1^{\ 2}$.

Here, $v_1 = (GM_{earth}/R_{earth})^{1/2} = 7.89\,km/\sec = 1^{\text{st}}$ cosmic speed [R_{earth} = the Earth radius], whereas

$\widetilde{K}$ is a dimensionless constant. For the USRSE in Figure 1(b), we set $\widetilde{K} = 1/4$, corresponding, by Eq. (8''),

to the USRSE tensile stress $p = \widetilde{K}\rho\, v_1^2 = 20.24\ GPa$ if the USRSE is made of carbon nano-tubes (CNT)

with $\rho \approx 1,300\ kg/m^3$. Thus, pleasingly, the tensile stress p of this USRSE is smaller than the tensile

strengths $p_{max} \approx 60\ GPa$ of single-wall CNT, and $p_{max} \approx 150\ GPa$ of multi-wall CNT (Yu et al., 2000a,

Yu et al., 2000b). *So, this* USRSE *is technologically achievable with modern day materials.* By Eq. (8) with
$K(s) = K = const.$, and by

$$- \vec{a}_{net} \cdot \hat{t} = (\partial\Phi_{eff}/\partial\vec{R}) \cdot (\partial\vec{R}/\partial s) = \partial\Phi_{eff}\ (\vec{R}(s))/\partial s,$$

we find that the USRSE magical mass line density obeys the equation

$$\ln\left[\frac{\mu(s)}{\mu(s_1)}\right] = \ln\left[\frac{A(s)}{A(s_1)}\right] = \frac{\Phi_{eff}\ (\vec{R}(s)) - \Phi_{eff}\ (\vec{R}(s_1))}{K}, \tag{8''''}$$

for any (s, s_1). It is depicted in lower panel of Figure 3 for the USRSE in Figure 1(b) [with T_{RSE}=4.22 min, and

$\widetilde{K} = 1/4$], with $s_1 = 0$ corresponding to the USRSE bottom at the Earth. This line density profile can be

technologically achieved by using tapered cable having the cross-sectional area $A(s)$ given by our Eq. (8''''). The USRSE in Figure 1(b) is actually attached to a Linear Space Elevator (LSE) which can also be designed to have a uniform stress $< p_{max}$ (Pearson, 1975). The LSE line mass density has a discontinuity at the junction between the USRSE and the LSE (to balance the USRSE tension force pulling down the LSE along the R_1-axis). Away from the junction, the uniform stress LSE line mass density obeys Eq. (8'''') with $\vec{R}(s) = (R_1 = s, R_2 = 0, R_3 = 0)$ (Pearson, 1975). We note that unlike the technologically achievable USRSE in Figure 1(b) with $p < p_{max}$ (for CNT), the Earth based elliptic RSE in Figure 1(a) has a *non*-uniform stress that actually exceeds the CNT tensile strength in the midsection of the RSE. However, elliptic RSEs built on dwarf planets such as asteroids Ceres and Vesta would have a tensile stress $p < p_{max}$ (for CNT). They are thus technologically achievable.

Periodic like motion of sliding climbers (seen in the simulations in Figure 2) goes on along nearly constant shape RSE strings. This climber motion can be understood by means of Eq. (6). For a time-independent RSE shape $\vec{R}(s)$, the Eq. (1'), in combination with Eq. (6), reduces to

$$\ddot{s} = -\hat{t}(s) \cdot \frac{\partial\Phi_{eff}}{\partial\vec{R}(s)} = -\frac{\partial\vec{R}(s)}{\partial s} \cdot \frac{\partial\Phi_{eff}}{\partial\vec{R}(s)} = -\frac{\partial\Phi_{eff}\ (\vec{R}(s))}{\partial s},$$

yielding the conservation law

$$\frac{1}{2}\left(\frac{ds(t)}{dt}\right)^2 + \Phi_{eff}\ [\vec{R}(s(t))] = const. \tag{9}$$

Eq. (9) is isomorphic to the familiar conservation law describing oscillatory 1-d motion of a particle with the position $s(t)$ in the potential $U(s) = \Phi_{eff}(\vec{R}(s))$. Strikingly, in this potential, sliding climbers simply oscillate between two turning points, one of which is close to the Earth (starting point) whereas the other one is in outer space; see Figs. 1 and 2. In fact, the RSE bottom (the point $s = 0$) becomes a local maximum of the potential $U(s)$ (seen by sliding climbers) provided the RSE angular frequency Ω_{RSE} is bigger than the minimal frequency,

$$\Omega_{min} \cong \frac{v_1}{R_{earth}}(1 + R_{earth}\ |C(s = 0)|)^{1/2} = \frac{v_1}{R_{earth}}(1 + \frac{1}{\widetilde{K}(s = 0)})^{1/2}, \tag{10}$$

with $\widetilde{K}(s) = K(s)/v_1^2$ (Golubovic & Knudsen, 2009; Knudsen & Golubovic, 2014). Due to this, for

$\Omega_{RSE} > \Omega_{min}$, i.e., $T_{RSE} < T_{max} = 2\pi/\Omega_{min}$, a climber initially at rest will start moving up no matter how

close is its initial position to the RSE bottom [$R_1 = R_{earth}$, $R_2 = R_3 = 0$)] in Figs. 1(a) and (b), at $s = 0$. For the elliptical RSE in Figure 1(a) with the semi-axes $b = 0.5\ R_{earth}$, $a = 3.2107\ R_{earth}$, the climbing threshold RSE

period $T_{max} = 22.71\ min$. This is bigger than the T_{RSE} of $10.83\ min$ of the elliptic RSE in Figure 1(a), yielding the oscillatory sliding climber dynamics seen in the simulations in Figure 2, upper panel. For an USRSE

with $\widetilde{K}(s) = 1/4$, by Eq. (10), $T_{max} = 37.78\ min$. This is bigger than the T_{RSE} of $4.22\ min$ of the USRSE

in Figure 1(b), yielding the oscillatory sliding climber dynamics seen in the simulations in Figure 2, lower panel. We note that the USRSE point having the maximum distance R_2 away from the $R_1 - axis$ in Figure 1(b) has the speed $\Omega_{RSE} \cdot (R_2)_{max} \cong v_1 = 1^{st}$ cosmic speed (for the USRSE with $T_{RSE}=4.22\ min$). Thus, the USRSE loop in Figure 1(b) can be used for launching satellites. We will discuss potential applications of the RSE in Sec. 6. It is significant to note that [by using differential Eq. (8''')] the USRSE loops can be designed with their bottoms anywhere above the Earth surface (e.g., above the dense atmospheric layer, to avoid air-resistance).

5. Shape Stability of RSE and Crumpling Transition

Numerical simulations of RSEs reveal an interesting morphological phase transition of the RSE strings that occurs with changing the (initial) RSE angular frequency Ω_{RSE}, i.e., its period T_{RSE} (Golubovic & Knudsen, 2009; Knudsen & Golubovic, 2014). This transition was seen both in the USRSE and the elliptical RSE. E.g., for the elliptical RSE in Fig 1(a), it occurs at a critical value for the RSE period $T_{crit} \approx 17\ min$. For $T_{RSE} < T_{crit}$, the tension field $T(s,t)$ remains everywhere positive. It exhibits only small oscillations around $T(s,t = 0) > 0$ given by Eqs. (8) and (8'). However, for $T_{RSE} > T_{crit}$, the RSE string (in both the elliptical RSE and the USRSE) undergoes a dramatic shape change and chaotic long time dynamics: Tension field develops a noise like behavior in which $T(s,t)$ assumes both positive and negative values. In effect, the RSE string *crumples* due to the buckling of the string sections that are under locally negative $T(s,t)$. Macroscopically, the string crumpling triggers a narrowing of the RSE initial shape, displayed in Figure 4(d) for the elliptical RSE with $T_{RSE} = 21.66\ min$. The narrowing eventually turns initially elliptic RSE into two nearly independently fluctuating linear type space elevators connecting the massive elevator top and bottom regions. Chaotic dynamics of the two elevator's branches reflects an ergodic-like (thermal equilibrium like) string state similar to that of the directed polymers (Kardar, 2007) stretched between the RSE top and bottom. Related to the RSE narrowing is the dynamics of the RSE angular momentum L_1 about the R_1-axis (in the frame rotating with the Earth as in Figure 1); see Figure 4(c) for the elliptical RSE with $T_{RSE} = 21.66\ min$: The L_1 decays to zero over a two week period. [In contrast to this, for $T_{RSE} > T_{crit}$, the USRSE narrows and loses its L_1 only partially.] As seen in Figs. 4 (a) and (b) at $T_{RSE} = 21.66\ min$, these phenomena destabilize the position of the elliptical RSE top. It drifts away from its initial position (at 0.8 Earth radii above geostationary level) to a new slightly higher position around which the RSE top continues to chaotically oscillate.

In drastic contrast to this, for $T_{RSE} < T_{crit}$, the string shapes of both the elliptic RSE and the USRSE remain

nearly the same as in their initial configuration in Figs. 1(a) and (b), i.e., no RSE narrowing occurs. Related to this, as evidenced in Figure 4(c) for the elliptical RSE with $T_{RSE} = 10.83\ min$, the RSE angular momentum L_1 is nearly constant in time, whereas the RSE top exhibits only very small oscillations around its initial position; see Figure 4(a) and (b).

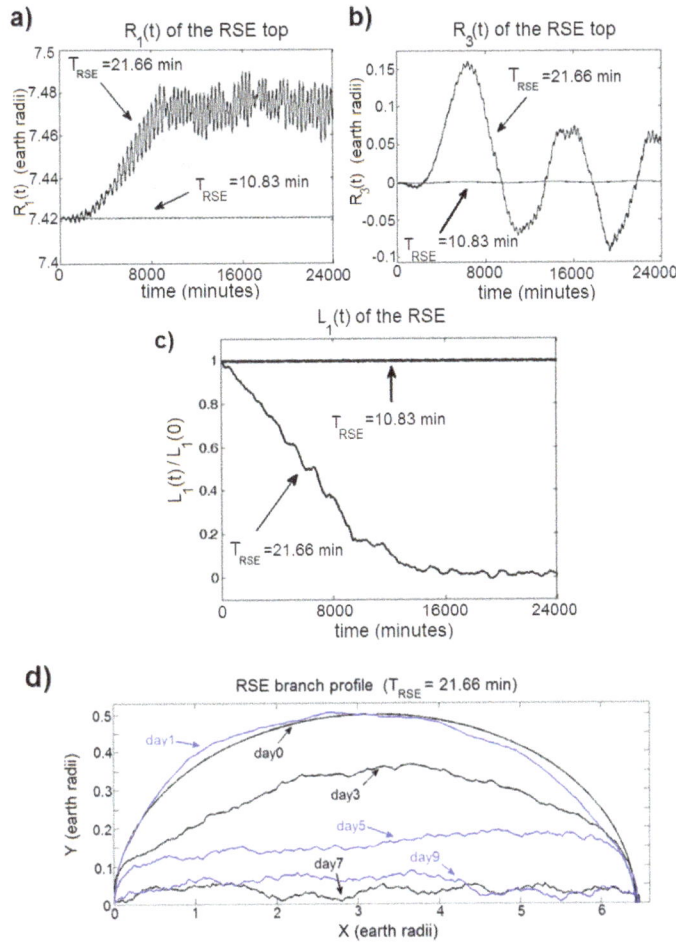

Figure 4. From the simulations of Knudsen and Golubovic (2014): For the elliptic RSE in Figure 1(a), the RSE top coordinates $R_1(t)$ in (a), $R_3(t)$ in (b), and, in (c), the evolution of the RSE angular momentum L_1 about the R_1-axis (in the frame rotating with the Earth), for $T_{RSE} = 10.83$ min and $T_{RSE} = 21.66$ min . In (d), the evolution of the RSE profile (of one of its two branches), for $T_{RSE} = 21.66$ min over the first ten days. Here, for any RSE point P, the Y is its distance away from the (instantaneous) axis A connecting the RSE bottom and top, and X is the distance between the normal projection of P onto the axis A and the RSE bottom point   (at $$X = 0)$$

6. Using RSEs to Launch Spaceships and Satellites

By the discussions of Sec. 4, RSE is a rapid extraterrestrial transportation system which requires no internal engines for the climbers sliding along the elevator strings. Climbers motion is naturally facilitated by employing basic natural phenomena, the inertial forces due to the internal RSE rotation around the R_1-axis in Figure 1. As noted before in Sec. 4, RSEs can be used to launch satellites and interplanetary spaceships (Knudsen & Golubovic, 2014). In this section we will discuss this RSE capability in more detail. It is also significant to note that the RSE strings can be used to host space stations and research posts. Sliding climbers can be then used to transfer useful loads and humans from the Earth to these extraterrestrial locations.

Satellites and spaceships transported by sliding climbers can be released (launched) along RSEs. Let us look at a climber that has started its motion at near rest at the RSE tying position with the Earth in Figure 1 [there, $|\vec{R}(s = 0)| = R_{earth}$]. Let us then look at an object released from the climber when it reaches the RSE position $\vec{R}(s) = [R_1(s), R_2(s), R_3(s) = 0]$ in the DRF. The released object's speed in the DRF, that is its speed along the tangent at $\vec{R}(s)$ is ds/dt. This tangential velocity is obtained by Eq. (9) yielding,

$$\frac{1}{2}\left(\frac{ds(t)}{dt}\right)^2 = \Phi_{eff}[\vec{R}(s(0) = 0)] - \Phi_{eff}[\vec{R}(s(t))] \cdot \qquad (11)$$

Using here the Eqs. (5) and (7), we find,

$$\frac{1}{2}\left(\frac{ds(t)}{dt}\right)^2 = -G\frac{M_{earth}}{R_{earth}} + G\frac{M_{earth}}{|\vec{R}(s)|} + \frac{1}{2}\Omega^2_{earth}[R_1(s)]^2 + \frac{1}{2}(\Omega^2_{RSE} + \frac{1}{2}\Omega^2_{earth})[R_2(s)]^2. \quad (12)$$

Of the practical interest is the released object speed observed in the *inertial frame*. Consider, for example, the situation in which the object is released when the elevator loop is in the plane of Figure 1(a, upper panel) or Figure1 (b) (then the R_2 axis of DRF points along the north-south direction). Let the object be released from a climber which is on the *lower branch* of RSE in Figure 1. The RSE velocity at this point is in the direction of the Earth rotation and has the magnitude

$$v_{RSE} = \Omega_{RSE}|R_2(s)| + \Omega_{earth}R_1(s), \quad (13)$$

in the inertial frame. This velocity points into the plane of Figure 1. In addition to this velocity, the released object also has the tangential velocity *ds/dt* which is in the plane of Figure 1. Thus, by Pythagorean Theorem, the total released object speed in the inertial frame, $v_{released}$ satisfies

$$v_{released} = \sqrt{(v_{RSE})^2 + (ds/dt)^2}. \quad (14)$$

By Eqs. (12) through (14), the speed of the released object in the inertial frame satisfies the equation

$$\frac{1}{2}(v_{released})^2$$
$$= -G\frac{M_{earth}}{R_{earth}} + G\frac{M_{earth}}{|\vec{R}(s)|} + (\Omega^2_{RSE} + \frac{1}{4}\Omega^2_{earth})[R_2(s)]^2 \quad (15)$$
$$+ \Omega_{RSE}\Omega_{earth}|R_2(s)|R_1(s) + \Omega^2_{earth}[R_1(s)]^2,$$

if the object is released when the elevator loop is in the plane of Figure 1(a, upper panel) or Figure 1 (b), from a climber which is on the lower branch of RSE in Figure 1. If this speed is large enough, the released object will unbind from the Earth, and approach infinity with the speed v_∞ ("escaping speed") that can be obtained from the mechanical energy conservation law,

$$\frac{1}{2}(v_\infty)^2 = \frac{1}{2}(v_{released})^2 - G\frac{M_{earth}}{|\vec{R}(s)|}. \quad (16)$$

By Eqs. (15) and (16),

$$\frac{1}{2}(v_\infty)^2$$
$$= -(v_1)^2 + [\Omega_{RSE}R_2(s)]^2\left[1 + \left(\frac{\Omega_{earth}}{2\Omega_{RSE}}\right)^2 + \frac{\Omega_{earth}R_1(s)}{\Omega_{RSE}R_2(s)} + \left(\frac{\Omega_{earth}R_1(s)}{\Omega_{RSE}R_2(s)}\right)^2\right]. \quad (17)$$

Above, we introduced the first cosmic speed $v_1 = (GM_{earth}/R_{earth})^{1/2} = 7.89\,km/sec$. Again we stress that the Eq. (17) applies if the object is released when the elevator loop is in the plane of our Figure 1(a, upper panel) of Figure 1(b), from a climber which is on the lower branch of RSE in Figure 1. This particular case is significant because the enhancement of the released object speed provided by the rotation of the Earth is at its maximum (for the RSE rotating around the R_1 axis in the direction indicated in Figure 1). For the interesting (for RSE systems) situations with $\Omega_{RSE} \gg \Omega_{earth}$ [see the discussions following Eq. (5)], the results in Eqs. (13), (15) and (17) reduce to approximate yet more illuminating results,

$$v_{RSE} \approx \Omega_{RSE}|R_2(s)|, \quad (13')$$

$$\frac{1}{2}(v_{released})^2 \approx -G\frac{M_{earth}}{R_{earth}} + G\frac{M_{earth}}{|\vec{R}(s)|} + \Omega^2_{RSE}[R_2(s)]^2, \quad (15')$$

while the escaping speed ("speed at infinity") approximately satisfies the simple equation,

$$\frac{1}{2}(v_\infty)^2 \approx -(v_1)^2 + [\Omega_{RSE}R_2(s)]^2. \quad (17')$$

The approximate results (13'), (15'), and (17') are equivalent to ignoring the Earth rotation (setting $\Omega_{earth}=0$) in the exact results in Eqs. (13), (15), and (17). Thus, the above approximate results are significant also because they (approximately) apply to the objects released from the RSE at any orientation of the rotating RSE plane relative the R_1-R_2 plane in Figure 1. [We stress that $|R_2(s)|$ in the above equations is the distance between the release point and the R_1 axis.]

The approximate results (13'), (15'), and (17') offer a better insight into the launching actions of the RSE. By Eq. (17'), the released object (spaceship) will escape to an interplanetary travel $[(v_\infty)^2 > 0]$ if the RSE speed of the release point of $(=\Omega_{RSE}|R_2(s)|)$ is bigger than the first cosmic speed v_1. By the Eq. (17'), the highest possible escaping speed v_∞ is achieved if the object is released from a sliding climber at the RSE point with the maximum value of $|R_2|$. For example, for the elliptic RSE in the figure 1(a), this point is the midpoint of the RSE, with $|R_2|_{max}=b=0.5R_{earth}$ at $R_1=a+R_{earth}=4.2107\ R_{earth}$. At this point, with $T_{RSE}=10.83\ min$, one has $\Omega_{RSE}|R_2(s)|=3.9v_1$. With this value, the Eq. (17') predicts the value of the highest possible escaping speed (speed at infinity) from this RSE to be $(v_\infty)_{max} \approx 5.3310\,v_1$ which is only slightly smaller than $v_\infty = 5.5200\,v_1$ as obtained by using the exact Eq. (17). We note that for the marginal case with $\Omega_{RSE}|R_2(s)|=v_1$, whence the approximation Eq. (17') predicts $v_\infty = 0$, the exact equation (17) yields an $(v_\infty)^2 > 0$, meaning that the object still unbinds with a small escape velocity at infinity. A situation like this is (incidentally) realized in the USRSE in the Figure 1(b), with $T_{RSE}=4.22$ min. For its point with the maximum value of $|R_2|$ ($|R_2|_{max}=0.052R_{earth}$ at $R_1 =1.02R_{earth}$) we find $\Omega_{RSE}|R_2(s)|=1.04v_1$ (which is only slightly above v_1). The approximate Eq. (17') would then yield $v_\infty \approx 0.4040\,v_1$, whereas the exact Eq. (17) gives $v_\infty = 0.54303\,v_1$ for this case [if the object is released when the elevator loop is in the plane of our Figure 1, from a climber which is on the lower branch of USRSE in Figure 1(b)].

The shape of a USRSE loop is determined by solving the differential equation (8'''), and thus it depends on the value of $T_{RSE}=2\pi/\Omega_{RSE}$. In Figure 5 (upper panel), we display the USRSE shapes for several different values of T_{RSE}, all for the same value of the parameter $\widetilde{K}=1/4$ (corresponding to the string tensile stress $p = 20.24\,GPa$ if the USRSE is made of carbon nano-tubes, see Sec. 4). In Figure 5 (lower panel), we plot, versus T_{RSE}, the USRSE speed $\Omega_{RSE}|R_2(s)|_{max}$ as well as the speed at infinity v_∞ of an object released from a climber at $|R_2(s)|_{max}$ on the lower branch of USRSE in Figure 1(b) when this branch is in the plane of our Figure 1. With a known RSE shape, this speed can be calculated from Eq. (17). Note that v_∞ vanishes at a characteristic value of T_{RSE} of about 6 min.

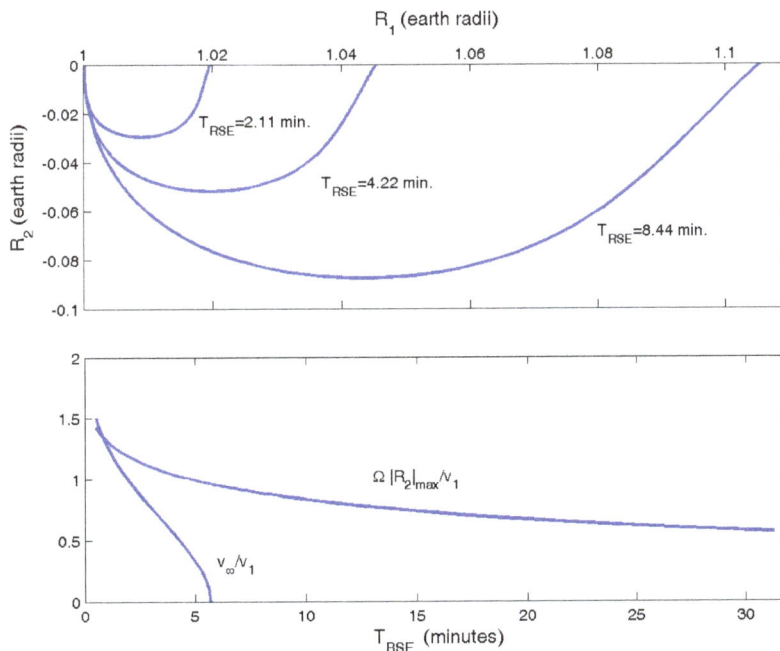

Figure 5. In the upper panel, we plot the USRSE loop shapes obtained for several different values of T_{RSE}, all for the same value of the parameter $\widetilde{K}=1/4$. In the lower panel, we plot, versus T_{RSE}, the USRSE speed $\Omega_{RSE}|R_2(s)|_{max}$ as well as the speed at infinity v_∞ of an object released from a sliding climber at $|R_2(s)|_{max}$ on the lower branch of USRSE in Figure 1(b) when this branch is in the plane of our Figure 1. It is obtained from Eq. (17). The two speeds are given in units of the first cosmic speed v_1. Adapted from Knudsen and Golubovic (2014)

Thus far we have discussed the case with no sliding friction between the climber and the RSE string. Only then Eq. (9) applies. The friction may be significantly depressed, e.g., by magnetic levitation. With some friction present, climbers would eventually stop near the RSE point minimizing the effective potential seen by the climber in DRF, $U(s) = \Phi_{eff}(\vec{R}(s))$. From the equipotentials displayed in Figure 1, this point occurs very close to the RSE point maximizing its R_2 coordinate in Figure 1. Interestingly, as noted above, the USRSE point having the maximum distance $|R_2|$ away from the R_1 – axis in Figure 1(b) has the speed $\Omega_{RSE} \cdot |R_2|_{max} \cong v_1 = 1^{st}$ cosmic speed (for the particular USRSE shown in Figure 1(b), with T_{RSE}=4.22 min and $\tilde{K} = 1/4$; see Sec. 4)   Thus, the USRSE loop in Figure 1(b) can be used for launching satellites carried from the Earth by a sliding climber. Indeed, due to the friction, the climber would eventually stop close to the USRSE point having the maximum distance away from the R_1 –axis in Figure 1(b). At this point, the stopped climber rotates with the RSE with the speed $\cong v_1 = 1^{st}$ cosmic speed. Thus, the climber can directly release the carried satellite into a nearly circular low Earth orbit.

7. Physics of Untied RSE

What will happen if one *unties* the elliptic RSE (ERSE) in Figure 1 from the Earth? This interesting question is investigated by Knudsen and Golubovic (2015). Interestingly, it was found that the tying may not be needed at all to achieve the stable double rotating motion of ERSE. In fact, the magical mass distribution $\mu(s)$ in Eq. (8) does *not* assume that the loopy ERSE is tied. Thus, it is in principle possible that an untied ERSE exhibits persistent shape and everlasting double rotating motion much like the tied ERSE. This intriguing possibility was explored by studying the dynamics of the untied elliptical RSE (Knudsen & Golubovic, 2015). The actual untied ERSE behavior was found to depend on the length of its long semi-axis a (along the R_1-axis in Figure 1).   The study shows that there are two characteristic values of a, called $a_{hopping}$ and $a_{unbinding}$. If $a < a_{hopping}$, the untied ERSE exhibits nearly the same dynamics as a tied ERSE. That is, its bottom and top points execute only very small oscillations about their initial positions. Thus, strikingly, for $a < a_{hopping}$, untied ERSE bottom effectively remains quasi-tied to the Earth. It remains close to the Earth as if the ERSE were tied to the Earth. On the other side, if $a > a_{hopping}$, the untied ERSE as a whole hops away and then it falls back to the Earth. The amplitude of this hopping (maximum height reached by the RSE bottom) increases with increasing a and it diverges as a approaches the $a_{unbinding}$. In this limit, as well as for any $a > a_{unbinding}$, the ERSE unbinds from the Earth much like an object with a speed above the second cosmic speed.

8. Summary

In summary, the RSEs are rapid outer space transportation systems that require no internal engines for the climbers sliding along the elevator strings. RSE strings exhibit interesting nonlinear dynamics and statistical physics phenomena.   RSEs' action fundamentally employs truly basic natural phenomena -- gravitation and inertial forces. Satellites and space-crafts carried by sliding climbers can be released (launched) along RSEs. RSE strings can host space stations and research posts. Sliding climbers can be then used to transport useful loads and humans from the Earth to these outer space locations. The RSE exhibits a variety of interesting dynamical phenomena explored by numerical simulations. Thanks to its special design aided by its magical mass distribution, the RSE exhibits persistent shape and enduring double rotating motion. Under some conditions however the RSE may undergo a morphological transition to a chaotic state reminiscent of fluctuating directed polymers encountered in the statistical physics of strings and membranes.

Acknowledgments

The authors thank WVHTC and NASA for the sub-grant WVHTC-W-NASA-IR-07-1367 of the Grant NNX06AE557: Innovative Research for Next Generation Space Exploration.

References

Artsutanov, Y. V. (1960, July 31). V Kosmos na Elektrovoze (To the Cosmos by Electric Train). *Komsomolskaya Pravda* (in Russian), front page of Sunday Supplement. English translation retrieved from http://www.spaceward.org/documents/Artsutanov_Pravda_SE.pdf

Clarke, A. C. (1978). *The Fountains of Paradise*. New York, NY: Harcourt Brace Jovanovich.

Clarke, A. C. (1981). The Space Elevator: 'Thought Experiment' or Key to the Universe?. *Advances in Earth Oriented Applied Space Technology Vol. 1*, 39-48. London, UK: Pergamon Press. Retrieved from http://www.islandone.org/LEOBiblio/CLARK1.HTM

Edwards, B. C., & Westling, E. A. (2003). *The Space Elevator: A Revolutionary Earth-to-Space Transportation System*. USA: BC Edwards.

Fitzgibbons, T. C., Guthrie, M., En-shi Xu E., Crespi, V. H., Davidowski, S. K., ... Badding, J. V. (2014), Benzene-derived carbon nanothreads, *Nature Materials, 14*, 43-47. http://dx.doi.org/10.1038/nmat4088

Golubović, L., & Knudsen, S. (2009). Classical and statistical mechanics of celestial scale spinning strings: Rotating space elevators. *Europhysics Letters, 86(3),* 34001. http://dx.doi.org/10.1209/0295-5075/86/34001

Kardar, M. (2007). *Statistical Physics of Fields.* Cambridge, UK: Cambridge University Press.

Knudsen, S., & Golubović, L. (2014). Rotating space elevators: Physics of celestial scale spinning strings. *European Physical Journal Plus, 129*, 242. http://dx.doi.org/10.1140/epjp/i2014-14242-8

Knudsen, S., & Golubović, L. (2015). Physics of untied rotating space elevators. *European Physical Journal Plus, 130*, 243. http://dx.doi.org/10.1140/epjp/i2015-15243-9

Landau, L. D., & Lifshitz, E. M. (1976). *Mechanics, Course in Theoretical Physics, Vol. 1.* London, UK: Pergamon Press.

Lvov, V. (1967). Sky-Hook: old idea. *Science, 158,* 946-947. http://dx.doi.org/10.1126/science.158.3803.946

Nelson, D. R. (2002). *Defects and Geometry in Condensed Matter Physics.* Cambridge, UK: Cambridge University Press.

Nelson, D. R., Piran, T., & Weinberg, S. (Eds.). (1988). *Statistical Mechanics of Membranes and Surfaces.* Singapore: World Scientific.

Pearson, J. (1975). The Orbital Tower: A Spacecraft Launcher Using the Earth's Rotational Energy. *Acta Astronautica, 2, 785–799.* http://dx.doi.org/10.1016/0094-5765(75)90021-1

Yu, M. F., Files, B. S., Arepalli, S., & Ruoff, R. S. (2000a). Tensile Loading of Ropes of Single Wall Carbon Nanotubes and their Mechanical Properties. *Physical Review Letters, 84,* 5552-5555; http://dx.doi.org/10.1103/PhysRevLett.84.5552

Yu, M. F., Lourie O., Dyer M. J., Moloni, K., Kelly, T. F., & Ruoff. R. S. (2000b). Strength and breaking mechanism of multiwalled carbon nanotubes under tensile load, *Science 287*, 637-640. http://dx.doi.org/10.1126/science.287.5453.637

20

The Effects of Pressure and Temperature on the Magnetic Susceptibility of Semiconductor Quantum Dot in A Magnetic Field

Faten BZOUR[1], Mohammad K. ELSAID[1] & Ayham SHAER[1]

[1] Physics Department, Faculty of Science, An- Najah National University, Nablus, West Bank, Palestine

Correspondence: Mohammad K. ELSAID, Physics Department, Faculty of Science, An- Najah National University, Nablus, West Bank, Palestine. E-mail: mkelsaid@najah.edu

Abstract

In this work, we present a theoretical study of the magnetic susceptibility (χ) of two-electron GaAs parabolic quantum dot (QD) under the combined effects of external pressure, temperature and magnetic field. We used the exact diagonalization method to obtain the eigenenergies by solving the two electron quantum dot Hamiltonian taking into account the dependence of the effective mass and dielectric constant on the hydrostatic pressure and temperature. The pressure and temperature show significant effects on the calculated QD spectra. Next, we investigate the behavior of the magnetization of a quantum dot as a function of external pressure, temperature, confining frequency and magnetic field. The singlet-triplet transitions in the ground state of the quantum dot spectra and the corresponding jumps in the magnetic susceptibility spectra have been shown. The comparison shows that our results are in very good agreement with the reported works.

Keywords: Pressure; temperature; magnetic susceptibility; quantum dot, magnetic field; exact diagonalization

1. Introduction

Recent nanofabrication methods have it possible to design different types of quantum dots with the flexibility of controlling the size, shape, and number of electrons. These controllable physical properties of the zero-dimensional nanostructure makes it promising candidate for a wide range of device applications like quantum dot lasers, solar cells, single electron transistors and quantum computers (Ashoori et al., 1993; Ciftja, 2013; Kastner, 1992; Loss & DiVincenzo, 1998; Burkard, Loss, & DiVincenzo, 1999).

Different approaches had been used to solve the two interacting electrons QD- Hamiltonian, including the presence of an applied magnetic field, and had obtained the eigenenergies and eigenstates of the QD-system as a function of magnetic field strength (Wagner, Merkt, & Chaplik, 1992; Taut, 1994; Ciftja & Kumar, 2004; Kouwenhoven, Austing, & Tarucha, 2001; Sanjeev Kumar, Mukhopadhyay, & Chatterjee, 2016; Kandemir, 2005; El-Said, 1995; El-Said, 1998; El-Said, 2000; Elsaid, Al-Naafa, & Zugail, 2008; Maksym & Chakraborty, 1990; De Groote, Hornos, & Chaplik, 1992). The energy spectra shows spin-singlet (S) and spin-triplet (T) ground state oscillations. These spin oscillations show themselves as transition peaks in the spectra of magnetic and thermodynamic quantities like magnetization (M), magnetic susceptibility (χ) and heat capacity (C_v) (Nguyen & Peeters, 2008; Nammas, Sandouqa, Ghassib, & Al-Sugheir, 2011; Boyacioglu & Chatterjee, 2012; Helle, Harju, & Nieminen, 2005; Schwarz et al., 2002; Räsänen et al., 2003; Climente, Planelles, & Movilla, 2004; Nguyen & Sarma, 2011; Rezaei & Kish, 2012; Dybalski & Hawrylak, 2005; Avetisyan, Chakraborty, & Pietiläinen, 2016).

The aim of this work, is to investigate the magnetic susceptibility of two interacting electrons confined in a parabolic quantum dot which is presented in a magnetic field. The applied magnetic field is uniform and its direction is taken to be along z-axis that is perpendicular to the x-y plane of the QD. In addition, we consider the effects of the external pressure and temperature on the magnetic susceptibility curve. We initially applied the exact diagonalization method to solve the QD Hamiltonian and obtain the eigenenergies for various values of physical QD parameters. Secondly, we investigate the dependence of the QD magnetic susceptibility, as a thermodynamic quantity, on the pressure, temperature confining frequency and magnetic field strength.

The rest of this paper is organized as follows: section II presents the Hamiltonian theory, computation diagonalization technique and the statistical thermodynamic relations of magnetic susceptibility for two interacting electrons in the quantum dot. In section III, we give the numerical results for the energy spectra and the magnetic susceptibility of the QD. We devoted the final section for conclusions.

2. Theory

In this section we describe in detail the main two parts of the theory, namely: quantum dot Hamiltonian and exact diagonalization method and the magnetic susceptibility of GaAs quantum dot.

2.1 Quantum Dot Hamiltonian

The effective mass Hamiltonian for two interacting electrons confined in a QD by a parabolic potential in a uniform magnetic field $\vec{B} = B\,\hat{k}$ can be written in a separable form as:

$$\hat{H} = \hat{H}_{CM} + \hat{H}_r \tag{1}$$

$$\hat{H}_{CM} = \frac{1}{2M}\left[\vec{P}_R + \frac{Q}{c}\vec{A}(\vec{R})\right]^2 + \frac{1}{2}M\omega_0^2 R^2 \tag{2}$$

$$\hat{H}_r = \frac{1}{2\mu}\left[\vec{p}_r + \frac{q}{c}\vec{A}(\vec{r})\right]^2 + \frac{1}{2}\mu\omega_0^2 r^2 + \frac{e^2}{\epsilon|\vec{r}|} \tag{3}$$

Where ω_0 is the confining frequency, $\mu = \frac{m^*}{2}$ is the reduced mass, M=2m is the total mass, $q = \frac{e}{2}$ is the reduced charge, Q=2e is the total charged and ϵ is the dielectric constant for the GaAs medium. $\vec{R} = \frac{\vec{r_1}+\vec{r_2}}{2}$ and $\vec{r} = \vec{r_2} - \vec{r_1}$ are the center of mass and relative coordinates, respectively. $\omega_c = \frac{eB}{m^*}$ is the cyclotron frequency and $\mathbf{A} = \frac{1}{2}\mathbf{B} \times \mathbf{r}$ is the vector potential.

The corresponding energy of this Hamiltonian equation (1) is:

$$E_{total} = E_{CM} + E_r \tag{4}$$

The center of mass Hamiltonian given by equation (2) is a harmonic oscillator type with well-known eigenenergies:

$$E_{cm} = E_{n_{cm},m_{cm}} = (2n_{cm} + |m_{cm}| + 1)\hbar\,\omega + m_{cm}\frac{\hbar\omega_c}{2} \tag{5}$$

where n_{cm}, m_{cm} and $\omega = \sqrt{\frac{\omega_c^2}{4} + \omega_0^2}$ are the radial, angular quantum numbers and effective confining frequency, respectively.

However, the relative motion Hamiltonian part (H_r), given by equation (3) does not have an analytical solution for all ranges of ω_0 and ω_c. In this work, we applied the exact diagonalization method to solve the relative part of the Hamiltonian and obtain the corresponding eigenenergies E_r.

2.2 Exact Diagonalization Method and Magnetic Susceptibility

For non-interacting case the relative Hamiltonian in equation (3) is a single particle problem with eigenstates $|n_r m_r >$ (Kouwenhoven, Austing, & Tarucha, 2001),

$$|n_r m_r\rangle = N_{n_r m_r}\frac{e^{im_r\phi}}{\sqrt{2\pi}}(\alpha r)^{|m_r|}e^{-\alpha^2 r^2/2}L_{n_r}^{|m_r|}(\alpha^2 r^2) \tag{6}$$

where the functions $L_{n_r}^{|m_r|}(\alpha^2 r^2)$ are the standard associated Laguerre polynomials . We calculated the normalization constant $N_{n_r m_r}$ from the normalization condition of the basis, $< n_r m_r|\ n_r m_r >= 1$, which resulted in

$$N_{n_r m_r} = \sqrt{\frac{2n_r!\alpha^2}{(n_r + |m_r|)!}} \tag{7}$$

We used α as a constant which has the dimensionality of an inverse length

$$\alpha = \sqrt{\frac{m\omega}{h}} \tag{8}$$

The eigenenergies of the QD Hamiltonian which are given by equation (4) consist of the sum of the energies for the center of mass Hamiltonian (E_{cm}) and the eigenenergies(E_r)which are obtained by direct diagonalization to the relative Hamiltonian part. For interacting case, we applied the diagonalization method to solve equation (3) and find the corresponding exact eigenenergies for arbitrary values of ω_c and ω_0.

We can write the matrix element of the relative Hamiltonian part using the basis $|n_r m_r >$ as,

$$h_{nn'} = \langle n_r, m_r | \hat{H}_r | n'_r, m_r \rangle = < n_r m_r | \; -\frac{\hbar^2}{2\mu}\nabla^2 + \frac{1}{2} \; \mu \, \omega^2 r^2 \; |n'_r m_r > +$$

$$< n_r m_r | \; \frac{e^2}{\epsilon r} \; | n'_r m_r >. \tag{9}$$

The first term in the right side of equation (9) is diagonalized as,

$$[(2n + |m_z| + 1) \; \sqrt{(1 + \frac{\gamma^2}{4})} - \frac{\gamma}{2} | \; m_z \; | \;] \; \delta_{nn'} \tag{10}$$

Where the coulomb matrix energy can be given as

$$\frac{\lambda}{\sqrt{2}} \sqrt{\frac{n'! n!}{(n'+|m_z|)! \, (n+|m_z|)!}} \; \times I_{nn'} \; |m_z \; | \tag{11}$$

where $\gamma = \frac{\omega_c}{\omega_0}$ and $\lambda = \frac{e^2 \alpha}{\hbar\omega}$ are dimensionless parameters while $\omega^2 = 1 + \frac{\gamma^2}{4}$is the effective confining frequency.

By changing the coordinate transformation to t-variable by direct substitution of $r = \frac{\sqrt{t}}{\alpha}$ in the integration $I_{nn'} = I_{n_r n'_r}$, we can express the coulomb energy matrix element into the integral form:

$$< n_r m_r | \; \frac{e^2}{\epsilon r} \; |n'_r m_r > \propto I_{nn'|m_z|} = \int_0^\infty dt \; t^{|m_z|} e^{-t} L_n^{|m_z|}(t) L_{n'}^{|m_z|}(t) \frac{1}{\sqrt{t}} \quad . \tag{12}$$

We evaluated the above coulomb energy matrix element in a closed form by using the Laguerre relation given in the appendix A (Nguyen & Sarma, 2011).

This closed form result of the coulomb energy reduces greatly the computation time needed in the diagonalization process.

In our calculation, we have used the basis $| \; n_r m_r >$ defined by equation (6) to diagonalize the relative QD Hamiltonian and obtain its corresponding eigenenergies E_r. The exact diagonalization method is used in spanning the total Hamiltonian for the selected single electron basis and extract the lowest eigenvalues (eigenenergies) of the matrix. The procedure of increasing the number of linearly independent eigenstates is converging to the exact results. In each step the new energy results are compared with previous results from a smaller apace, until satisfactory convergence is achieved.

Next step, we have calculated the magnetic susceptibility (χ) from the mean energy $\langle E\,(T)\,\rangle$ and the magnetization (M) of the two-electron quantum dot using the statistical thermodynamic relations:

$$M = \frac{\partial \langle E(T) \rangle}{\partial B} \quad \text{and} \quad \chi = \frac{\partial \, M}{\partial \, B} \quad , \tag{13}$$

where, the statistical energy $\langle E(T) \rangle$ is obtained by the help of the partition function.

To include the effect of the pressure (P) and temperature (T) on the QD energy states and the magnetization we replace the dielectric constant ϵ with $\epsilon_r(P,T)$ and the effective mass m^* with $m(P,T)$ in the QD Hamiltonian as defined by Equations 2 and 3, where $\epsilon_r(P,T)$ and $m^*(P,T)$ are the pressure and temperature dependent dielectric constant and electron effective mass, respectively. These pressure and temperature dependent mass parameters should be included in the energy spectrum Eq.4 and the wave functions basis eq.6 of the Hamiltonian. For quantum dot made of GaAs the dependency of $\epsilon_r(P,T)$ and $m^*(P,T)$ are given in appendix B (Rezaei & Kish, 2012).

The pressure and temperature effective Rydberg $(R_y^*(P,T))$ is used as the energy unit and given as follows:

$$R_y^*(P,T) = \frac{e^2}{2\epsilon(P,T)a_B^*(P,T)} \tag{14}$$

Where $a_B^*(P,T)$ is the effective Bohr radius and given as:

$$a_B^*(P,T) = \epsilon(P,T)\hbar^2/(m^*(P,T)e^2) \tag{15}$$

So the effective Rydberg can be written as:

$$R_y^*(P,T) = \frac{e^4 m^*(P,T)}{2(\epsilon(P,T))^2 \hbar^2} \tag{16}$$

We change the pressure and temperature values to see their effects on the ground state energy of the QD Hamiltonian in both cases: zero ($\omega_c = 0$) and finite magnetic field (ω_c). Eventually, the ground state energies of the two electron-quantum dot system will be calculated as function of temperature (T), pressure (P), confining frequency (ω_0) and magnetic field cyclotron frequency (ω_c) .The obtained numerical results are displayed in the next section.

3. Results and Discussions

We present the effects of pressure, temperature, confining frequency and magnetic field cyclotron frequency on the magnetic susceptibility of two interacting electrons in a quantum dot made from GaAs material (effective Rydberg R*= 5.825 meV) in Figures 1 to 5 and Table 1. To achieve our aim, it is essential, as a first step, to investigate the dependence of the QD energy levels on the pressure and temperature. In Figure 1, we display the dependence of the QD energy states (m=0, 1, 2, 3 and 4) on the magnetic field, ω_c, for pressure P=10Kbar and temperature T=0.0K. We found that the overall shape of the spectra of the QD remains the same while the eigenenergies are enhanced under the effect of external pressure. For zero magnetic field and zero pressure case, we have tested in Table 1, the computed numerical results against the corresponding ones taken from the work of Ciftja and Kumar (Ciftja & Kumar, 2004). The comparisons give excellent agreement between the energy spectra of two-electron QD Hamiltonian solved by both exact numerical and variational methods (Ciftja & Kumar, 2004; Ciftja & Kumar, 2004; Kandemir, 2005; Dybalski & Hawrylak, 2005). The QD spectra shows transitions in the ground state angular momentum (m) as the magnetic field increases. For example, we observed the first transitions in the angular momentum of the ground state of the QD system, from m=0 to m=1, occurs at $\omega_c \approx 0.8\ R^*$ while the second transition (from m=1 to m=2) occurs at $\omega_c \approx 1.2\ R^*$. These transitions show themselves as cusps in the presented QD-magnetization curves. Figure 2, displays the energies of the quantum dot state (m=0) against the magnetic field for various pressure values: P =0, 10, 20 and 30 Kbar and Temperature T=0.0. For fixed particular value of the magnetic field, the figure clearly shows an enhancement in the energy level as the pressure increases. As the pressure increases the dielectric constant decreases leading to a significant electron-electron coulomb energy enhancement. Next we show the dependence of the magnetic susceptibility of the QD on the pressure, temperature and magnetic field. Figures 3a and 3b show the effects of the pressure on the dependence of the magnetic susceptibility on magnetic field. The magnetic susceptibility plotting is given for no external pressure (P=0.0) and for pressure value, P=30.0 Kbar, at temperature T=.01 K and confining energy frequency $\omega_0 = 0.5\ R^*$. Figures 3a and 3b show that as the pressure increases: P=0.0 Kbar to 30 Kbar, the absolute value of the magnetic susceptibility ($|\chi|$) at $\omega_c = 0.0$ also increases. In addition the heights of the peaks enhance as the magnetic field increases .For example, we can observe the changes in the height of peak number five which is located at $\omega_c \approx 3.8R^*$. This magnetic susceptibility behavior, under the effect of pressure, can be attributed to the enhancement of the electron-electron repulsive coulomb interaction energy in the quantum dot. As the pressure increases the dielectric constant,$\epsilon(p)$, decreases and thus the coulomb interaction energy term, $v_c \approx \frac{1}{\epsilon(p)r}$ in the QD-Hamiltonian, enhances. This leads to a significant energy increase in the total energy state of the electron making the electron unstable. The electron in this case makes a jump to another state with higher angular momentum (m) in order to reduce its coulomb energy, and in effective its total energy decreases, to become more stable. Figure 4 shows the effect of temperature on the QD magnetic susceptibility plot by changing the temperature from T = 0.01K (Figure 3a) to T=1K (Figure 4) for no external pressure (P=0.0 Kbar) and the same confining frequency, $\omega_c = 0.5R^*$. The magnetic susceptibility plot shows again a significant temperature dependence. We can clearly see the significant reduction in the height of magnetic susceptibility peaks as the temperature increases from T=0.01K to T=1K .For example,only the first peak located at $\omega_c = 0.4\ R^*$ remains, while the rest of the peaks at high magnetic field strength almost disappear due to the thermal fluctuations and the magnetic susceptibility becomes smooth. Furthermore, we study, in Figure 5, the effect of confining frequency ω_0 on the shape of the magnetic susceptibility spectra. We consider different values of confining frequencies:

$\omega_0 = 0.5 \, and \, 0.8 \, R^*$ keeping the temperature and the pressure parameters both are unchanged: P=0.0 kbar and T=0.01 K. The magnetic susceptibility spectra calculated at $\omega_0 = 0.5R^*$ and $\omega_0 = 0.8R^*$ are given in Figure 3a and Figure 5, respectively. The comparison of the spectra in both figures clearly shows a significant confining frequency dependence ω_0, which results in shifting the peak positions of the magnetic susceptibility spectra as clearly shown. These peaks- shift behavior can be understood in this way. As we increase the confining frequency, ω_0, the electron in the quantum dot becomes more confined and this electron energy enhancement shifts significantly the location of the energy level crossings to a higher magnetic field strength. For example, our confining frequencies: $\omega_0 = 0.5 \, and \, 0.8 \, R^*$ corresponds to transition magnetic field values: $\omega_c = 0.4 \, and \, 0.9 \, R^*$, respectively. The eigenenergies are obtained by diagonalizing the full QD-Hamiltonian matrix given by equation (9). In all steps of calculations, the numerical convergence is achieved. For example, the ground state energy calculated at $\lambda = 3$ in Table 1, converges to E_r =4.324, 4.320, 4.319 and 4.319 meV as the number of basis in the matrix elements S_p = 5, 20, 40 and 50, respectively.

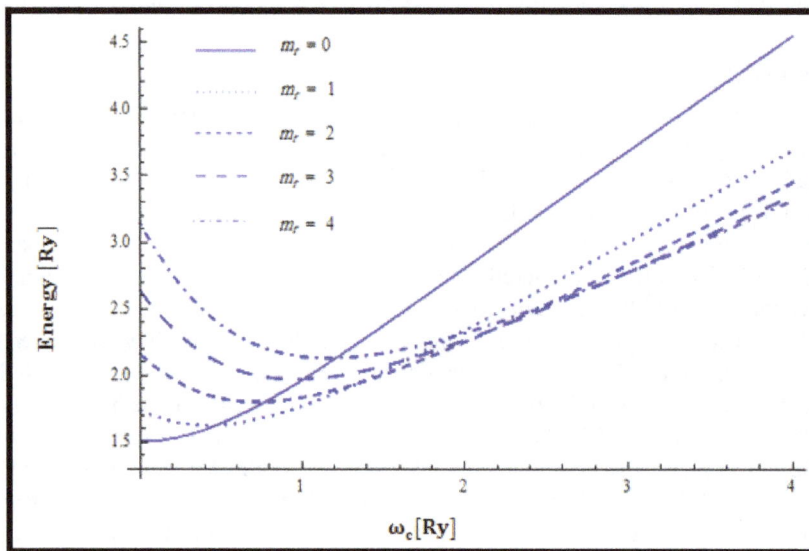

Figure 1. The computed energy spectra of quantum dot versus the strength of the magnetic field for $\omega_0 = 0.5R^*$, T=0K, P=10Kbar and angular momentum m = 0,1,2,3,4

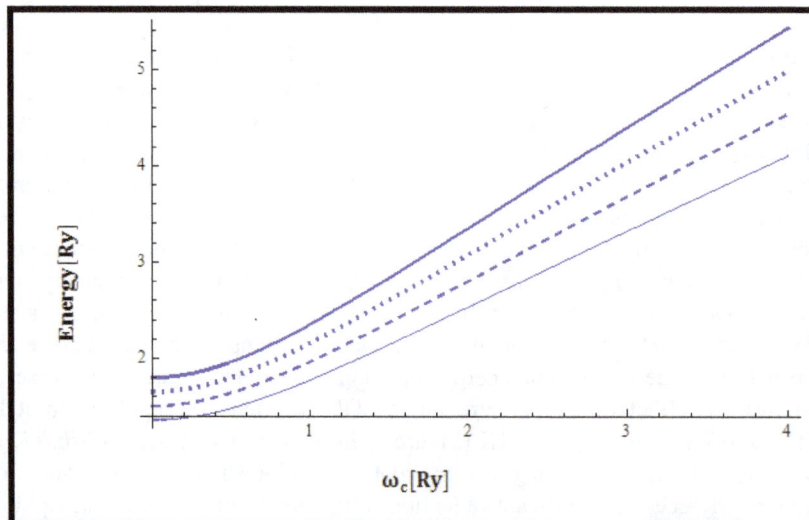

Figure 2. The computed energy spectra of quantum dot versus the strength of the magnetic field for $\boldsymbol{\omega_0 = 0.5R^*}$, T=0K, m=0 and various pressures (P=0 Kbar solid; P=10Kbar dashed; P=20Kbar dotted and P=30Kbar thick)

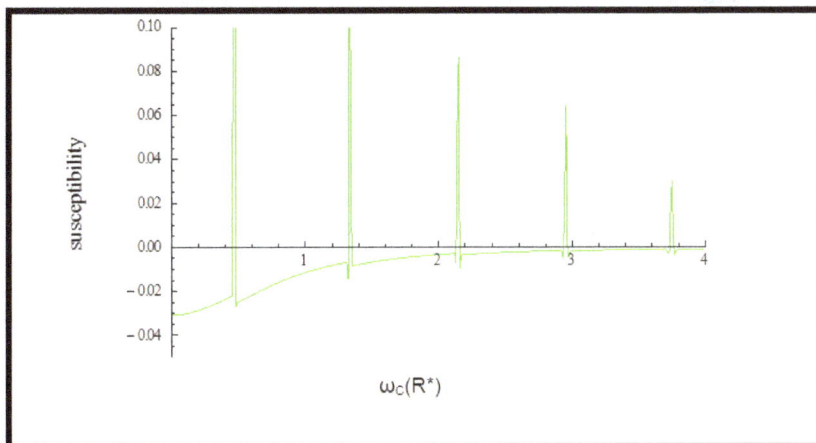

Figure 3a

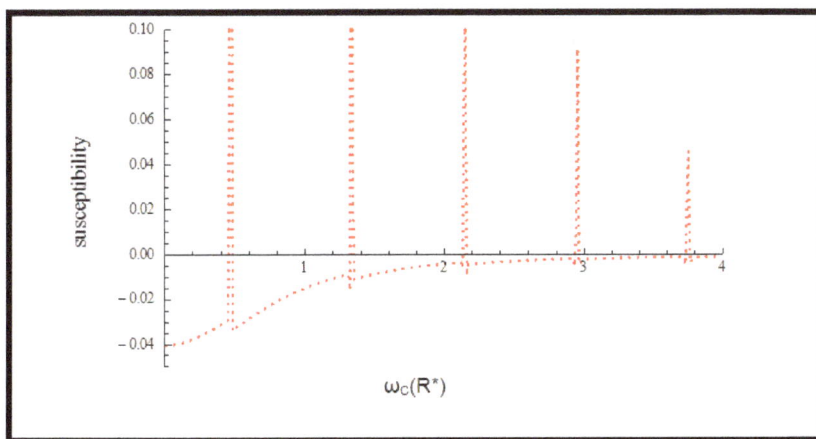

Figure 3b

Figure 3. Magnetic susceptibility (χ) of the two electrons quantum dot as function of magnetic field strength, calculated at fixed confining frequency $\omega_0 = 0.5R^*$, pressure T=0.01K and different temperatures : a) P=0.0 Kbar   and b) P=3.0 Kbar

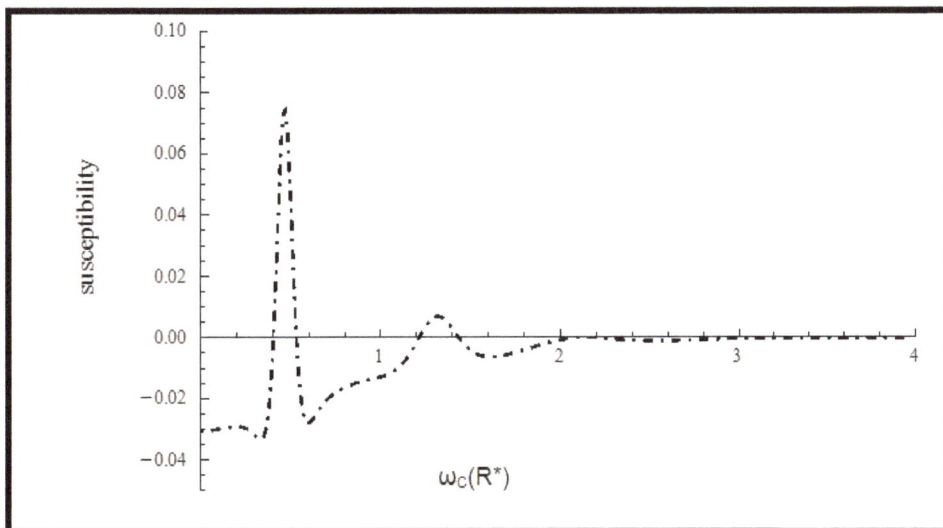

Figure 4. Magnetic susceptibility (χ) of the quantum dot as function of magnetic field strength at fixed confining frequency $\omega_0 = 0.5R^*$, Temperature T=1 K and pressure P= 0 Kbar

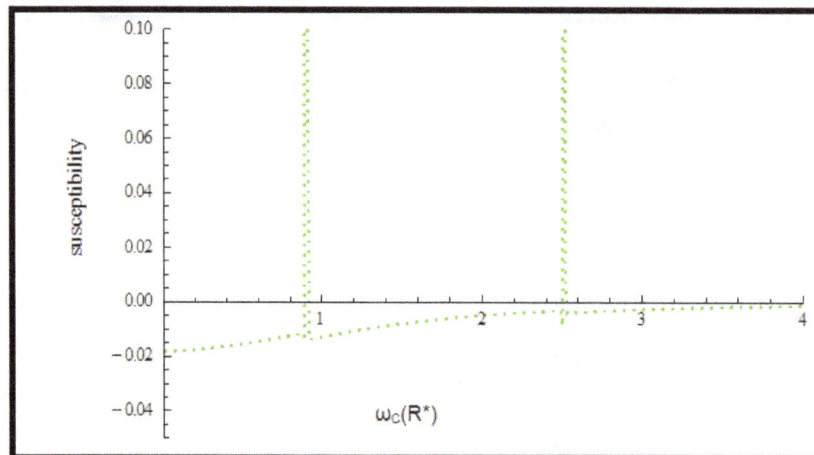

Figure 5. Magnetic susceptibility (χ) of the quantum dot as function of magnetic field strength at fixed confining frequency $\omega_0 = 0.8R^*$, Temperature T=0.01 K and pressure P= 0 Kbar

Table 1. The ground state energies of QD (in R^*) as a function of dimensionless coulomb coupling parameter λ obtained from exact diagonalization method (second column) compared with reported work (third column) Reference (Ciftja & Kumar, 2004)

λ	E (Present work)	E(Ciftja & Kumar, 2004)
0	2.00000	2.00000
1	3.000969	3.00097
2	3.721433	3.72143
3	4.318718	4.31872
4	4.847800	4.84780
5	5.332238	5.33224
6	5.784291	5.78429
7	6.211285	6.21129
8	6.618042	6.61804
9	7.007949	7.00795
10	7.383507	7.38351

4. Conclusions

The electronic energy levels and the magnetic susceptibility of two interacting electrons in the quantum dot have been calculated as a function of pressure, temperature confining frequency and magnetic field. The magnetic susceptibility spectra shows some peak structures which correspond to the energy levels and the associated spin-singlet-triplet transitions in the ground state of the quantum dot .The calculations show that the absolute value of the magnetic susceptibility ($| \chi |$) and the height of the transition peaks in the susceptibility spectra enhances also as the pressure increases, while keeping the values of temperature and magnetic field strength unchanged. In addition, the magnetic susceptibility spectra shows a temperature dependent behavior. The height and the number of the transition peaks in the magnetic susceptibility spectra changes significantly due to the thermal fluctuations. Furthermore, the confining frequency has a significant effect on the magnetic susceptibility spectra. We have observed that, as the confining frequency increases, the energy of the electrons increases and thus the positions of the energy level crossings are shifted towards a higher transition field strength. This in turn leads to changes in locations of the peaks spectra in the magnetic susceptibility spectra of the quantum dot.

Appendix A: Properties of the Laguerre polynomials

The following Laguerre relation was used to evaluate the coulomb energy matrix element in a closed form (Nguyen & Sarma, 2011):

$$\int_0^\infty t^{\alpha-1} e^{-pt} L_m^\lambda(at) L_n^\beta(bt) dt =$$

$$\frac{\Gamma(\alpha)(\lambda+1)_m(\beta+1)_n p^{-\alpha}}{m!n!} \sum_{j=0}^m \frac{(-m)_j(\alpha)_j}{(\lambda+1)_j\, j!} \left(\frac{a}{p}\right)^j \sum_{k=0}^n \frac{(-n)_k(\alpha+j)_k}{(\beta+1)_k\, k!} \left(\frac{b}{p}\right)^k \qquad (A\,1)$$

Appendix B: The pressure and temperature dependent dielectric constant and electron effective mass.

$$\epsilon_r \, (P,T) = \begin{cases} 12.74 \exp(-1.73 \times 10^{-3}P) \exp[9.4 \times 10^{-5}(T - 75.6)] \text{ for T} < 200 \text{ K} \\ 13.18 \exp(-1.73 \times 10^{-3}P) \exp[20.4 \times 10^{-5}(T - 300)] \text{ for T} \geq 200 \text{ K} \end{cases} \tag{B 1}$$

$$m^*(P,T) = \left[1 + 7.51 \left(\frac{2}{E_g^r(P,T)} + \frac{1}{E_g^r(P,T)+0.341}\right)\right]^{-1} m_0 \tag{B 2}$$

$$E_g^r(P,T) = \left[1.519 - 5.405 \times 10^{-4} \frac{T^2}{T+204}\right] + bP + cP^2 \tag{B 3}$$

Where m_0 is the free electron mass, $E_g^r(P,T)$ is the pressure and temperature dependent energy band gap for GaAs quantum dots at Γ point, b= 1.26×10^{-1} eV GPa^{-1} and c = -3.77$\times 10^{-3}$ eV GPa^{-2} (Rezaei & Kish, 2012).

References

Ashoori, R. C., Stormer, H. L., Weiner, J. S., Pfeiffer, L. N., Baldwin, K. W., & West, K. W. (1993). N-electron ground state energies of a quantum dot in magnetic field. *Physical review letters*, *71*(4), 613.

Avetisyan, S., Chakraborty, T., & Pietiläinen, P. (2016). Magnetization of interacting electrons in anisotropic quantum dots with Rashba spin–orbit interaction. *Physica E: Low-dimensional Systems and Nanostructures*, *81*, 334-338.

Boyacioglu, B., & Chatterjee, A. (2012). Heat capacity and entropy of a GaAs quantum dot with Gaussian confinement. *Journal of applied physics*, *112*(8), 083514.

Burkard, G., Loss, D., & DiVincenzo, D. P. (1999). Coupled quantum dots as quantum gates. *Physical Review B*, *59*(3), 2070.

Ciftja, O. (2013). Understanding electronic systems in semiconductor quantum dots. *Physica Scripta*, *88*(5), 058302.

Ciftja, O., & Kumar, A. A. (2004). Ground state of two-dimensional quantum-dot helium in zero magnetic field: Perturbation, diagonalization, and variational theory. *Physical Review B*, *70*(20), 205326.

Climente, J. I., Planelles, J., & Movilla, J. L. (2004). Magnetization of nanoscopic quantum rings and dots. *Physical Review B*, *70*(8), 081301.

De Groote, J. J. S., Hornos, J. E. M., & Chaplik, A. V. (1992). Thermodynamic properties of quantum dots in a magnetic field. *Physical Review B*, *46*(19), 12773.

Dybalski, W., & Hawrylak, P. (2005). Two electrons in a strongly coupled double quantum dot: From an artificial helium atom to a hydrogen molecule. *Physical Review B*, *72*(20), 205432.

El-Said, M. (1995). Two-electron quantum dots in a magnetic field. *Semiconductor science and technology*, *10*(10), 1310.

El-Said, M. (1998). The energy level ordering in two-electron quantum dot spectra. *Superlattices and microstructures*, *23*(6), 1237-1243.

El-Said, M. (2000). Spectroscopic structure of two interacting electrons in a quantum dot by the shifted 1/N expansion method. *Physical Review B*, *61*(19), 13026.

Elsaid, M. K., Al-Naafa, M. A., & Zugail, S. (2008). Spin Singlet-Triplet Energy Splitting in the Ground State of a Quantum Dot with a Magnetic Field: Effect of Dimensionality. *Journal of Computational and Theoretical Nanoscience*, *5*(4), 677-680.

Helle, M., Harju, A., & Nieminen, R. M. (2005). Two-electron lateral quantum-dot molecules in a magnetic field. *Physical Review B*, *72*(20), 205329.

Kandemir, B. S. (2005). Variational study of two-electron quantum dots. *Physical Review B*, *72*(16), 165350.

Kastner, M. A. (1992). The single-electron transistor. *Reviews of Modern Physics*, *64*(3), 849.

Kouwenhoven, L. P., Austing, D. G., & Tarucha, S. (2001). Few-electron quantum dots. *Reports on Progress in Physics*, *64*(6), 701.

Loss, D., & DiVincenzo, D. P. (1998). Quantum computation with quantum dots. *Physical Review A*, *57*(1), 120.

Maksym, P. A., & Chakraborty, T. (1990). Quantum dots in a magnetic field: Role of electron-electron interactions. *Physical review letters*, *65*(1), 108.

Nammas, F. S., Sandouqa, A. S., Ghassib, H. B., & Al-Sugheir, M. K. (2011). Thermodynamic properties of two-dimensional few-electrons quantum dot using the static fluctuation approximation (SFA). *Physica B: Condensed Matter*, *406*(24), 4671-4677.

Nguyen, N. T., & Peeters, F. M. (2008). Magnetic field dependence of the many-electron states in a magnetic quantum dot: The ferromagnetic-antiferromagnetic transition. *Physical Review B*, *78*(4), 045321.

Nguyen, N. T., & Sarma, S. D. (2011). Impurity effects on semiconductor quantum bits in coupled quantum dots. *Physical Review B*, *83*(23), 235322.

Räsänen, E., Saarikoski, H., Stavrou, V. N., Harju, A., Puska, M. J., & Nieminen, R. M. (2003). Electronic structure of rectangular quantum dots. *Physical Review B*, *67*(23), 235307.

Rezaei, G., & Kish, S. S. (2012). Effects of external electric and magnetic fields, hydrostatic pressure and temperature on the binding energy of a hydrogenic impurity confined in a two-dimensional quantum dot. *Physica E: Low-dimensional Systems and Nanostructures*, *45*, 56-60.

Sanjeev Kumar, D, Mukhopadhyay, S., & Chatterjee, A. (2016). Magnetization and susceptibility of a parabolic InAs quantum dot with electron–electron and spin–orbit interactions in the presence of a magnetic field at finite temperature. *Journal of Magnetisim and Magnetic Materials, 418*, 169.

Schwarz, M. P., Grundler, D., Wilde, M., Heyn, C., & Heitmann, D. (2002). Magnetization of semiconductor quantum dots. *Journal of applied physics*, *91*(10), 6875-6877.

Taut, M. (1994). Two electrons in a homogeneous magnetic field: particular analytical solutions. *Journal of Physics A: Mathematical and General*, *27*(3), 1045.

Wagner, M., Merkt, U., & Chaplik, A. V. (1992). Spin-singlet–spin-triplet oscillations in quantum dots. *Physical Review B*, *45*(4), 1951.

Permissions

List of Contributors

Olivier K. Bagui and Jeremie T. Zoueu
Laboratoire d'Instrumentation Image et Spectroscopie, Institut National Polytechnique Felix Houphouet-Boigny, BP 1093 Yamoussoukro, Cote d'Ivoire

Kenneth A. Kaduki
Laser Physics and Spectroscopy Group, Department of Physics, University of Nairobi, P. O. Box 30197-00100, Nairobi, Kenya

Edouard Berrocal
Department of Physics, Division of Combustion Physics, Lund Institute of Technology, Box 118, Lund 221 00, Sweden

Rob Langley
Norwich, UK

Carolyne M. M. Songa
Natural Science Department (Physics), Faculty of Science, The Catholic University of Eastern Africa, Nairobi, Kenya
Physics Department, Jomo Kenyatta University of Agriculture and Technology, Nairobi, Kenya

Jared H. O. Ndeda
Physics Department, Jomo Kenyatta University of Agriculture and Technology, Nairobi, Kenya

Gilbert Ouma
Meteorological Department, University of Nairobi, Kenya

David Zareski
I.A.I, Yehud, Israel

Adama Penetjiligue Soro, Emma Georgina Zoro-Diama, Guy Euloge Bany, Yvon Bibila Mayaya Bisseyou and Adjo Viviane Adohi-Krou
Laboratoire de Cristallographie et Physique Moléculaire (LaCPM), Université Félix Houphouët Boigny, Côte d'Ivoire

Kédro Sidiki Diomande
Centre National de Recherche Agronomique (CNRA), Côte d'Ivoire

Jiří Stávek
Bazovského 1228, 163 00 Prague, Czech republic

Eyal Brodet
Michaeslon St, Jerusalem, 93707, Israel

Karam Hamdan, Ghassan Younes and Mahmoud Korek
Faculty of Science, Beirut Arab University, P. O. Box 11-5020 Riad El Solh, Beirut 1107 2809, Lebanon

Philip J. Tattersall
Private researcher, 8 Lenborough St, Beauty Point, Tasmania, Australia

Hassan H. Mohammed
Department of Computer Engineering Technology, Iraq University College, Alestiqlal Street, Basrah, Iraq

Salwan K. J. AL-Ani
Department of Physics, College of Science, Al-Mustansirya University, Baghdad, Iraq
Correspondence: Hassan H. Mohammed, Department of Computer Engineering Technology, Iraq University

Jérémie T. Zoueu
Laboratoire d'Instrumentation Image et Spectroscopie, Unité Mixte de Recherche et d'Innovation en Electronique et Electricité Appliquée, Ecole Supérieure d'Industrie, Institut National Polytechnique Houphouët Boigny BP 1093 ESI Yamoussoukro, Côte d'Ivoire

Jocelyne Bosson
Laboratoire de Physique Fondamentale et Appliquée, Unité de Formation et de Recherche des Sciences Fondamentales et Appliquées, Université Nangui Abrogoua, 02 BP 801 Abidjan 02, Côte d'Ivoire

Wilfried G. Dibi
Laboratoire d'Instrumentation Image et Spectroscopie, Unité Mixte de Recherche et d'Innovation en Electronique et Electricité Appliquée, Ecole Supérieure d'Industrie, Institut National Polytechnique Houphouët Boigny BP 1093 ESI Yamoussoukro, Côte d'Ivoire
Laboratoire de Physique Fondamentale et Appliquée, Unité de Formation et de Recherche des Sciences Fondamentales et Appliquées, Université Nangui Abrogoua, 02 BP 801 Abidjan 02, Côte d'Ivoire

Beaulys Fotso, Casimir Y. Brou and Adolphe Zeze
Laboratoire des Biotechnologies Végétale et Microbienne, Unité Mixte de Recherche et d'Innovation en Sciences Agronomiques et Génie Rural, Ecole Supérieure d'Agronomie, Institut National Polytechnique Houphouët-Boigny BP 1313 ESA Yamoussoukro, Côte d'Ivoire

Anna C.M. Backerra
Gualtherus Sylvanusstraat 2, 7412 DM Deventer, The Netherlands

Masoud Asadi-Zeydabadi and Clyde Zaidins
Department of physics, University of Colorado Denver, Denver, CO 80217, USA

Dirk J. Pons
Department of Mechanical Engineering, University of Canterbury, New Zealand

Arion D. Pons
University of Canterbury, Christchurch, New Zealand

Aiden J. Pons
Rangiora New Life School, Rangiora, New Zealand

Jack Jia-Sheng Huang
Source Photonics, 8521 Fallbrook Avenue, Suite 200, West Hills, CA 91304, USA
Source Photonics, No.46, Park Avenue 2nd Rd., Science-Based Industrial Park, Hsinchu, Taiwan, R.O.C.

Yu-Heng Jan
Source Photonics, No.46, Park Avenue 2nd Rd., Science-Based Industrial Park, Hsinchu, Taiwan, R.O.C.
Source Photonics, 8521 Fallbrook Avenue, Suite 200, West Hills, CA 91304, USA

Jack Jia-Sheng Huang
Source Photonics, 8521 Fallbrook Avenue, Suite 200, West Hills, CA 91304, USA
Source Photonics, No.46, Park Avenue 2nd Rd., Science-Based Industrial Park, Hsinchu, Taiwan, R.O.C.

Yu-Heng Jan
Source Photonics, No.46, Park Avenue 2nd Rd., Science-Based Industrial Park, Hsinchu, Taiwan, R.O.C
Source Photonics, 8521 Fallbrook Avenue, Suite 200, West Hills, CA 91304, USA

H. S. Chen, H. S. Chang, C. J. Ni and Emin Chou
Source Photonics, No.46, Park Avenue 2nd Rd., Science-Based Industrial Park, Hsinchu, Taiwan, R.O.C.

Koshun Suto
Chudaiji Buddhist Temple, Isesaki, Japan

Sergio P. F
Independent Researcher, graduated from Computer Science, Madrid, Spain

Leonardo Golubovic and Steven Knudsen
Physics and Astronomy Department, West Virginia University, Morgantown, WV 26506-6515, USA

Faten Bzour, Mohammad K. Elsaid and Ayham Shaer
Physics Department, Faculty of Science, An- Najah National University, Nablus, West Bank, Palestine

Index